1977

SMOKE, DUST
AND
HAZE

FUNDAMENTALS OF
AEROSOL BEHAVIOR

An 80 m^3 2 mil Teflon bag used in the study of aerosol dynamics (p. 257-8).

SMOKE, DUST AND HAZE

FUNDAMENTALS OF AEROSOL BEHAVIOR

S. K. FRIEDLANDER
California Institute of Technology

A WILEY-INTERSCIENCE PUBLICATION

JOHN WILEY & SONS

New York · London · Sydney · Toronto

Library of Congress Cataloging in Publication Data:

Friedlander, Sheldon Kay.
Smoke, dust, and haze.

"A Wiley-Interscience publication."
Includes bibliographies and index.
1. Aerosols. I. Title.
TD884.5.F76 628.5′3′2 76-26928

ISBN 0-471-01468-0

Printed in the United States of America

10 9 8 7 6 5 4 3 2 1

Preface

Over the last ten years I have taught a course in particulate pollution to seniors and first-year graduate students in environmental and chemical engineering. A course in this field has now become essential to the training of engineers and applied scientists working in the field of air pollution. The subject matter is sufficiently distinctive to require separate coverage; at the same time, it is inadequately treated in most courses in engineering or chemistry. Although there are a few good reference works covering different parts of the field, I have felt the need for a text; this one is based on my own course notes.

There are three main types of practical problems to which the contents of this book can be applied: How are aerosols formed at pollution sources? How can we remove particles from gaseous emissions to prevent them from becoming an air pollution problem? How can we relate air quality to emission sources and thereby devise effective pollution control strategies? The fundamentals of aerosol behavior necessary to deal with these problems are developed in this text. Although fundamentals are stressed, examples of practical problems are included throughout.

The treatment that I have given the subject assumes some background in fluid mechanics and physical chemistry. A student with good preparation in either of these fields should, with diligence, be able to master the fundamentals of the subject. This has been my experience in teaching first-year graduate students with undergraduate majors in almost all branches of engineering, chemistry, and physics.

The first half of the text is concerned primarily with the transport of particles and their optical properties. It is this part of the field that, until recently, had been the most developed. Particle transport theory has application to the design of gas-cleaning devices, such as filters and electrical precipitators, and this is pointed out in the text. However, I have not dealt with the details of equipment design; in most cases, direct application of the theory to design is difficult because of the complexity of gas-cleaning equipment. This leads to the use of methods that are more empirical than otherwise employed in this text. With a good understanding of particle transport, the student will be able to read the specialized works on equipment design intelligently and critically.

Once the student has mastered the concepts of particle transport and optical behavior, he will also find it easy to understand aerosol measurement methods. A chapter on this subject ends the first half of the text on an experimental note; progress in aerosol science is heavily dependent on experimental advances, and it is important to get this across to the student early in his studies. Indeed, throughout the text, theory and experiment are closely linked.

In the second half of the book, the dynamics of the size distribution function are discussed. It is this theory that gives the field of small particle behavior its distinctive theoretical character. The organization of this material is completely new, so far as coverage in book form is concerned. It begins with a chapter on coagulation, and is followed by chapters on thermodynamics and gas-to-particle conversion. Next, the derivation of the general dynamic equation for the size distribution function and its application to emission sources and plumes are discussed. This leads to the final chapter on the relationship of air quality to emission sources for particulate pollution. This chapter is based in part on the preceding theory. However, the power of the theory has not yet been fully exploited, and the next few years should see significant advances.

One of my goals in writing the book was to introduce the use of the equation for the dynamics of the particle size distribution function at the level of advanced undergraduate and introductory graduate instruction. This equation is relatively new in applied science, but has many applications in air and water pollution and the atmospheric sciences.

I have also taken a step toward the linking of aerosol physics and chemistry in the last few chapters. Chemistry enters into the general dynamic equation through the term for gas-to-particle conversion. This turns out to have many important air pollution applications as shown in the last four chapters.

To keep the subject matter to manageable proportions, I have omitted interesting problems of a specialized nature, such as photophoresis and diffusiophoresis, which are seldom of controlling importance in applied problems. Details of the kinetic theory of aerosols have also been omitted. Although of major importance, they usually enter fully developed, so to speak, in applications. Besides, their derivation is covered in other books on aerosol science.

The resuspension of particles from surfaces and the break-up of agglomerates, important practical problems, are not well understood; the methods of calculation are largely empirical and not conveniently subsumed into the broad categories covered in the book.

Before I began writing, I considered the possibility of a general text covering small particle behavior in both gases and liquids. Much of the theory of physical behavior is the same or very similar for both aerosols and hydrosols, almost as much as in the fluid mechanics of air and water. The differences include double layer theory in the case of aqueous solutions and mean free path effects in gases. There are other important, specifically chemical differences.

After some thought, I decided against the general approach. Since I wanted to write a book closely linked to applications, I thought it best to limit it to the air field in which I can claim expertise. Including topics from water pollution would have unduly lengthened the book. However, the students who take my course are often interested in water pollution, and I frequently point out both similarities and differences between the air and water fields.

Special thanks are due C. I. Davidson and P. H. McMurry who served as teaching assistants and helped prepare some of the figures and tables. D. L. Roberts assisted in reviewing the manuscript for clarity and consistency. Professor R. B. Husar of Washington University made a number of useful suggestions on the text.

S. K. FRIEDLANDER

Pasadena, California
November 1976

Contents

Common Symbols

(Defining equation in parentheses)

a	radius, cm
A	aerosol surface area per unit volume of gas, cm^{-1} (1·11)
b	extinction coefficient, cm^{-1}
c	mass fraction, dimensionless
$\vec{c}$	migration velocity in the presence of an external force field, cm/sec
c_s	Stokes terminal settling velocity, cm/sec
C	slip correction, dimensionless (2·16)
d_p	particle diameter, cm
D	diffusion coefficient, cm^2/sec
E	electric field intensity, V/cm
f	friction coefficient, g/sec, or Fanning friction factor, dimensionless
g	size-composition probability density function, dimensionless (1·18)
I	intensity of light, erg/cm^2 sec
I_λ	intensity distribution function, erg/cm^3 sec
J	flux in one dimension, $cm^{-2}sec^{-1}$ or light source function, erg/cm^3 sec
J_x, J_y, J_z	flux components, $cm^{-2}sec^{-1}$
k	mass transfer coefficient, cm/sec (3·13)
K	light scattering efficiency, dimensionless (5·2)
l	mean free path, cm (1·1)
m	mass of particle or molecule, g
M	molecular weight, g/mole
n	concentration, cm^{-3}
n_∞	concentration at large distances from a surface, cm^{-3}
n_1	n/n_∞, dimensionless concentration (Section 3·2) or monomer concentration, cm^{-3}
$n(v)$	particle size distribution function with volume the distributed variable, cm^{-6} (1·2)
$n_a(a)$	particle size distribution function with radius the distributed variable, cm^{-4}

$n_d(d_p)$	particle size distribution function with diameter the distributed variable, cm^{-4} (1·3)
N_∞	total particle number concentration, cm^{-3}
p	pressure, $dyne/cm^2$
p_d	equilibrium vapor pressure above a drop of diameter d_p
p_s	equilibrium vapor pressure above a flat surface
Q	coupling constant for dispersion forces, erg (2·30)
r	radial coordinate, cm
$\vec{r}$	position vector, cm
s	particle stop distance, cm (4·19)
s^*	visual range, cm (5·32)
S	saturation ratio, dimensionless
t	time, sec
T	absolute temperature, °K
u, v, w	velocity components in x,y,z directions, respectively, cm/sec
u_f, v_f, w_f	gas velocity components in equations in which the particle velocity appears, cm/sec
U	mainstream or average velocity, cm/sec
v	particle volume, cm^3
v_m	molecular volume, cm^3
x,y,z	Cartesian coordinates
Z	electrical mobility, (cm/sec)/(V/cm), (2·23)
β	collision frequency function, cm^3/sec, (7·1) or monomer flux, $cm^{-2}sec^{-1}$ (Section 8·9)
ϵ	eddy diffusivity, cm^2/sec
ϵ_d	turbulent energy dissipation, cm^2/sec^3
ϵ_p	dielectric constant of particle
η	dimensionless particle volume, (Section 7·11)
η_R	single cylinder removal efficiency, dimensionless, (3·16)
θ	angular coordinate or dimensionless time
κ	thermal conductivity
λ	wavelength of light, cm
μ	viscosity, g/cm sec
ν	kinematic viscosity $=\mu/\rho$, cm^2/sec
ρ	gas density, g/cm^3, or aerosol mass per unit volume of gas. g/cm^3
ρ_p	particle density, g/cm^3
σ	surface tension, dyne/cm
ϕ	electrostatic potential
Φ	potential energy, erg
ψ	self-preserving size distribution function, dimensionless (7·26), or stream function (Chap. 3)

		Section
Kn	Knudsen	1·2
Le	Lewis	8·3
Pe	Péclet	3·2
R	Interception	3·2
Re	Reynolds	—
Sc	Schmidt	3·2, Table 2·1
Sh	Sherwood	3·2
St	Stokes	4·4

PHYSICAL CONSTANTS

Electronic charge	$e = 1.6 \times 10^{-19}$ coulomb
	$= 4.8 \times 10^{-10}$ statcoulomb
Gravitational acceleration	$g = 981$ cm/sec^2
Boltzmann constant	$k = 1.380 \times 10^{-16}$ erg/°K
Avogadro number	$N_{av} = 6.023 \times 10^{23}$ mole^{-1}
Gas constant	$R = 8.31 \times 10^7$ erg/°K mole

SMOKE, DUST AND HAZE

FUNDAMENTALS OF AEROSOL BEHAVIOR

CHAPTER 1

Introduction: Aerosol Characterization

Aerosols are formed either by the conversion of gases to particulate matter or by the disintegration of liquids or solids. They may also result from the resuspension of powdered material or the break-up of agglomerates. Formation from the gas phase tends to produce much finer particles than disintegration processes (except when condensation takes place directly on existing particles). Particles formed directly from the gas are usually smaller than 1 μm in diameter. (1 micron = 1 micrometer = 10^{-4} cm is designated by the symbol 1 μm).

The great variety of words that have been invented to describe particulate systems attests to their ubiquity and to the impression they have made on man from early times. Dust, smoke, fume, haze, and mist are all terms in common use with somewhat different popular meanings. Thus dust usually refers to solid particles produced by disintegration processes, while smoke and fume particles are generally smaller and formed from the gas phase. Mists are composed of liquid droplets. In this text, however, we will rarely employ these special terms because of the difficulty of exact definition and the complexity of many real systems composed of mixtures of particles. Instead, we employ the generic term aerosol to describe all such systems of small particles suspended in air or another gas.

As an example of the factors determining the chemical and physical characteristics of such systems, we consider the atmospheric aerosol. This is composed of a mixture of particles originating from various sources, together with products of gas-to-particle conversion processes. Material introduced into the atmosphere in particulate form is called *primary* while material resulting from gas-to-particle conversion is called *secondary*.

In the case of the Los Angeles aerosol, one of the most carefully studied, a pattern of the following type sometimes develops (Fig. 1·1): Marine air carrying sea salt droplets passes over industrial sources in the western portion of the air basin; products of the combustion of fuel oil from power plants and refineries, including sulfur dioxide and fuel oil flyash, are introduced into the air flow. Some of the sulfur dioxide is oxidized to sulfate and accumulates in the particulate phase, perhaps by displacement of chloride from sea salt droplets.

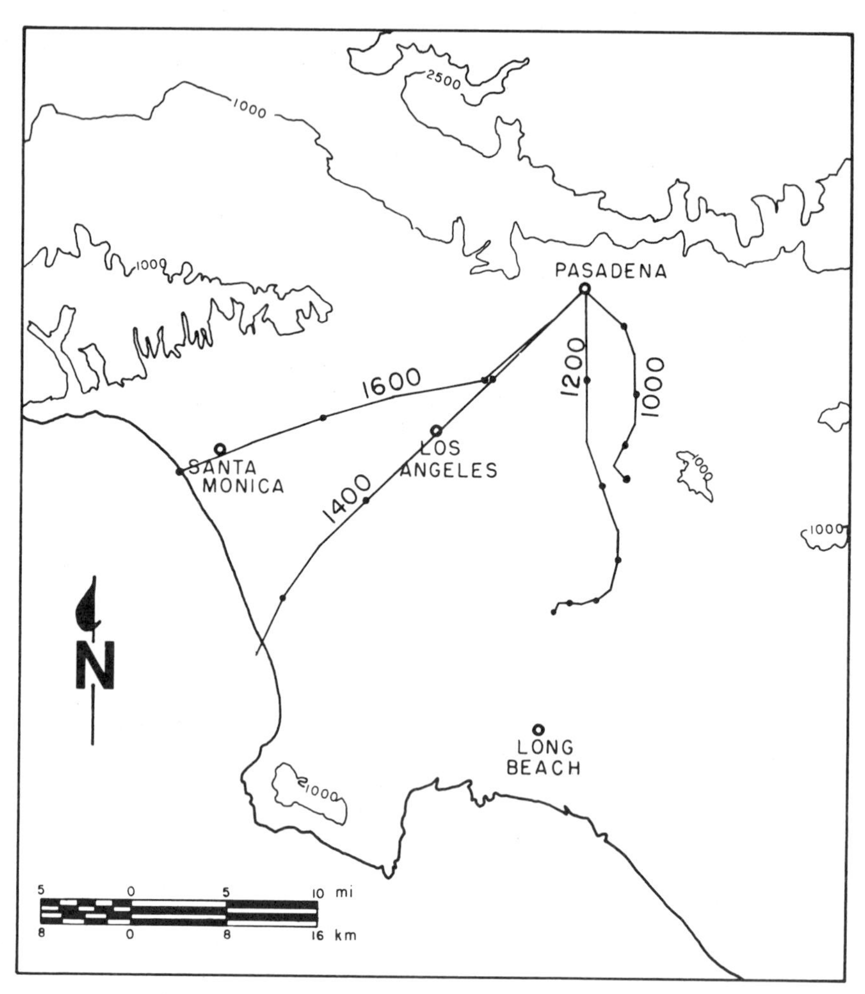

Fig. 1·1 Calculated air trajectories arriving in Pasadena on September 3, 1969. Numbers give time of arrival; dots denote hour intervals (White and Husar, 1976). Air arriving between 1200 and 1400 hours comes from the southwest part of the basin. This air carries industrial and power plant emissions, probably mixed with the marine aerosol; over downtown Los Angeles, automobile emission products are added in large volumes.

Increasing amounts of automobile exhaust are added to the air flow as it moves northeast after leaving the industrial areas. Automobile exhaust includes particulate matter composed of lead bromochloride and tarry organic matter. The gas phase includes nitrogen oxides, organic vapors, and sulfur dioxide. Small percentages of these gases, converted to particulate matter, add significantly to the total aerosol concentration and contribute significantly to visibility degradation. The conversion products include nitrates resulting from the oxidation of NO and NO_2, oxygenated organic compounds, and sulfates.

Also present is soil dust that becomes airborne as a result of the winds and surface activities associated with transportation, construction, and agriculture. Such particles are usually larger than 1 μm in diameter. Because of their large size, they tend to settle out but are continually reintroduced by surface activities.

Methods of estimating the contributions of different emission sources to the atmospheric burden are discussed in Chap. 11.

1·1 Factors Influencing Aerosol Behavior

Aerosols may affect human health, visibility, and climate. Particle size, concentration, and chemical composition are usually the most important factors determining such effects. For example, from experimental studies on humans inhaling aerosols of controlled particle size, the fraction of the particles that deposit in the lung can be determined. The fraction deposited varies from individual to individual because of differences in lung geometry and air flow patterns; a typical curve of deposition as a function of particle size is shown in Fig. 1·2 for spherical particles of a given density. Very small particles are removed by Brownian motion, whereas the larger ones settle out or deposit because their inertia does not permit them to follow the turns in the air flow. The minimum in the efficiency curve results because the intensity of the Brownian motion increases for the smaller particles, while removal by sedimentation and inertial deposition increase with particle size. The physiological effect of the deposited particles is determined by their chemical composition.

Light scattering by small particles is also a sensitive function of size. In Fig. 1·3 the light scattered per unit mass of aerosol is shown as a function of particle diameter for wavelengths in the visible range. The maximum scattering efficiency corresponds to the particle size of the same order as the wavelength of the incident light. Processes leading to the accumulation of particles in that size range produce the most severe visibility degradation. Such processes depend on the chemical composition of the system, which also determines the refractive index, another important factor in light scattering and absorption.

We can express the nucleating efficiency of small particles, that is, their ability to form clouds, in terms of the rate of condensation per unit mass of aerosol. For particles of a given material growing by diffusion from the gas phase, the nucleating efficiency is shown in Fig. 1·4. Particles smaller than the critical size determined by the Kelvin effect

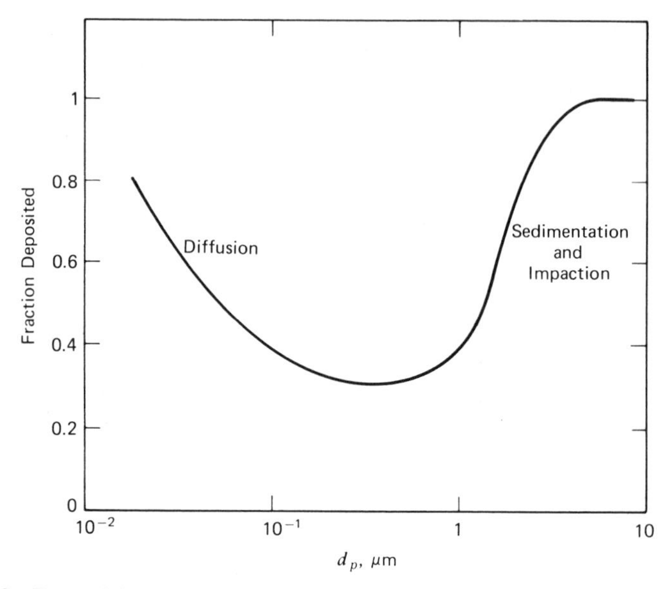

Fig. 1·2 Form of the efficiency curve for particle deposition in the lung. Particles in the size range between 0.1 and 1 μm are too large to diffuse rapidly and too small to settle rapidly or deposit because of inertial effects.

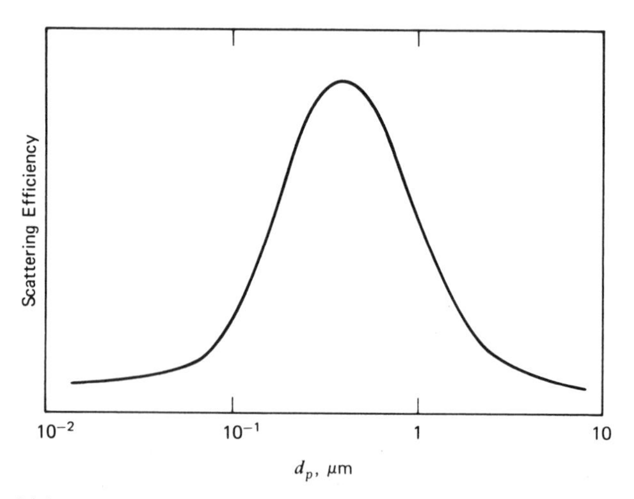

Fig. 1·3 Light scattering efficiency, expressed as light scattered per unit mass of particles. The peak occurs in the range corresponding to the wavelength of the scattered light in the visible range.

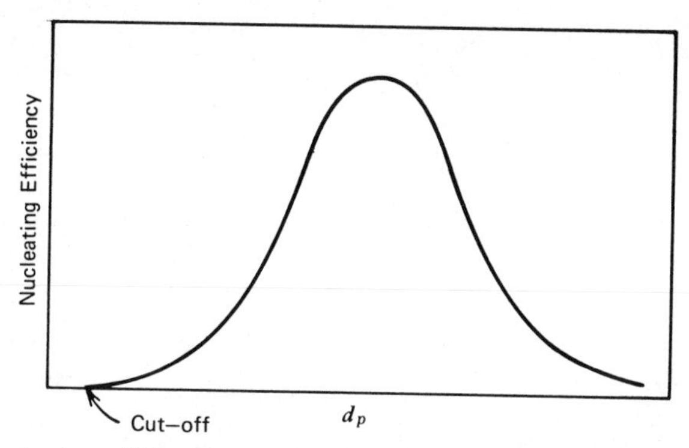

Fig. 1·4 Nucleating efficiency expressed as rate of condensation per unit mass of aerosol. Particles smaller than the cut-off do not grow because of the increased vapor pressure over a small particle (Kelvin effect). Large particles are inefficient because they have less surface area per unit mass.

do not grow. Large particles have a relatively small amount of surface area per unit mass, which accounts for their reduced nucleating efficiency.

Particle emissions can be controlled by gas-cleaning devices added to the source. The efficiency of filters, scrubbers, and other such devices primarily depends on particle size. As shown in Chap. 5, a minimum is often found when the efficiency of particle removal is plotted as a function of particle size. The performance of electrical precipitators and scrubbers is also strongly affected by the chemical nature of the particles and gases. In the case of the precipitators, the electrical field at the collector electrode depends on the conductivity of the deposited layer of particles.

In the following sections, mathematical methods of characterizing aerosol size and chemical nature are discussed. These are primarily accounting methods. They provide no direct information on mechanisms of aerosol formation, dynamic behavior, or effect on visibility. These and other subjects are covered in later chapters.

Fortunately, aerosol characterization theory is not merely a paper-and-pencil exercise; major advances in experimental techniques now permit some, but by no means all, of the most important theoretical parameters to be measured (Chap. 6).

1·2 Particle Size

The particle sizes of interest in aerosol behavior range from molecular clusters of 10 Å to fog droplets and dust particles as large as 100 μm.

This represents a variation of 10^5 in size and of 10^{15} in mass. For spherical liquid droplets, the diameter (d_p) is an unequivocal measure of particle size. Spherical particles are frequently encountered in polluted atmospheres because of the growth of nuclei by condensation of liquid from the gas phase.

For nonspherical particles such as crystal fragments, fibers, or agglomerates, a characteristic size is more difficult to define. In the special case of geometric similarity, particles of different size have the same shape so that a single length parameter serves to characterize any size class. An ellipsoid with fixed ratio of major to minor axes is an example. Sizes of irregular particles are often defined in terms of the method of measurement (Chap. 6). In the case of the cascade impactor, for example, the aerodynamic diameter of a particle of arbitrary shape and density refers to the size of a spherical particle of unit density that would deposit on a given impactor stage.

Particle behavior often depends on the ratio of particle size to some other characteristic length. The mechanisms of heat, mass, and momentum transfer between particle and carrier gas depend on the Knudsen number, $2l/d_p$, where l is the mean free path of the gas. The mean free path or mean distance traveled by a molecule between successive collisions can be calculated from the kinetic theory of gases. As a good approximation for a gas composed of molecules that act like rigid elastic spheres:

$$l = \nu\left(\frac{\pi m}{2kT}\right)^{1/2} \tag{1·1}$$

where ν is the kinematic viscosity of the gas, m is the molecular mass, k is Boltzmann's constant, and T is the absolute temperature. For normal temperatures and pressures, the mean free path in air is about 0.065 μm.

Consider the case of a spherical particle moving at constant velocity through a gas. When the particle diameter is much smaller than the mean free path ($l/d_p \gg 1$), molecules bouncing from the surface rarely collide with the mainstream molecules until far from the sphere. Most of the molecules striking the surface of the sphere come from the main body of the gas and are essentially unaffected by the presence of the sphere. The exchange of heat, mass, and momentum between particle and gas can be calculated from molecular collision theory. Appropriately, this range ($l/d_p \gg 1$) is known as the free molecule range.

When the particle diameter is much greater than the mean free path ($l/d_p \ll 1$), molecules striking the surface are strongly affected by those leaving. It is found experimentally that the gas behaves as a continuum with a zero velocity boundary condition at the surface. The drag, which can be calculated from the Navier–Stokes equations of fluid motion, is much greater at the same relative velocity than for the range of free molecule flow. The transition between the continuum and free molecule ranges takes place continuously but the transition theory poses formidable difficulties.

Light scattering by small particles depends on the ratio of diameter to the wavelength of the incident light, d_p/λ. For $d_p/\lambda \gg 1$, geometric optics apply and the scattering cross section is proportional to the cross-sectional area of the particle. When $d_p/\lambda \ll 1$, the scattering is calculated from the theory of the oscillation of a dipole in an oscillating electric field. For light in the visible range, λ is of the order of 0.5 μm and the transition between the two scattering regimes takes place over the range from 0.05 to 5 μm. Light scattering in the transition range can be calculated from classical electromagnetic theory, but the computations are complex (Chap. 5).

1·3 Concentration

Particle concentration at a point—expressed as the number of particles per unit volume of gas—is defined in a manner similar to gas density: Let δN be the number of particles in an initially rather large volume δV surrounding the point P in the gas (Fig. 1·5). Then the ratio $\delta N/\delta V$ is called the average (number) concentration of the particles within the volume δV. As the volume δV shrinks toward the point P, the average concentration can either increase or decrease depending on the concentration gradient in δV; in general, however, it will approach a constant value over a range of values of δV in which the gradient is small but many particles are still present (Fig. 1·6). This constant value is the particle concentration at the point P. As the volume continues to

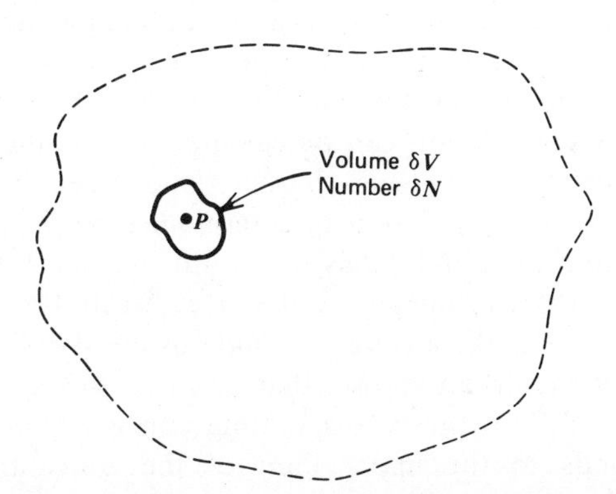

Fig. 1·5 The volume δV containing δN particles shrinks toward point P.

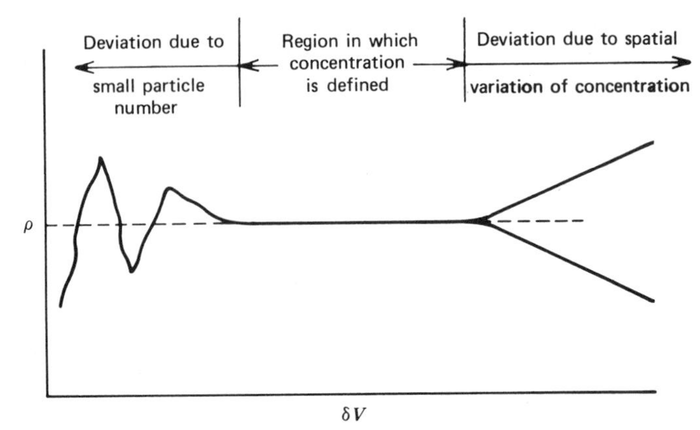

Fig. 1·6 Variation of the average concentration with size of region. For large values of δV, the average concentration may be larger or smaller than ρ depending on the gradient. For small values of δV, large positive or negative deviations from the average may occur because small numbers of particles are involved.

shrink, the number of particles becomes so small that the average concentration fluctuates markedly as shown in Fig. 1·6.

Particle concentrations in a polluted urban atmosphere may be of the order of 10^5 cm^{-3}, with source concentrations several orders of magnitude larger. (Excluded from consideration are the molecular clusters that play an important role in the theory of homogeneous condensation.) In polluted air, most of the particles by number fall in the size range between 0.01 and 0.1 μm. Concentrations of particles larger than 1 μm fall to only a few—less than ten per cubic centimeter. This leads to difficulty in defining the instantaneous concentrations for such particles.

Suspensions of small particles in gases at high concentrations are unstable; the particles collide and coagulate as a result of the Brownian motion. The time to reduce the particle concentration to one-tenth its original value by coagulation can be calculated from theory (Chap. 7). Table 1·1 shows values of this characteristic time as a function of concentration for the coagulation of a monodisperse aerosol with $d_p = 0.1$ μm. This time changes relatively little with particle size for monodisperse systems, but does depend on the size distribution function. We can use Table 1·1 to make a rough estimate of the stability with respect to coagulation of the urban aerosol that has a residence time of several hours. For $t_{1/10} = 3.5$ hr, the corresponding concentration is 10^6 cm^{-3}, which corresponds to the upper limit of the concentration usually observed in polluted atmospheres.

TABLE 1·1

TIME TO REDUCE THE CONCENTRATION OF A MONODISPERSE AEROSOL TO ONE-TENTH THE ORIGINAL VALUE, N_0 ($d_p = 0.1\ \mu m$, $T = 20°C$) (see also Section 7·3)

N_0 cm^{-3}	$t_{1/10}$ (approximate)
10^{10}	1.2 sec
10^9	12 sec
10^8	2 min
10^7	20 min
10^6	3.5 hr
10^5	35 hr

The average volume accessible to a particle is equal to the inverse of the concentration. A number concentration of 10^5 cm^{-3} (a fairly high atomospheric concentration) corresponds to 10^{-5} cm^3/particle or a sphere with a diameter of about 300 μm. Since this is so much larger than the particle diameter, particle–particle interactions can be neglected when calculating heat, mass, and momentum exchange between the particles and the surrounding gas.

Aerosol *mass* concentrations are usually determined by filtering a known volume of gas and weighing the collected particulate matter. The mass concentration averaged over the measurement time is found by dividing the measured mass by the volume of gas filtered. Mass concentrations for particulate matter in polluted atmospheres are usually well below 1000 μg/m^3 (1 μg/m^3 = 10^{-6} g/m^3).

Example. At a certain location, the mass concentration of particles in air (20°C and 1 atm) is 240 μg/m^3. Determine the corresponding mass ratio of particles to gas. If the local SO_2 concentration is 0.1 ppm on a volume basis, determine the mass ratio of particles to SO_2.

SOLUTION. The density of dry air at 20°C is 1.2 g/liter. Hence the mass ratio of particles to gas is $240 \times 10^{-6}/1.2 \times 10^3 = 2 \times 10^{-5}$. Despite the small value of this ratio, an atmospheric concentration of 240 μg/m^3 usually results in severe visibility degradation.

The mass ratio of particles to SO_2 present at a concentration of 0.1 ppm is given by

$$\frac{240 \times 10^{-6} \times 10^6(29)}{0.1(1.2) \times 10^3(64)} = 0.9$$

Particle mass concentrations in process gases may be many orders of magnitude higher than in atmospheric air. Such concentrations are often expressed in the engineering literature in grains per standard cubic foot (gr/SCF); process gas concentrations in these units may range from 1 to 10. Standard conditions refer to 0°C and 1 atm. Since 1 gr/SCF = 2.288 g/m^3, the ratio of particles to gas for a mass loading of 10 gr/SCF is about 2×10^{-3}.

At these low mass ratios, the effects of the particles on the flow of the gas can be neglected. The gas velocity distribution can be calculated from the equations of fluid motion or taken from measurements made in the absence of particles.

1·4 The Size Distribution Function

By taking special precautions, it is possible to generate aerosols composed of particles, which are all the same size. These are called *monodisperse* or homogeneous aerosols. In most practical cases both at emission sources and in the atmosphere, aerosols are composed of particles of many different sizes. These are called *polydisperse* or *heterodisperse* aerosols.

The most important physical characteristic of polydisperse systems is their particle size distribution. This distribution can take two forms: The first is the *discrete distribution* in which only certain "allowed" particle sizes are considered. Consider a suspension that is built up of unitary particles by some type of coagulation process. All particles will then be composed of integral numbers of these unitary particles, and the size distribution can be defined by the quantity n_g, where $g = 1, 2, \ldots, k$ represents the number of unitary particles composing the aggregates. As a limiting process, all distributions can be considered discrete in the sense that particles are composed ultimately of molecules that can be looked upon as unitary particles.

Usually more useful, however, is the concept of the *continuous size distribution*. Let dN be the number of particles per unit volume of fluid at a given position in space and at a given time in the particle size volume range v to $v + dv$:

$$dN = n(v, \mathbf{r}, t)\, dv \qquad (1\cdot 2)$$

This expression defines the particle size distribution function $n(v, \mathbf{r}, t)$. The volume is used here as the characteristic size parameter because it is relatively easy to define even for irregular particles, compared to the diameter or the surface area. Moreover, it is the volume that is conserved when two particles collide and adhere as in coagulation processes.

For spherical particles, the size is given unequivocally by the particle diameter. The corresponding particle size distribution function is then defined by the expression

$$dN = n_d(d_p, \mathbf{r}, t)\,d(d_p) \tag{1·3}$$

where dN is the concentration of particles in the particle size range d_p to $d_p + d(d_p)$ and $n_d(d_p)$ is the particle size distribution function with diameter the distributed variable instead of volume. The diameter may also be the aerodynamic diameter or some other equivalent size parameter.

The two distribution functions n and n_d are not equal but are related as follows: For a spherical particle, the volume and particle diameter are related by the expression:

$$v = \frac{\pi d_p^{\,3}}{6} \tag{1·4}$$

Hence

$$dv = \frac{\pi d_p^{\,2}}{2}\,d(d_p)$$

Substituting in (1·2)

$$dN = \frac{\pi d_p^{\,2} n\,d(d_p)}{2}$$

Hence by (1·3) for the same increment of radius,

$$n_d = \frac{\pi d_p^{\,2} n}{2} \tag{1·5}$$

In a similar way, a particle size distribution function can be defined with surface area as the distributed variable.

Since particle sizes cover such a wide range, it is convenient to employ a log scale. One common method of presenting data is to plot $\log n_d$ versus $\log d_p$. Another is to present the data as $dV/d\log d_p$ versus $\log d_p$, where $V = \int_0^v nv\,dv$. The area under curves plotted in this way is proportional to the mass of aerosol over a given size range provided that the particle density is a constant, independent of size. The following relationship holds with $v = \pi d_p^{\,3}/6$:

$$\frac{dV}{d\log d_p} = \frac{2.3\pi^2 d_p^{\,6} n(v)}{12} \tag{1·6}$$

The cumulative number distribution is defined by the expression

$$N(d_p) = \int_0^{d_p} n_d(d_d)\,d(d_p) \tag{1·7}$$

and is the number of particles of diameter less than or equal to d_p. Clearly $n_d = dN(d_p)/d(d_p)$. Hence the distribution function can be determined by differentiating the cumulative function with, however, the inaccuracies often attendant upon differentiation of experimental data.

When several disperse systems are mixed, the resulting size distribution, in the absence of coagulation, is given by the volume average:

$$n = \frac{\Sigma_i n_i Q_i}{\Sigma_i Q_i}$$

where n_i is the size distribution and Q_i is the volume of gas corresponding to the ith aerosol.

The time average distribution function at a fixed point is given by

$$\bar{n}_d(d_p, \mathbf{r}) = \frac{1}{t}\int_0^t n_d(d_p, \mathbf{r}, t')\,dt'$$

where the average is taken over the interval from 0 to t.

Example. During the combustion of pulverized coal, the coal particles often shatter. Assume that particles in a given size range break up to form p times as many new particles, where p is independent of particle size. What is the relation of the new size distribution function $n'(v')$ to the original distribution $n(v)$?

SOLUTION. In the size range v to $v + dv$, there are originally $n(v)dv$ particles. As a result of combustion, $pndv$ form from this original group. Only a fraction α of the original volume of the particle remains (as flyash). Thus the volume of the final group of particles is $\alpha v/p$. The number of particles in the new size range, $dv' = \alpha\, dv/p$, is given by

$$n'(v')\,dv' = pn(v)\,dv$$

Substituting for dv',

$$n'(v') = \frac{p^2}{\alpha} n(v)$$

1·5 Dynamics of the Distribution Function

The aerosol size distribution function changes with time and position as air passes through a city or a smokestack. Consider a small volume element of gas whose motion we follow. The size distribution function of the aerosol contained in the element is modified by a number of processes. Cooling or chemical reaction leads to the conversion of molecules originally in the gas phase to the aerosol phase. In the atmosphere, gases that react to form aerosols include SO_2, NO_2, olefins, and NH_3. Conversion may take place by the formation of many tiny new particles, less than 100 Å in diameter, or by condensation on existing nuclei. The formation of new particles from the gas phase is called *homogeneous nucleation*.

Molecular accretion from the gas phase can be considered a continuous growth process, equivalent to a continuous movement through v space. Particle collisions result from the Brownian motion and lead to clumping of particles by *coagulation*. This process is most rapid near aerosol sources, where concentrations are high since the rate of coagulation is proportional to the product of the concentrations of the colliding particles.

Sedimentation is important for the larger particles but not for those in the submicron range. Finally, particles diffuse across the walls of the elemental volume. Small particles diffuse most rapidly because their Brownian motion is most violent.

The physical and chemical processes shaping the size distribution are summarized in Fig. 1·7. The change in $n(v, \vec{r}, t)$ with time in the volume element whose motion we follow can be expressed as the sum of

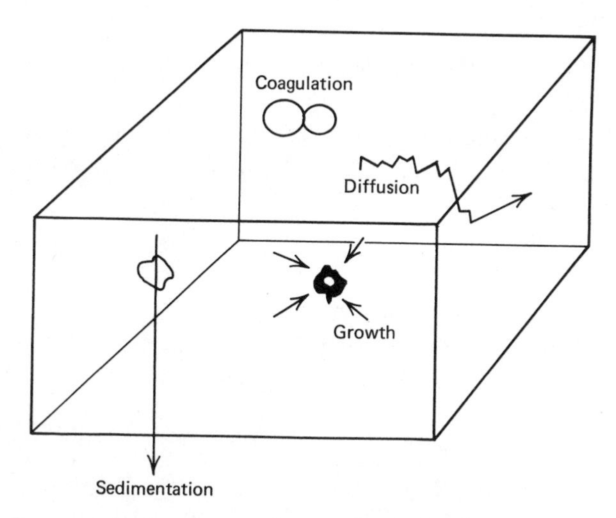

Fig. 1·7 Processes shaping the particle size distribution function in a small volume element of gas. Diffusion and sedimentation involve transport across the walls of the element. Coagulation and growth take place within the element.

two terms:

$$\frac{dn}{dt} = \left[\frac{dn}{dt}\right]_i + \left[\frac{dn}{dt}\right]_e$$

The first term on the right represents processes which occur inside the element including gas-to-particle conversion and coagulation. The second term represents particle transport across the boundaries of the element by diffusion and sedimentation. Transport processes are discussed in the first half of the text; coagulation and gas-to-particle conversion are covered in the second half. A general dynamic equation for $n(v, \vec{r}, t)$ incorporating both internal and external processes is set up and discussed in Chap. 10.

1·6 Moments of the Distribution Function

It is instructive to consider the various moments of the particle size distribution function $n_d(d_p, \mathbf{r}, t)$, since they have great physical importance. The general moment can be defined by the expression

$$M_\nu(\mathbf{r}, t) = \int_0^\infty n_d d_p^{\,\nu} \, d(d_p)$$

where ν represents the order of the moment.

The *zeroth* moment,

$$M_0 = \int_0^\infty n_d \, d(d_p) = N_\infty(\mathbf{r}, t) \tag{1·8}$$

is the total concentration of particles in the suspension at a given point and time.

The *first* moment,

$$M_1 = \int_0^\infty n_d d_p \, d(d_p) \tag{1·9}$$

when divided by the zeroth moment gives the number average particle diameter,

$$\bar{d}_p = \frac{\int_0^\infty n_d d_p \, d(d_p)}{\int_0^\infty n_d \, d(d_p)} = \frac{M_1}{M_0} \tag{1·10}$$

The *second* moment is proportional to the surface area of the particles composing the disperse system:

$$\pi M_2 = \pi \int_0^\infty n_d d_p^2 d(d_p) = A \tag{1·11}$$

where A is the total surface area per unit volume of fluid in the disperse system. The average surface area per particle is given by $\pi M_2 / M_0 = A / N_\infty$.

The *third* moment is proportional to the total volume of the particles:

$$\frac{\pi M_3}{6} = \frac{\pi}{6} \int_0^\infty n_d d_p^3 d(d_p) = V \tag{1·12}$$

where V is the volume fraction of dispersed material in the fluid, cubic centimeters of material per cubic centimeters fluid, for example. If the particle density is independent of size, the third moment is proportional to the mass concentration of particulate matter. The average volume of a particle is defined by

$$\bar{v} = \frac{V}{N_\infty} = \frac{\pi M_3}{6 M_0} \tag{1·13}$$

The *fourth* moment is proportional to the total surface area of the material sedimenting from a stationary fluid. For spherical particles larger than 1 μm, the terminal settling velocity is given to a good approximation by

$$c_s = \frac{\rho_p d_p^2 g}{18\mu} \tag{1·14}$$

where ρ_p is the particle density, g is the gravitational constant, and μ is the fluid viscosity. Hence the rate at which a horizontal surface is covered by settling particulate matter is given by

$$\int_0^\infty \left(\frac{\pi d_p^2}{4} \right) \left(\frac{\rho_p d_p^2 g}{18\mu} \right) n_d(d_p) d(d_p) = \frac{\pi \rho_p g}{72\mu} M_4 \tag{1·15}$$

with c.g.s. units of $\sec^{-1}$.

The *fifth* moment is proportional to the mass flux of material sedimenting from a fluid (g/cm^2 sec in c.g.s. units):

$$\int_0^\infty n_d c_s \frac{\pi d_p^3}{6} d(d_p) = \frac{\pi \rho_p^2 g}{108\mu} M_5 \tag{1·16}$$

The *sixth* moment is proportional to the total light scattering by particles when they are much smaller than the wavelength of the incident light. The scattering efficiency of a small, single spherical particle or fraction of the incident light scattered is (Chap. 5)

$$K_{scat} = \frac{8}{3}\left(\frac{\pi d_p}{\lambda}\right)^4 \left|\frac{m^2-1}{m^2+2}\right|^2 \qquad (d_p << \lambda)$$

where m is the refractive index, λ is the wavelength of incident light, and I is the intensity of incident light. Then the total scattering by an aerosol composed of very small particles ($d_p << \lambda$) is given by

$$b_{scat} = \frac{2}{3}\frac{\pi^5}{\lambda^4}\left(\frac{m^2-1}{m^2+2}\right)^2 I \int_0^\infty n_d d_p^6 d(d_p) \sim M_6 \tag{1·17}$$

Rayleigh scattering usually does not contribute significantly to light scattering by small particles in the atmosphere. Most of the scattering

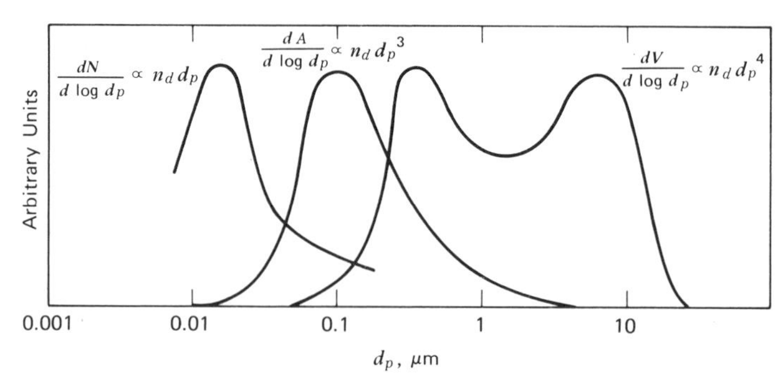

Fig. 1·8 The main contributions to different moments of the particle size distribution function come from different particle size ranges. For example, the area under the curve representing $dN/d\log d_p$ is proportional to the total number concentration; most of the contribution comes from the size range near 0.01 μm. The volume distribution, which is proportional to $n_d d_p^4$ by (1·5) and (1·6) sometimes shows several modes with peaks above and below 1 μm. The lower mode results from gas-to-particle conversion and the upper mode from the break-up of solids or the resuspension of dusts.

by atmospheric particles occurs in the size range $d_p \sim \lambda$, for which much more complex scattering laws (Mie theory) must be used (Chap. 5).

Different parts of the particle size distribution function make controlling contributions to the various moments. In a polluted urban atmosphere, the number concentration or zeroth moment is often dominated by the 0.01 to 0.1 μm size range, and the surface area, by the 0.1 to 1.0 μm range; contributions to the volumetric concentration come from both the 0.1 to 1.0 and 1.0 to 10 μm size ranges. An example is given in Fig. 1·8.

Moments of fractional order appear in the theory of particle diffusion to surfaces (Chap. 3).

1·7 Chemical Composition: Size–Composition Probability Density Function

A single aerosol particle may be composed of many chemical compounds, but chemical analysis is difficult because the amount of material involved is so small. Some success in measuring the elemental composition of individual particles has been achieved with the electron microprobe (Chap. 6), but identification of chemical compounds in individual particles is usually beyond our capabilities.

Aerosol particles in the same size range may have different chemical compositions because of variations in temperature or concentration at the emission source or as a result of mixing of aerosols from different sources.

To take into account variations in chemical composition from particle to particle, the particle size distribution function must be generalized, and for that purpose the *size–composition probability density function* has been introduced. Let dN be the number of particles per unit volume of gas containing molar quantities of each chemical species in the range between n_i and $n_i + dn_i$ with $i = 1, 2, \ldots, k$ where k is the total number of chemical species. It is assumed that in each size range v to $v + dv$, the chemical composition of the particles is distributed continuously. Then the size–composition probability density function (p.d.f.), g, is defined by the relation (Friedlander, 1970, 1971):

$$dN = N_\infty g(v, n_2 \cdots n_k, \mathbf{r}, t)\, dv\, dn_2 \cdots dn_k \qquad (1\cdot18)$$

It is not necessary to include n_1 as one of the independent variables because of the relationship between v and n_i:

$$v = \sum_i n_i \bar{v}_i$$

where $\bar{v}_i$ is the partial molar volume of species i. This description is adequate so far as chemical composition is concerned. However, it does not account for structural effects, such as particle surface layers and the morphological characteristics of agglomerates.

Since the integral of dN over all v and n_i is equal to N_∞,

$$\int_v \cdots \int_{n_k} g(v, n_2 \cdots n_k, \mathbf{r}, t)\, dv\, dn_2 \cdots dn_k = 1$$

Also, the size distribution function can be found from g by integrating over all of the chemical constituents of the aerosol:

$$n(v, \mathbf{r}, t) = N_\infty \int_{n_2} \cdots \int_{n_k} g(v, n_2 \cdots n_k, \mathbf{r}, t)\, dn_2 \cdots dn_k$$

1·8 Average Chemical Concentrations

Filtration is the most common method of sampling particulate matter for the determination of chemical composition. If the filter is perfectly efficient, it provides information on the chemical composition of the aerosol averaged over all particle sizes and over the time interval of sampling. For a constant gas-sampling rate, the concentration of species i averaged over particle size and time is related to the size–composition probability density function as follows:

$$\bar{\rho}_i = \frac{M_i}{t} \int_0^t N_\infty \int_0^\infty \left[\int \cdots \int g n_i\, dn_2 \cdots dn_i \cdots dn_k \right] dv\, dt' \quad (1 \cdot 19)$$

where $\bar{\rho}_i$ is the mass of species i per unit volume of gas averaged over time, and M_i is the molecular weight of species i. The concentration of species i in the particulate phase is given by the mass fraction, usually defined as follows:

$$\bar{\bar{c}}_i = \frac{\bar{\rho}_i}{\bar{\rho}}$$

where $\bar{\rho}$ is the total mass of particulate matter per unit volume of air. As an example, over a certain time period the average concentration of particulate sulfur in the Los Angeles atmosphere is 15 $\mu g/m^3$, reported as sulfate, and the total mass of particulate matter is 125 $\mu g/m^3$. Thus $\bar{\bar{c}}_{SO_4} \approx 0.12$. Such information is reported on a routine basis for many chemical constituents of the atmospheric aerosol (EPA, 1971).

1·9 Chemical Composition: Distribution with Respect to Particle Size

It is currently not possible to obtain information on a routine basis concerning the composition of individual particles, sufficient to determine the size–composition p.d.f. However, the average composition of the particle in a discrete size interval between v_1 and v_2 can be measured at a fixed point over a finite time interval, by using a cascade impactor (Chapter 6). The concentration measured in this way is related to $g(v, n_2, \ldots, n_k, \mathbf{r}, t)$ as follows:

$$\overline{\Delta\rho_i} = \frac{M_i}{t}\int_0^t N_\infty \int_{v_1}^{v_2}\left[\int \cdots \int g n_i \, dn_2 \cdots dn_i \cdots dn_k\right] dv\, dt' \quad (1\cdot 20)$$

where $\overline{\Delta\rho_i}$ is the mass of species i per unit volume of air in the size range v_1 to v_2. For $v_1 \to 0$ and $v_2 \to \infty$, $\overline{\Delta\rho_i}$ becomes the concentration that would be measured by a total filter.

The discrete mass distribution can be plotted in the form $\overline{\Delta\rho_i}/\Delta \log d_p$ versus $\log d_p$. An example of a distribution of this type is shown in Fig. 1·9 for particulate sulfur, reported as sulfate. The area

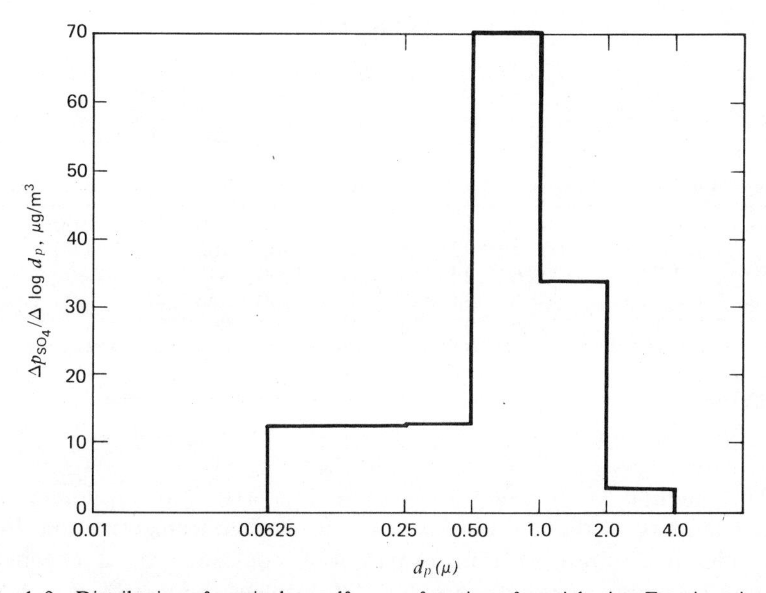

Fig. 1·9 Distribution of particulate sulfur as a function of particle size. Fractionation by cascade impactor followed by vaporization analysis using flame photometry. Measurements made at California State University, Dominguez Hills, Calif., October 5, 1973, 0840–0940 PST, downwind from petroleum refinery and chemical plant complex. Arbitrary lower cut-off set at 0.0625 μm.

under the histogram between two values of d_p is proportional to the mass of the species in that size range.

Data on chemical composition as a function of particle size can be reported in other ways. A common method is to plot the fraction of the total mass of chemical species larger (or smaller) than a given size, often on log-probability paper, in the hope that the substance will be log-normally distributed and representable by a straight line. From a plot of this kind, it is possible to determine the chemical species mass median diameter, that is, the particle size for which half the mass of the species is larger and half smaller.

If the total mass or volume of particulate matter in a given size range is known, together with the corresponding mass of a chemical species, the concentration of the species (mass fraction or mass per unit volume) can be reported as a function of size. In Table 1·2, mass fractions for certain trace metals present in flyash are shown as a function of particle size. This group of metals tends to concentrate in the smaller size range as shown in the table. Other metals (not shown) showed less tendency to accumulate, or none at all.

TABLE 1·2

AIRBORNE FLYASH ELEMENTS SHOWING PRONOUNCED CONCENTRATION TRENDS WITH RESPECT TO PARTICLE SIZE (Davison et al., 1974)

Particle diameter (μm)	Pb	Ti	Sb	Cd	Se	As	Ni	Cr	Zn	S
	μg/g									wt.%
>11.3	1100	29	17	13	13	680	460	740	8100	8.3
7.3–11.3	1200	40	27	15	11	800	400	290	9000	—
4.7–7.3	1500	62	34	18	16	1000	440	460	6600	7.9
3.3–4.7	1550	67	34	22	16	900	540	470	3800	—
2.1–3.3	1500	65	37	26	19	1200	900	1500	15,000	25.0
1.1–2.1	1600	76	53	35	59	1700	1600	3300	13,000	—
0.65–1.1	—	—	—	—	—	—	—	—	—	48.8

This method of presentation is sometimes useful in developing an understanding of the mechanism of gas-to-particle transformation. For example, if the rate of transformation is controlled by a chemical reation occurring in polydisperse aerosol droplets, the rate of growth of each droplet will be proportional to the droplet volume; the concentration of the chemical species converted will be independent of particle size.

If the dispersed phase is a binary solution, such as sulfuric acid and water, with the solvent in equilibrium with the gas phase, all droplets will have the same composition regardless of how sulfuric acid is distributed with respect to size. This is not true for very small droplets when the Kelvin effect becomes important.

The distribution of chemical composition with respect to particle size determines the pattern of deposition of different chemical species in the lung, hence health effects. The nucleation of water vapor in the atmosphere depends on both size and chemical nature.

Average concentrations may give a misleading impression of particulate effects, however. In the case of lung deposition, for example, the local dosage to tissue may depend on the composition of individual particles. The nucleating properties of a particle for the condensation of water and other vapors also depends on the chemical nature of the individual particles.

Problems

1. Derive an expression for the rate at which the particles of a polydisperse aerosol settle on a horizontal plate from a stagnant gas. Dimensions are number per unit time per unit area. Assume that Stokes law holds for the terminal settling velocity, and express your answer in terms of the appropriate moments.

2. An aerosol initially has a number distribution function $n_0(d_p)$.

(*a*) This aerosol is mixed with a volume of filtered air q times the original volume. What is the distribution function of the resulting aerosol?

(*b*) The original aerosol is composed of sea salt droplets that are 3.5% salt by weight. After mixing with equal volumes of dry, filtered air, all of the water evaporates. What is the new distribution function? (Neglect the gas volume of the water evaporated.)

(*c*) The original aerosol loses particles from each size range at a rate proportional to the particle surface area. Derive an expression for the distribution function as a function of time and particle diameter.

In all cases, express your answer in terms of the initial distribution and any other variables that may be necessary. Define all such variables.

3. The size distribution of atmospheric aerosols sometimes follow a power law of the form:

$$n_d(d_p) = \text{const}\frac{V}{d_p^4}$$

where V is the volume fraction of dispersed material (volume aerosol per unit volume air). When applicable, this form usually holds for particles ranging in size from about 0.1 to 5 μm. Show that when aerosols that follow this law are mixed, the resulting size distribution obeys a law of the same form. Before mixing, the aerosols have different volume fractions.

4. A certain chemical species is adsorbed by the particles of an aerosol. The amount adsorbed is proportional to the surface area of the particle. Derive an expression for the distribution of the species with respect to particle size expressed as mass of the species per unit volume of gas in the size range v to $v + dv$. Express your answer in terms of v and $n(v)$. Define any constants you introduce.

5. The size distribution function $n_d(d_p) \approx \Delta N/\Delta d_p$ for the Pasadena aerosol averaged over the measurements in August and September 1969 is shown in the attached table.

(*a*) Determine the mean particle diameter in micrometers.

(*b*) Determine the mass median particle diameter, that is, the diameter for which the mass of the larger particles is equal to the mass of smaller particles.

(*c*) Estimate the total surface area of particulate matter per unit volume of air corresponding to this distribution.

PARTICLE SIZE DISTRIBUTION FUNCTIONS
AVERAGED OVER MEASUREMENTS MADE IN PASADENA,
AUGUST TO SEPTEMBER 1969
(Whitby, Husar, and Liu, 1972)

d_p (μm)	$\Delta N/\Delta d_p$ (No./cm^3 μm)	$\Delta V/\Delta \log d_p$ (μm^3/cm^3)
0.00875	1.57×10^7	0.110
0.0125	5.78×10^6	0.168
0.0175	2.58×10^6	0.289
0.0250	1.15×10^6	0.536
0.0350	6.01×10^5	1.08
0.0500	2.87×10^5	2.14
0.0700	1.39×10^5	3.99
0.0900	8.90×10^4	7.01
0.112	7.02×10^4	13.5
0.137	4.03×10^4	17.3
0.175	2.57×10^4	28.9
0.250	9.61×10^3	44.7
0.350	2.15×10^3	38.6
0.440	9.33×10^2	42.0
0.550	2.66×10^2	29.2
0.660	1.08×10^2	24.7
0.770	5.17×10^1	21.9
0.880	2.80×10^1	16.1
1.05	1.36×10^1	22.7
1.27	5.82	18.6
1.48	2.88	13.6
1.82	1.25	19.7
2.22	4.80×10^{-1}	13.4
2.75	2.17×10^{-1}	15.2
3.30	1.18×10^{-1}	13.7
4.12	6.27×10^{-2}	25.3
5.22	3.03×10^{-2}	26.9

Total number of particles $= 1.14 \times 10^5$/cm^3.
Total volume of particles $= 58.1$ μm^3/cm^3.

(*d*) Many of the constituents of photochemical smog, such as free radicals, are highly reactive. Assume that such species are destroyed on striking a surface. Estimate the time for the concentration of such species to decay to one-tenth of their original value in the atmosphere if they are destroyed and not replaced on striking the aerosol surfaces. Assume that the rate at which molecules in a gas collide with a surface is given by the kinetic theory expression,

$$\beta = c\left(\frac{RT}{2\pi M}\right)^{1/2}$$

where M is the molecular weight of colliding species $= 100\,\mathrm{g/}$ mole (assumed), R is the gas constant (8.3×10^7 ergs/°K mole), T is the absolute temperature (300 °K, assumed), and c is the concentration of reactive species.

(*e*) Determine the sulfur concentration of the particles as a function of size using the data of Fig. 1·9 for the distribution of sulfur with respect to particle size. Report your result in grams of sulfur per cubic centimeter of aerosol as a function of particle size. Note that the data do not correspond to the same aerosol samples, so the calculation must be regarded as approximate.

(*f*) Show that the power law form of Problem 3 applies approximately to the data and evaluate the constant.

References

Cramer, H. (1955) *The Elements of Probability Theory*, Wiley, New York. A good introductory discussion of the mathematical properties of distribution functions.

Davison, R. L., Natusch, D. F. S., Wallace, J. R., and Evans, C. A., Jr. (1974) *Environ. Sci. Technol.*, **8**, 1107.

Environmental Protection Agency (1971) Air Quality Data for 1967, APTD 0741 Research Triangle Park, N.C.

Friedlander, S. K. (1970) *J. Aerosol Sci.*, **1**, 295.

Friedlander, S. K. (1971) *J. Aerosol Sci.*, **2**, 331.

Heichel, G. H., and Hankin, L. (1973) *Environ. Sci. Technol.*, **6**, 1121.

Roberts, P. T. (1975) Gas-To-Particle Conversion: Sulfur Dioxide in a Photochemically Reactive System, Ph.D. thesis, California Institute of Technology.

Whitby, K. T., Husar, R. B., and Liu, B.Y.H. (1972) *J. Colloid Interface Sci.*, **39**, 177.

White, W. H., and Husar, R. B. (1976) *J. Air Poll. Control Assoc.*, **26**, 32.

CHAPTER 2

Transport Properties

An understanding of particle transport or movement from one point to another in a gas is basic to the design of gas cleaning equipment and aerosol sampling instruments. The scavenging of particulate matter from the atmosphere is also determined by particle transport processes.

The classical problems of particle transport were studied by well-known physicists in the late nineteenth and early twentieth centuries. Stokes, Einstein, and Millikan investigated the motion of small spherical particles under applied forces primarily because of application to (at that time) unsolved problems in physics. They derived relatively simple relationships for spherical particles that can be considered as ideal cases; irregular particles are usually discussed in terms of their deviations from spherical behavior.

In this chapter, we consider Brownian diffusion, sedimentation, migration in an electric field, and thermophoresis. The last refers to particle movement produced by a temperature gradient in the gas. We consider also the London–van der Waals forces that are important when particles approach surfaces. Turbulent diffusion and migration resulting from fluid acceleration are discussed in later chapters.

The rate of transport of particles across a surface at a point, expressed as number per unit time per unit area, is called the *flux* at the point. Common dimensions for the flux are particles/cm^2 sec. Expressions for the diffusion flux and diffusion coefficient are derived from first principles in this chapter. The presence of an external force field acting on the particles leads to an additional term in the flux. Expressions for these terms are given without derivation in most cases but with a discussion of the conditions under which they apply and references to the derivations. For practical applications, the flux expressions are usually introduced in the equations of conservation for the particles or the equation of particle motion, in both cases for moving fluids.

2·1 Equation of Diffusion

Small particles suspended in a fluid exhibit a haphazard dancing motion resulting from the fluctuating forces exerted on them by the surrounding molecules. The motion was reported in 1827 by the botanist Robert Brown who made the first detailed studies. Brown first observed the phenomenon in aqueous suspensions of pollen and then with particles of mineral origin, showing that the motion was a general property of matter independent of its origin—organic or not. As a result of their random motion, there is a net migration of particles from regions of high to low concentration, a process known as diffusion. An equation that describes the rate of diffusion can be derived in the following way:

A fluid, which is not flowing, contains small particles in Brownian motion. Gradients exist in the concentration of particles, all of which are of the same size. Concentrations are small, however, so that any small flows that accompany diffusion can be neglected. A balance can be carried out on the number of particles in an elemental volume of fluid $\delta x \, \delta y \, \delta z$, fixed in space (Fig. 2·1), as follows: The rate at which particles enter the elemental volume across the face $ABCD$ is

$$\left[J_x - \frac{\partial J_x}{\partial x} \frac{\partial x}{2} \right] \delta y \, \delta z$$

where J_x is the flux of particles across the face with c.g.s. units of particles/cm² sec and $\partial J_x / \partial x$ is the gradient of J_n in the x direction at the centroid of the elemental volume. The rate at which particles leave

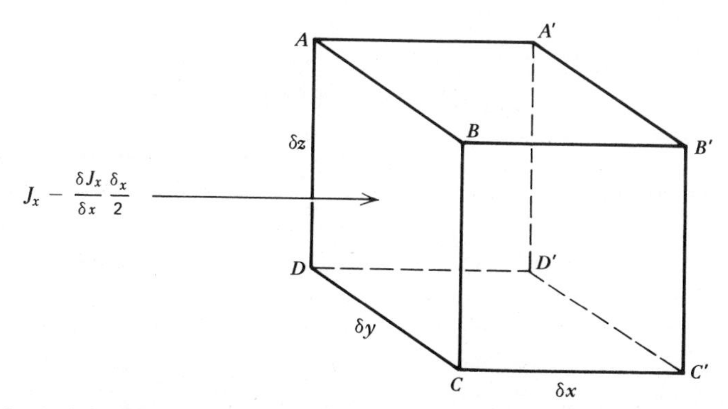

Fig. 2·1 An elemental volume of fluid, fixed in space, through which diffusion is occurring in all directions. At the centroid of the element, the particle flux in the x direction is given by J_x. The flux across one face, $ABCD$, is shown.

the elemental volume across the opposite face $A'B'C'D'$ is

$$\left[J_x + \frac{\partial J_x}{\partial x}\frac{\partial x}{2}\right]\delta y\,\delta z$$

The net rate of transport into the element for these two faces is obtained by summing the two previous expressions to give

$$-\,\delta x\,\delta y\,\delta z\frac{\partial J_x}{\partial x}$$

For the $\delta x\,\delta z$ faces, the corresponding term is

$$-\,\delta x\,\delta y\,\delta z\frac{\partial J_y}{\partial y}$$

and for the $\delta x\,\delta y$ faces

$$-\,\delta x\,\delta y\,\delta z\frac{\partial J_z}{\partial z}$$

The rate of change of the number of particles in the elemental volume $\delta x\,\delta y\,\delta z$ is given by

$$\frac{\partial n\,\delta x\,\delta y\,\delta z}{\partial t} = -\left[\frac{\partial J_x}{\partial x} + \frac{\partial J_y}{\partial y} + \frac{\partial J_z}{\partial z}\right]\delta x\,\delta y\,\delta z$$

Dividing both sides by $\delta x\,\delta y\,\delta z$, the result is

$$\frac{\partial n}{\partial t} = -\left[\frac{\partial J_x}{\partial x} + \frac{\partial J_y}{\partial y} + \frac{\partial J_z}{\partial z}\right] = -\nabla\cdot\mathbf{J} \tag{2·1}$$

which is the equation of conservation of species in terms of the flux vector, **J**, with components J_x, J_y, J_z.

The relationship between the flux and the concentration gradient depends on an experimental observation: A one-dimensional gradient in the particle concentration is set up in a fluid by fixing the concentration at two parallel planes. The fluid is isothermal and stationary. It is observed that the rate at which particles are transported from the high to the low concentration (particles/cm^2 sec) is proportional to the local concentration gradient, $\partial n/\partial x$:

$$J_x = -\,D\frac{\partial n}{\partial x} \tag{2·2}$$

where D, the diffusion coefficient, is a proportionality factor. In general, the diffusion coefficient is a variable depending on the particle size, temperature and concentration; its concentration dependence can often be neglected.

If the properties of the fluid are the same in all directions, it is said to be isotropic. This is the usual case, and D then has the same value for diffusion in any direction.

$$J_y = -D\frac{\partial n}{\partial y} \tag{2·3}$$

$$J_z = -D\frac{\partial n}{\partial z} \tag{2·4}$$

Substituting in (2·1), the result for constant D is

$$\frac{\partial n}{\partial t} = D\left[\frac{\partial^2 n}{\partial x^2} + \frac{\partial^2 n}{\partial y^2} + \frac{\partial^2 n}{\partial z^2}\right] = D\nabla^2 n \tag{2·5}$$

which is known as Fick's second law of diffusion, the first law being the linear relationship between flux and gradient.

As shown in later sections, the coefficient of diffusion is a function of particle size, small particles diffusing more rapidly than larger ones. For a polydisperse aerosol, the concentration variable can be set equal to $n_d(d_p,\mathbf{r},t)d(d_p)$ or $n(v,\mathbf{r},t)dv$, and both sides of (2·5) divided by $d(d_p)$ or dv. Hence the equation of diffusion describes the changes in the particle size *distribution* with time and position as a result of diffusive processes. Solutions to the diffusion equation for many different boundary conditions in the absence of flow have been collected by Carslaw and Jaeger (1959).

2·2 Coefficient of Diffusion

The coefficient of diffusion, D, with dimensions of square centimeters per second is one of the important transport properties of the particles in an aerosol. An expression for D can be derived as a function of the size of the particle and the properties of the gas.

We consider diffusion in one dimension alone. Suppose that a cloud of fine particles, all the same size, is released over a narrow region around the plane corresponding to $x=0$. The concentration everywhere else in the gas is zero. With increasing time, the particles diffuse as a result of the Brownian motion. The spread around the plane $x=0$ is

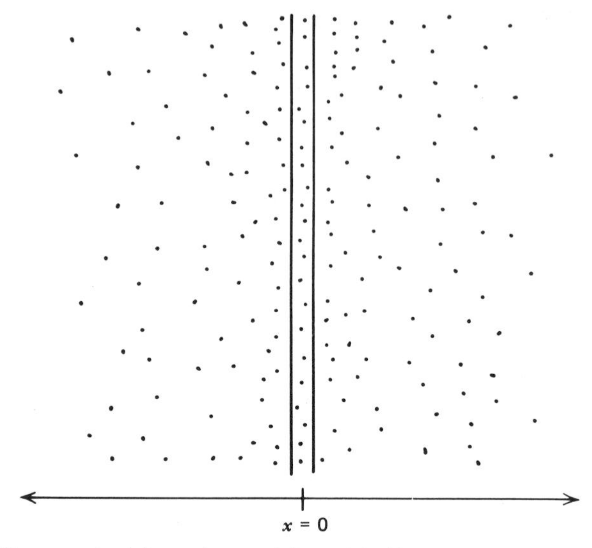

Fig. 2·2 The spread of Brownian particles originally concentrated at the differential element around $x=0$.

symmetrical (Fig. 2·2) in the absence of an external force field acting on the particles.

The spread of the particles with time can be determined by solving the one-dimensional equation of diffusion,

$$\frac{\partial n}{\partial t} = D\frac{\partial^2 n}{\partial x^2}$$

The solution for the concentration distribution is given by the Gaussian form

$$n(x,t) = \frac{N_0}{2(\pi D t)^{1/2}} \exp\frac{-x^2}{4Dt} \tag{2·6}$$

where N_0 is the number of particles released at $x=0$ per unit cross-sectional area. The mean square displacement of the particles from $x=0$ at time t is given by

$$\overline{x^2} = \frac{1}{N_0}\int_{-\infty}^{\infty} x^2 n(x,t)\,dx \tag{2·7}$$

Substituting (2·6) in (2·7), the result is

$$\overline{x^2} = 2Dt \tag{2·8}$$

Thus the mean square displacement of the diffusing particles is proportional to the elapsed time.

An expression for $\overline{x^2}$ can also be derived from a force balance on a particle in Brownian motion, which for one dimension takes the form:

$$m\frac{du}{dt} = -fu + F(t) \tag{2·9}$$

where m is the particle mass, u is the velocity, and t is the time. According to (2·9), the force acting on the particle is divided into two parts. The first term on the right is the frictional resistance of the fluid and is assumed proportional to the particle velocity. For particles much larger than the mean free path of the gas, the friction coefficient f is usually based on Stokes law:

$$f = 3\pi\mu d_p \tag{2·10}$$

Other forms for the friction coefficient are discussed in the next section.

The second term on the right of (2·9) represents a fluctuating force resulting from the thermal motion of molecules of the ambient fluid. The fluctuating force $F(t)$ is assumed to be independent of u and its mean value, $\overline{F}(t)$, over a large number of similar but independent particles vanishes at any given time. Finally, it is assumed that $F(t)$ fluctuates much more rapidly with time than u. Thus over some interval, Δt, u will be practically unchanged while there will be practically no correlation between the values of $F(t)$ at the beginning and end of the interval. These are rather drastic assumptions, but they have been justified by resort to models based on molecular theory (see the review by Green, 1952, p. 151 ff.). The conceptual difficulties attendant upon the use of (2·9) are discussed by Chandrasekhar (1943).

We now consider the group of small particles originally located near the plane $x=0$ at $t=0$ (Fig. 2·2). At a later time, these particles have wandered off as a result of the Brownian motion to form a cloud, symmetrical around $x=0$, as shown in the figure. Letting $A(t)=F(t)/m$ and multiplying both sides of (2·9) by x, the displacement from the plane $x=0$, the result for a single particle, is:

$$x\frac{du}{dt} + \beta ux = xA$$

where $\beta = f/m$. Rearranging,

$$\frac{du\,x}{dt} + \beta ux = u^2 + xA$$

since $dx/dt = u$. Integrating between $t=0$ and t, the result is

$$ux = e^{-\beta t}\int_0^t u^2 e^{\beta t'}\,dt' + e^{-\beta t}\int_0^t Axe^{\beta t'}\,dt'$$

where t' is a variable of integration that represents the time. We have made use of the fact that $ux=0$ at $t=0$. Averaging over all of the particles in the field, one finds

$$\overline{ux} = \frac{\overline{u^2}}{\beta}\left(1-e^{-\beta t}\right) \tag{2·11}$$

since $\overline{u^2}$ is constant by assumption and $\overline{Ax}=0$ since there is no correlation between the instantaneous force and the particle displacement. The following relationships also hold:

$$\overline{ux} = \overline{\frac{x\,dx}{dt}} = \overline{\frac{dx^2}{2dt}} = \frac{1}{2}\frac{d\left[\overline{x^2}\right]}{dt} \tag{2·12}$$

since the derivative of the mean over particles with respect to time is equal to the mean of the derivative. Equating (2·11) and (2·12) and integrating once more from $t=0$ to t gives

$$\frac{\overline{x^2}}{2} = \frac{\overline{u^2}\,t}{\beta} + \frac{\overline{u^2}}{\beta^2}\left(e^{-\beta t}-1\right)$$

For $t \gg 1/\beta$, this becomes

$$\frac{\overline{x^2}}{2} = \frac{\overline{u^2}\,t}{\beta} \tag{2·13}$$

We now introduce an important physical assumption, first made by Einstein, that relates the observable Brownian movement of the small particles to the molecular motion of the gas molecules: Since the particles share the molecular–thermal motion of the fluid, the principle of equipartition of energy is assumed to apply to the translational energy of the particles:

$$\frac{m\overline{u^2}}{2} = \frac{kT}{2}$$

Combining the equipartition principle with (2·8) and (2·13), the result is (Einstein, 1905):

$$D = \frac{\overline{x^2}}{2t} = \frac{kT}{f} \tag{2·14}$$

This is the Stokes–Einstein expression for the coefficient of diffusion. It relates D to the properties of the fluid and the particle through the friction coefficient discussed in the next section.

A careful experimental test of the theory was carried out by Perrin (1910). Emulsions composed of droplets about 0.4 μm in diameter were observed by means of an optical microscope and the positions of the particles were noted at regular time intervals. The Stokes–Einstein equation was checked by writing it in the form

$$N_{av} = \frac{2tRT}{3\pi\mu d_p \overline{x^2}}$$

where R is the gas constant and Avogadro's number N_{av} was calculated from the quantities on the right-hand side of the equation all of which were measured or known. The average value of Avogadro's number calculated this way was about 7.0×10^{23} in good agreement with the values determined in other ways at that time. The accepted modern value, however, is 6.023×10^{23}.

2·3 Friction Coefficient

The friction coefficient is a quantity fundamental to most particle transport processes. The Stokes law form, $f = 3\pi\mu d_p$, holds for a rigid sphere that moves through a fluid at constant velocity with a Reynolds number, $d_p U/\nu$, much less than unity. Here U is the velocity, and ν is the kinematic viscosity. The particle must be many diameters away from a surface and much larger than the mean free path of the gas molecules, which is about 0.065 μm at 25°C.

As particle size is decreased to the point where $d_p \sim l$, the drag for a given velocity becomes less than predicted from Stokes law and continues to decrease with particle size. In the range $d_p \ll l$, the free molecule range (Chap. 1), an expression for the friction coefficient, can be derived from kinetic theory (Epstein, 1924):

$$f = \frac{2}{3} d_p^2 \rho \left(\frac{2\pi kT}{m} \right)^{1/2} \left[1 + \frac{\pi\alpha}{8} \right] \tag{2·15}$$

where ρ is the gas density and m is the molecular mass of the gas molecules.

The accomodation coefficient α represents the fraction of the gas molecules that leave the surface in equilibrium with the surface. The fraction $1-\alpha$ is specularly reflected such that the velocity normal to the surface is reversed. As in the case of Stokes law, the drag is proportional to the velocity of the spheres. However, for the free molecule range, the friction coefficient is proportional to $d_p^{\,2}$, whereas in the continuum regime ($d_p \gg l$), it is proportional to d_p.

The coefficient α must, in general, be evaluated experimentally but is usually near 0.9 for momentum transfer (values differ for heat and mass transfer). The friction coefficient calculated from (2·15) is only 1% of that from Stokes law for a 20 Å particle.

An interpolation formula is often used to cover the entire range of values of d_p/l from the continuum to the free molecule regimes. It is introduced as a correction to the Stokes friction coefficient (2·10):

$$f = \frac{3\pi\mu d_p}{C} \tag{2·16}$$

where the slip correction factor C is given by

$$C = 1 + \frac{2l}{d_p}\left(A_1 + A_2 \exp\frac{-A_3 d_p}{l}\right) \tag{2·17}$$

and A_1, A_2, and A_3 are constants. Although basically empirical, the correction factor shows the proper limiting forms: For $d_p \gg l$, $C \to 1$ and f approaches (2·10), whereas for $d_p \gg l$, f approaches the form of the kinetic theory expression (2·15).

Values of the constants A_1, A_2, and A_3 are based on experimental measurements of the drag on small particles. Such measurements were made by Millikan and his students in their oil droplet experiments carried out to determine the electronic charge. A later compilation of experimental data (Davies, 1945) led to the following result: $A_1 = 1.257$, $A_2 = 0.400$, $A_3 = 0.55$. This corresponds to a value of α in (2·15) of 0.84. Values for the diffusion coefficient and settling velocity of spherical particles can be calculated over the entire particle size range using (2·14), (2·16) and (2·17). They are shown in Table 2·1 together with values of the slip correction and Schmidt number discussed in Chap. 3.

As pointed out previously, Stokes law is derived for the steady-state resistance to the motion of a particle. Why should it apply to the Brownian motion in which the particle is continually accelerated? The explanation is that the acceleration is always very small so that at each instant, a quasi-steady state can be assumed to exist.

TABLE 2·1

AEROSOL TRANSPORT PROPERTIES:
SPHERICAL PARTICLES IN AIR AT 20°C, 1 ATM

d_p (μm)	C	D (cm²/sec)	Schmidt Number, ν/D	c_s (cm/sec) ($\rho_p = 1$ g/cm³)
0.001	216.	5.14×10^{-2}	2.92	
0.002	108.	1.29×10^{-2}	1.16×10^{1}	
0.005	43.6	2.07×10^{-3}	7.25×10^{1}	
0.01	22.2	5.24×10^{-4}	2.87×10^{2}	
0.02	11.4	1.34×10^{-4}	1.12×10^{3}	
0.05	4.95	2.35×10^{-5}	6.39×10^{3}	
0.1	2.85	6.75×10^{-6}	2.22×10^{4}	8.62×10^{-5}
0.2	1.865	2.22×10^{-6}	6.76×10^{4}	2.26×10^{-4}
0.5	1.326	6.32×10^{-7}	2.32×10^{5}	1.00×10^{-3}
1.0	1.164	2.77×10^{-7}	5.42×10^{5}	3.52×10^{-3}
2.0	1.082			1.31×10^{-2}
5.0	1.032			7.80×10^{-2}
10.0	1.016			3.07×10^{-1}
20.0	1.008			1.22
50.0	1.003			7.58
100.0	1.0016			30.3

For nonspherical particles, the drag depends on the orientation of the particle as it moves through the air. When $d_p \gg l$, the drag can be calculated by solving the Stokes or "creeping flow" form of the Navier–Stokes equations for bodies of various shapes. In calculating the diffusion coefficient, it is necessary to average over all possible orientations because of the stochastic nature of the Brownian motion. This calculation has been carried out by Perrin (1936) for ellipsoids of revolution. These are bodies generated by rotating an ellipse around one of its axes. We consider an ellipse with semiaxes a and b rotated around the a axis.

For $z = b/a < 1$, the diffusion coefficient is given by:

$$\frac{D}{D_0} = \frac{z^{2/3}}{(1-z^2)^{1/2}} \ln \frac{1+(1-z^2)^{1/2}}{z}$$

and for $z > 1$,

$$\frac{D}{D_0} = \frac{z^{2/3}}{(z^2-1)^{1/2}} \tan^{-1}(z^2-1)^{1/2}$$

where D_0 is the diffusion coefficient of a sphere of the same volume as the ellipsoid. If a_0 is the radius of the sphere, then

$$a_0 = az^{2/3}$$

The results of the calculation are shown graphically in Fig. 2·3. The diffusion coefficient of the ellipsoid is always less than that of a sphere of equal volume. However, over the range $10 > z > 0.1$, the coefficient for the ellipsoid is always greater than 60% of the value for the sphere. These results are not directly applicable to the diffusion of particles suspended in a shear field, because all orientations of the particle are no longer equally likely.

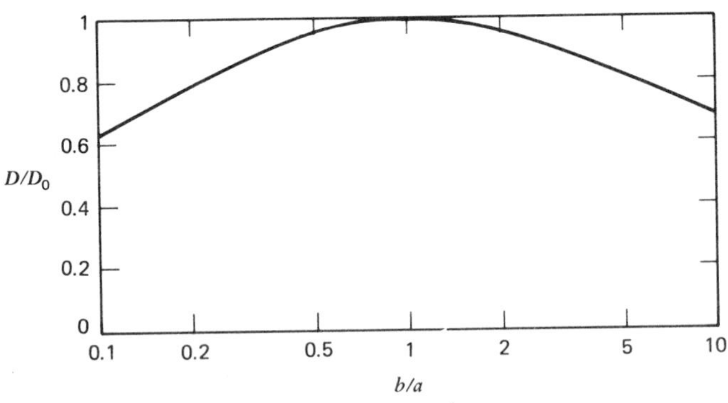

Fig. 2·3 The coefficient of diffusion for ellipsoids of revolution as a function of the ratio of the equatorial radius b to the radius of revolution a. D_0 is the coefficient of diffusion of a sphere of the same volume as the ellipsoid.

2·4 Migration in an External Force Field

The force fields of most interest in particle transport are gravitational, electrical, and thermal, the last produced by temperature gradients in the gas. If a balance exists locally in the gas between the force field and the drag on the particle, the two can be equated to give

$$\mathbf{c} = \frac{\mathbf{F}}{f} \tag{2·18}$$

where $\mathbf{c}$ is the migration or drift velocity in the field, $\mathbf{F}$ is the force, and f is the friction coefficient. For the gravitational field,

$$F = \frac{\pi d_p^3}{6}(\rho_p - \rho)\, g$$

where ρ and ρ_p are the gas and particle densities, respectively and g is the acceleration due to gravity; the migration velocity in this case is called the *terminal settling velocity*:

$$c_s = \frac{\rho_p g d_p^2}{18\mu} C\left[1 - \frac{\rho}{\rho_p}\right]$$

Usually ρ/ρ_p can be neglected in this equation. Values of c_s are given in Table 2·1.

The particle flux resulting from simultaneous diffusion and migration in an external force field can be obtained by summing the two effects to give

$$J_x = -D\frac{\partial n}{\partial x} + c_x n$$

for the one-dimensional case. In the vector form,

$$\mathbf{J} = -D\nabla n + \mathbf{c}n \tag{2·19}$$

The same result is obtained from the theory of the Brownian motion taking into account the external force field. Substituting in (2·1), the equation of conservation of species in the presence of an external force field becomes

$$\frac{\partial n}{\partial t} = \nabla\cdot D\nabla n - \nabla\cdot\mathbf{c}n \tag{2·20}$$

Solutions to this equation for constant D and c are given by Carslaw and Jaeger (1959) and Chandrasekhar (1943) for some special cases.

Example. Particles are transported through a thin layer of stationary gas above a horizontal flat plate. Derive an expression for the deposition rate in the steady state if diffusion and sedimentation are both operative.

SOLUTION. Let z be the distance from the plate measured from the surface. The one-dimensional particle transport rate is given by

$$J = -D\frac{dn}{dz} - c_s n$$

where c_s is given by (2·19). The negative sign appears because the flux, J, is positive in the direction of increasing z and c_s is positive. In the steady state, J is constant. Integrating the first-order linear equation across the gas film with the boundary conditions $n = n_b$ at $z = b$, the edge of the film, and $n = 0$ at $z = 0$, the result is

$$J = \frac{-c_s n_b}{1 - \exp(-c_s b/D)}$$

For $c_s b \gg D$, $J \rightarrow c_s n_b$, the flux is due to sedimentation alone. For $c_s b \ll D$, it is found that $J \rightarrow -Dn_b/b$, the flux due to diffusion alone, by expanding the exponential in the denominator.

In the sedimentation range, the flux increases with increasing particle size because the larger particles settle more rapidly. In the diffusion range, the flux increases with decreasing particle size because the smaller particles have a larger diffusion coefficient.

As a result, there is a minimum in the deposition flux at an intermediate particle size between the sedimentation and diffusion ranges. The particle size at which the minimum occurs can be found by setting $dJ/[d(d_p)] = 0$. As an approximation, the particle size at the minimum can be estimated by equating the sedimentation and diffusion fluxes and solving for d_p:

$$\frac{Dn_b}{b} = c_s n_b$$

We expect the minimum to occur approximately at the particle diameter for which $c_s b/D = 1$. For $b = 1$ mm, this occurs when $D/c_s = 1$ mm. From Table 2·1, $d_p \approx 0.1$ μm.

2·5 Electrical Migration: Field Charging

The force on a particle carrying n_e elementary units of charge in an electric field of intensity E is given by

$$F = n_e e E \tag{2·21}$$

where e is the electronic charge. When the electrical force is balanced by the drag, a steady migration velocity is obtained:

$$c_e = \frac{n_e e E}{f} \tag{2·22}$$

It is sometimes convenient to employ the electrical mobility,

$$Z = \frac{c_e}{E} = \frac{n_e e}{f} \tag{2·23}$$

which is the coefficient of proportionality between the migration velocity and the field intensity.

The charge acquired by a particle depends in a complex way on the ionic atmosphere, the electric field, and the particle size. In gas cleaning by electrical precipitation and in certain types of electrical mobility analyzers (Chap. 6), particles are charged by exposure to ions generated in a corona discharge. In industrial electrical precipitators, the corona is produced by discharge from a negatively charged wire. Electrons from the corona attach themselves to molecules of oxygen and other electronegative gases to form ions.

Charging by exposure to ions of one sign is called *unipolar charging*. In the atmosphere, both positive and negative ions are present, generated by cosmic rays and radioactive decay processes. Exposure to mixed ions leads to *bipolar charging* of particles. We consider only unipolar charging because of its importance in gas cleaning.

When a dielectric or conducting particle is placed in an electric field, the lines of force tend to concentrate in the neighborhood of the particle (Fig. 2·4). Particles become charged by collision with ions moving along lines of force that intersect the particle surface. This process is known as *field charging*.

As the particle becomes charged, it tends to repel additional ions of the same sign, and the distribution of electric field and equipotential lines changes. The field distribution can be calculated from electrostatic theory for the region surrounding a charged spherical particle. From the field distribution, the current flow toward the particle at any instant can be calculated. The number of electronic charges accumulated by the particle, n_e, found by integrating the current up to any time t, is given by

$$n_e = \left[\frac{\pi e Z_i n_{i\infty} t}{\pi e Z_i n_{i\infty} t + 1} \right] \left[1 + 2\frac{\epsilon_p - 1}{\epsilon_p + 2} \right] \frac{E d_p^2}{4e} \qquad (2 \cdot 24)$$

Here Z_i is the mobility of the ions, $n_{i\infty}$ is the ion concentration far from the particle, ϵ_p is the dielectric constant of the particle, and t is the time of exposure of the particle to the field.

For sufficiently long times, the charge on the particle approaches a saturation value:

$$n_{e\infty} = \left[1 + 2\frac{\epsilon_p - 1}{\epsilon_p + 2} \right] \frac{E d_p^2}{4e} \qquad (t \to \infty) \qquad (2 \cdot 25)$$

Under normal operating conditions, the limiting charge is approached after a time small compared with the time of gas treatment in a

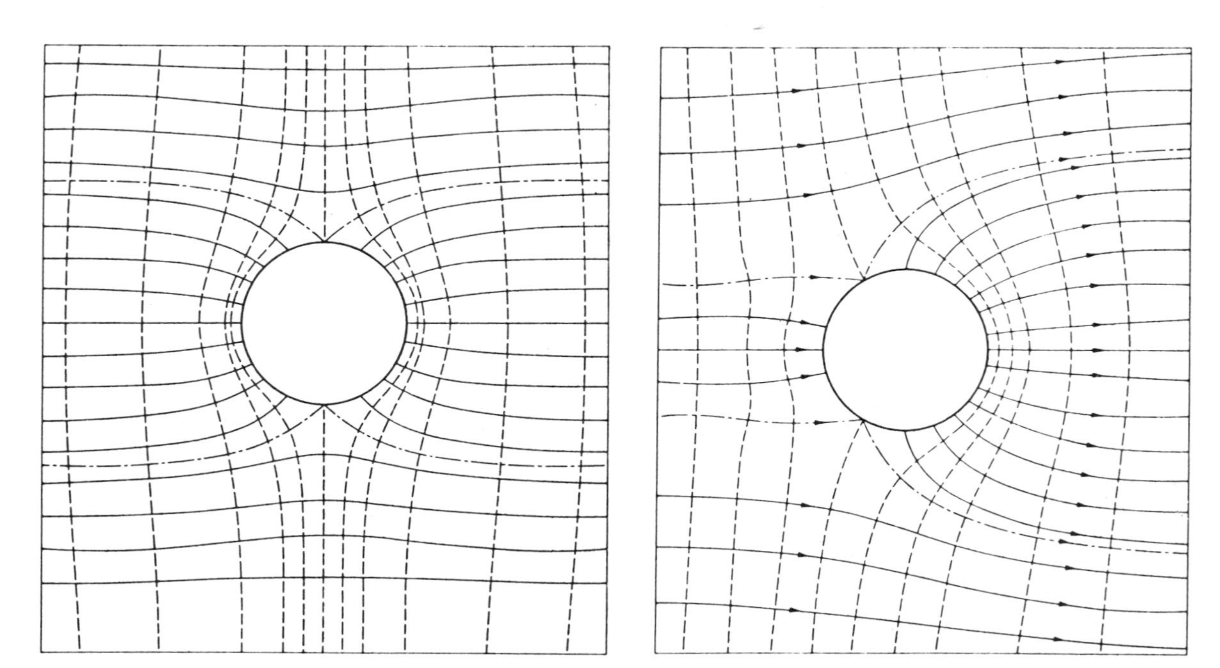

Fig. 2·4 Electric force (———) and equipotential (---------) lines around an uncharged conducting sphere in a uniform field (Fig. 2·4*a*) and around a partially charged conducting sphere (Fig. 2·4*b*) in a uniform field. Ions present in the field migrate along the electric force lines; those moving along lines that intersect the surface tend to collide with the sphere. As the particle becomes charged, the field lines become distorted in such a way that the charging process slows (White, 1963).

precipitator. Combining (2·25) with (2·22), it is found that the migration velocity for field charging increases linearly with particle size when f is given by Stokes law.

The factor $[1+2(\epsilon_p-1)/(\epsilon_p+2)]$ is a measure of the distortion of the electrostatic field produced by the particle. For $\epsilon_p=1$, there is no distortion, while for $\epsilon_p \to \infty$, the factor approaches 3, the value for conducting particles. For the usual dielectric materials, ϵ_p is less than 10, about 2.3 for benzene and 4.3 for quartz.

2·6 Electrical Migration: Diffusion Charging

In the previous section, we considered charging by the flow of ions along the lines of force in an electric field. Even in the absence of an applied electric field, particles exposed to an ion cloud become charged. Ion/particle collisions result from the thermal motion of the ions; the particle thermal motion can be neglected by comparison. This mechanism is called *diffusion charging*. An approximate expression for the charge developed by an initially uncharged particle by this mechanism can be derived in the following way: A charged aerosol particle in an ionic atmosphere produces a nonhomogeneous ion cloud in its neighborhood that is on the average spherically symmetrical. The average concentration of ions of type i, n_i, is a function of the radial distance from the center of the particle. If the charging process is not too rapid, the ion distribution can be approximately represented by the Boltzmann equilibrium law for the distribution of molecules in a force field:

$$n_i = n_{i\infty} \exp\left(\frac{-z_i e \phi}{kT} \right) \tag{2·26}$$

where z_i is the valence of the ion, ϕ is the electrostatic potential in the fluid surrounding the particle and $n_{i\infty}$ is the concentration of ions far from the particle in a region where $\phi=0$. Since the particle and ion cloud have the same charge, the ion concentration is reduced near the particle/gas interface and then rises to its maximum value $n_{i\infty}$.

The potential distribution surrounding the particle could, in principle, be determined by solving the Poisson electrostatic equation,

$$\nabla^2 \phi = -4\pi e \sum_i n_i z_i$$

with n_i given by (2·26). However, as an approximation, the effect of the charge density on ϕ is neglected. The potential at the surface is assumed to be given by the Coulomb relationship with a point charge at the center of the particle:

$$\phi = \frac{2n_e e}{d_p}$$

With (2·26), this expression determines the ionic concentration at the particle surface. The change in the particle charge is given by the rate of collision of ions with the surface of the particle; assuming that every ion that strikes the surface is captured,

$$\frac{dn_e}{dt} = n_{i\infty} \exp\left(\frac{-2e^2 n_e}{d_p kT}\right)\left(\frac{kT}{2\pi m_i}\right)^{1/2} \pi d_p^{\,2}$$

where m_i is the ionic mass and $(kT/2\pi m_i)^{1/2}$ is the mean velocity with which ions strike the particle surface, derived from kinetic theory. Integrating with the initial condition $n_e = 0$ at $t = 0$, the result is

$$n_e = \frac{d_p kT}{2e^2} \ln\left[1 + \left(\frac{2\pi}{m_i kT}\right)^{1/2} n_{i\infty} d_p e^2 t\right] \qquad (2\cdot27)$$

The charge acquired by a particle of given size depends on the product $n_{i\infty}t$. For $t \rightarrow \infty$, n_e approaches infinity logarithmically, according to (2·27). On physical grounds this cannot be true, because there are limits on the charge that a particle can carry. However, for values of $n_{i\infty}t$ encountered in practice ($\sim 10^8$ ion sec/cm^3), (2·27) gives results in qualitative agreement with the available experimental data.

Equation (2·27) holds best for particles much smaller than the mean free path of the gas molecules, but it is often used for larger particles. Equations more suitable for larger particles have been derived, (Whitby and Liu, 1966).

Example. Determine the migration velocity of a conducting 1 μm particle in an electric field with an intensity of 1 kV/cm. The ion concentration is 10^8 cm^{-3} and ion mobility 2(cm/sec)/(V/cm). These conditions approximate those in an electrical precipitator.

SOLUTION. The main point of interest are the mixed electrical and mechanical units. Following the customary practice in the field, (2·25) and (2·27) are based on electrostatic (esu) units. Thus when E is expressed in statvolts/cm and the mechanical parameters in cgs units, the charge $n_e e$ is in statcoulombs. Moreover,

$$1 \text{ coulomb} = 3 \times 10^9 \text{ statcoulombs}$$

$$300 \text{ V} = 1 \text{ statvolt}$$

The electronic charge is 1.6×10^{-19} coulomb $= 4.8 \times 10^{-10}$ statcoulomb. Substituting in (2·25) for field charging, $d_p = 10^{-4}$ cm, $E = 3.3$ statvolts/cm, $e = 4.8 \times 10^{-10}$ statcoulombs, it is found that $n_{e\infty} = 50$ electronic charges. Substituting in (2·27) for diffusion charging, $d_p = 10^{-4}$ cm, $k = 1.38 \times 10^{-16}$ ergs/°K, $T = 300$°K, $e = 4.8 \times 10^{-10}$ statcoulombs, $n_{i\infty} = 10^8$ cm^{-3} and $m_i = 5.3 \times 10^{-23}$ g (the mass of an oxygen molecule) the result is:

$$n_e = 9 \ln\left[1 + 3.9(10)^3 t\right] \text{ electronic charges}$$

For $t = 1$ sec, $n_e = 75$ electronic charges.

The migration velocity is given by (2·22). Substituting $e = 4.8 \times 10^{-10}$ statcoulombs, $E = 3.33$ statvolts/cm and $f = 3\pi\mu d_p / C$, with the appropriate value for n_e, it is found that

$$c_e \approx 0.5 \text{ cm/sec for field charging}$$

$$\approx 0.8 \text{ cm/sec for diffusion charging}$$

Field and diffusion charging are of comparable importance in this case. Theories which account for both effects simultaneously have been proposed, requiring numerical computations. Agreement of the calculated charge with some of the limited experimental data is fair (Smith and McDonald, 1975). Adding the charges calculated separately for field and diffusion charging gives somewhat poorer agreement. The total charge calculated by addition is usually less than the measured charge.

Particle migration velocities are shown in Fig. 2·5, based on (2·25) and (2·27) with (2·22). Field charging is the controlling mechanism for larger particles, whereas diffusion charging controls for smaller particles even in the presence of an applied field. For field charging the migration velocity increases linearly with particle diameter. For diffusion charging the mobility increases as particle diameter decreases because of the form of the slip correction C (2·17). The transition between the mechanisms usually occurs in the 0.1 to 1 μm range. Few experimental data are available under well-defined conditions, so these results must be regarded as semiquantitative.

For very small particles, diffusion charging theory breaks down; the number of charges per particle decreases to a value less than unity which is physically unacceptable. Only a fraction of such particles are charged, the fraction decreasing as particle size decreases. This value is usually determined experimentally.

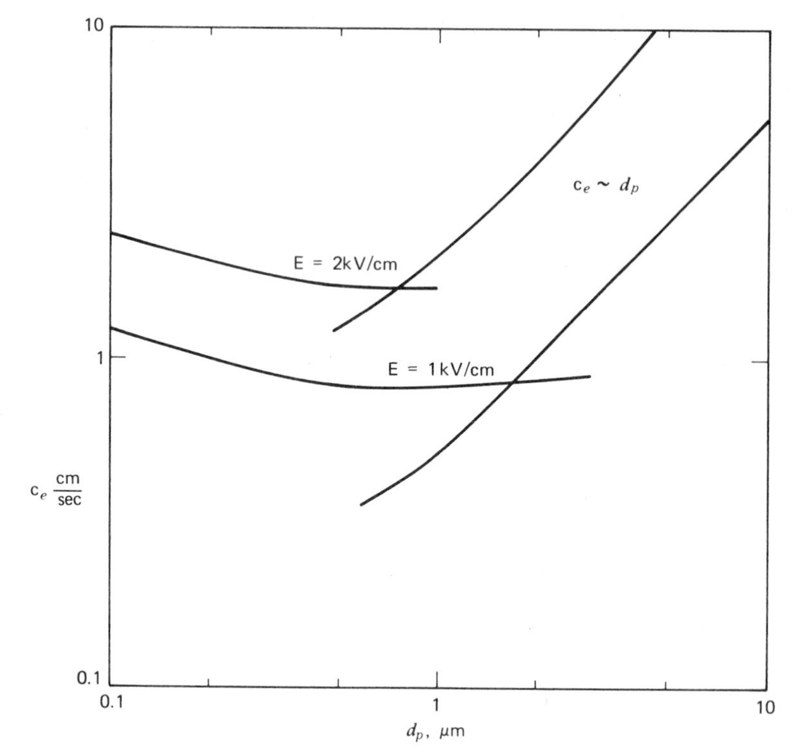

Fig. 2·5 Calculated migration velocities for $n_{i\infty}t = 10^8$ sec cm^{-3} and $T = 300°$K, based on (2·25) and (2·27). For an electric field intensity of 2 kV/cm, the transition from diffusion to field charging occurs near 0.75 μm under these conditions.

2·7 Thermophoresis

Small particles in a temperature gradient are driven from the high to low temperature regions. This effect was first observed in the nineteenth century when it was discovered that a dust-free or dark space surrounded a hot body, suitably illuminated. Particle transport in a temperature gradient has been given the name thermophoresis, literally, being carried by heat. Thermophoresis is closely related to the molecular phenomenon, thermal diffusion, separation produced by a temperature gradient in a multicomponent system.

Deposition by thermophoresis causes problems in process applications when hot gases containing small suspended particles flow over cool surfaces. For example, in petroleum refining, hot gases from fluidized beds carry particles produced by catalyst attrition and, perhaps, by condensation. When these gases pass through a heat exchanger, particles deposit on the cold surface causing scale formation and reduction of the heat transfer coefficient. Thermophoresis finds application in the sampling of small particles from gases. By choosing a proper flow geometry, the particles can be deposited on a surface for subsequent study.

For $d_p \ll l$ the mechanism of particle transport in a temperature gradient is easy to understand: Particles are bombarded by higher energy molecules on their "hot" side and thus driven toward the lower temperature zone. Their thermophoretic velocity can be calculated from the kinetic theory of gases (Waldmann and Schmitt, 1966)

$$\mathbf{c}_t = -\frac{3\nu \nabla T}{4(1+\pi\alpha/8)T} \qquad (d_p \ll l) \tag{2·28}$$

where the negative sign indicates that the motion is in the direction of decreasing temperature, ν is the kinematic viscosity, and α, the accommodation coefficient, is usually about 0.9 (Section 2·4). The thermophoretic velocity for $d_p \ll l$ is independent of particle size.

It is more difficult to explain the motion of particles that are larger than the mean free path. The explanation is based on the tangential slip velocity that develops at the surface of a particle in a temperature gradient (Kennard, 1938). This creep velocity is directed toward the high temperature side, propelling the particle in the direction of lower temperature. By solving the equation of fluid motion with a slip velocity boundary condition, Epstein derived the following equation for the thermophoretic velocity:

$$\mathbf{c}_t = -\frac{2\kappa\sigma}{2\kappa+\kappa_p}\left(\frac{\kappa}{p}\right)\nabla T \qquad (d_p \gg l) \tag{2·29}$$

where κ and κ_p are the thermal conductivities of the gas and particle, respectively, and σ is a dimensionless coefficient that appears in the relationship between the slip velocity at the particle surface and the temperature gradient in the gas, ∇T:

$$\mathbf{v}_{slip} = \sigma \frac{\kappa}{p} \nabla T$$

The value of σ, based on theory, is usually taken to be 1/5. The thermophoretic velocity (2·29) is also independent of particle size, although in general it differs from (2·28) for $d_p \ll l$.

Thermophoretic velocities have been measured for single particles suspended in a Millikan type cell with controlled electrical potential and temperature gradients. Particle diameters are usually larger than about 0.8 μm, for convenient optical observation. Data from experiments of this type are shown in Fig. 2·6, which illustrates how the transition takes place from the range of very small (2·28) to very large (2·29) particles. It is assumed that the dimensionless thermophoretic velocity, $-4Tc_t/15\nu(dT/dx)$, is a function only of d_p/l for a given particle–gas system.

Equation (2·29) is in satisfactory agreement with experiment when the ratio of particle-to-gas thermal conductivities is not too high, less than about 10, as for oil droplets. For particles of higher thermal conductivity, such as sodium chloride with $\kappa_p/\kappa \sim 100$, this expression predicts thermophoretic velocities that are much too small. For such particles, there is still uncertainty concerning the correct theoretical form of the thermophoretic velocity (Hidy and Brock, 1970).

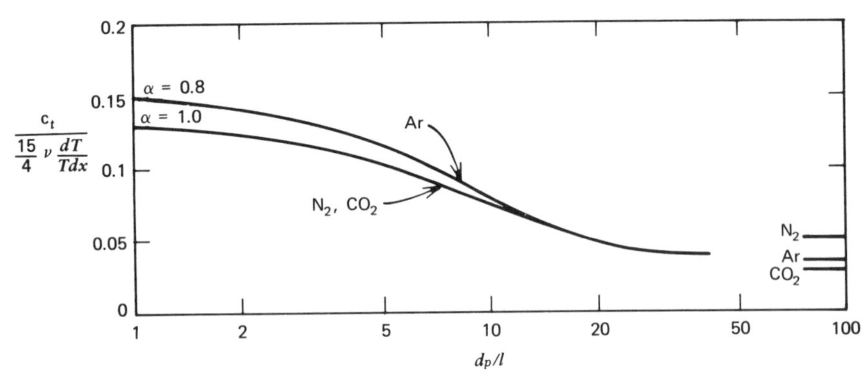

Fig. 2·6 Transition of the reduced thermophoretic velocity for silicone oil droplets from the free molecule to the continuum regions, based on the experimental results of Waldmann and Schmitt (1966). The lines are approximate representations of the data for different carrier gases. For $d_p/l \to 0$, the data approach free molecule theory (2·28); differences among the carrier gases can be explained in part by variation of the accommodation coefficient α. For $d_p/l \gg 1$, the data approach values calculated from continuum theory (2·29) but without the small variations expected among the different gases. Agreement between theory and experiment is considered good for these low-thermal conductivity droplets ($\kappa_p = 3.5 \times 10^{-4}$ cal/cm-sec-°K). Semitheoretical expressions for the variation of the reduced thermophoretic velocity in the transition zone have been derived (Hidy and Brock, 1970).

2·8 London–van der Waals Forces and the Boundary Conditions for Particle Diffusion

Gravitational, electrical, and thermophoretic forces act over distances large compared with particle size. London–van der Waals forces, which are attractive in nature, act over very short distances falling rapidly to zero away from a surface. These are the forces responsible for the effects of surface tension and for deviations from the ideal gas law (Chu, 1967). According to the theory of their origin, electrically neutral and symmetrical atoms (or molecules) have instantaneous dipoles resulting from fluctuations in the electron clouds surrounding the nucleus. These instantaneous dipoles induce dipoles in neighboring atoms or molecules. The resulting attractive forces can be calculated from quantum theory and are found to be proportional to r^{-7} with r the separation distance between the interacting atoms or molecules. As a good approximation, it can be assumed that the forces act independently of the presence of surrounding atoms and are, therefore, not influenced by the medium through which they are transmitted. Then the forces are additive and the energy of attraction between two volume elements dv_i and dv_j can be expressed as follows:

$$d\Phi = -\frac{Q\,dv_i\,dv_j}{r^6} \tag{2·30}$$

where the constant, Q, depends on the nature of the atoms or molecules involved and has the dimensions of an energy. This equation can be integrated for interacting bodies of various shapes.

For the case of a spherical particle in the neighborhood of an infinite mass bounded by a flat surface (Fig. 2·7) the interaction energy is given by

$$\Phi = -\frac{\pi^2 Q}{12}\left(\frac{1}{x} + \frac{1}{1+x} + 2\ln\frac{x}{1+x}\right) \tag{2·31}$$

where $x = s/d_p$, and s is the distance of closest approach of the particle to the surface. The negative sign indicates that the energy is attractive. As the particle approaches the surface, $x \to 0$ and the interaction energy becomes

$$\Phi = -\frac{\pi^2 Q d_p}{12 s} \qquad (s \to 0) \tag{2·32}$$

Thus the energy of attraction becomes infinite as the particle approaches the surface. For this reason, it is usually assumed that a surface acts as a perfect sink in the theory of aerosol diffusion; the boundary condition adopted is that the concentration vanishes at a distance of one particle radius from the wall.

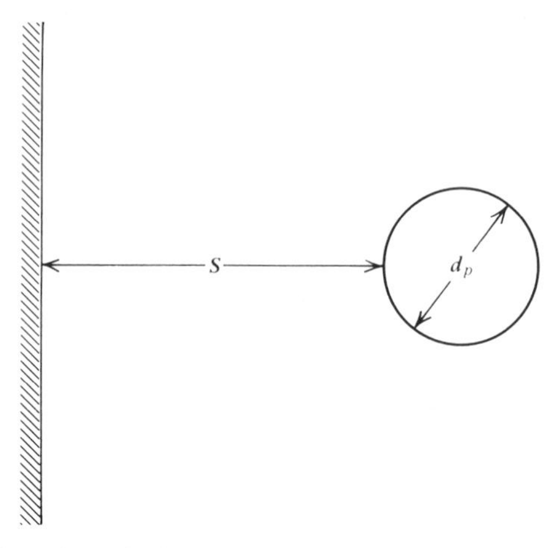

Fig. 2·7 Interaction of a spherical particle with an infinite mass bounded by a flat surface.

The range of operation of the dispersion forces can be estimated by comparing the ratio of the dispersion energy to the thermal energy, $|\Phi/kT|$. For $s/d_p \rightarrow 0$ by (2·32),

$$\left|\frac{\Phi}{kT}\right| = \left|\frac{\pi^2 Q d_p}{12kTs}\right|$$

Values of $\pi^2 Q$ often range between 10^{-13} and 10^{-12} ergs. For $T = 25°\text{C}$ and taking the smaller value of $\pi^2 Q$,

$$s < \frac{\pi^2 Q d_p}{12kT} = \frac{10^{-13} d_p}{12(1.38 \times 10^{-16})298} = 0.2 d_p$$

which satisfies the requirement $s/d_p \rightarrow 0$ on which (2·32) is based. For $d_p = 0.1$ μm, $s < 200$ Å.

The dispersion forces thus operate over a region very near the wall. Since the forces are attractive, they tend to increase the rate of particle diffusion to the surface. When the diffusion path is long compared with the range of operation of the dispersion forces, the attractive effects can be neglected (Fig. 2·8). The sink boundary condition is retained, however.

The particle flux can be calculated by solving the diffusion equation

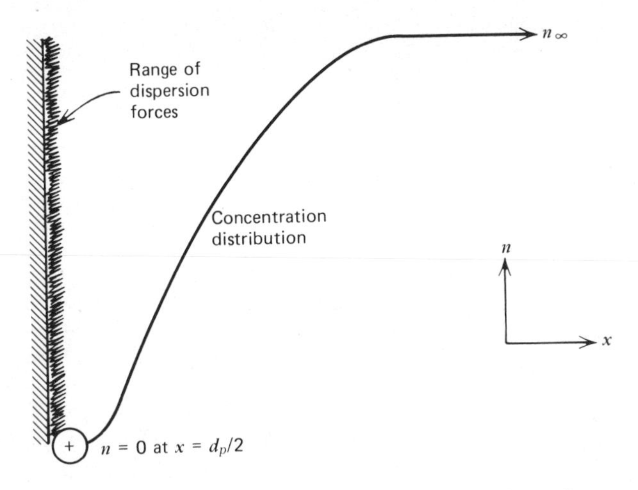

Fig. 2·8 Diffusion path or distance from mainstream of the gas is much greater than the range of the dispersion forces.

in the absence of an external force field with the condition

$$n = 0 \text{ at } x = \frac{d_p}{2}$$

The particle flux is given by

$$J = -D\left[\frac{\partial n}{\partial x}\right]_{x = d_p/2}$$

Finally, when d_p is also much smaller than the diffusion path, these conditions become

$$n = 0 \text{ at } x = 0 \quad \text{(on the wall)}$$

and the particle flux is given by

$$J = -D\left[\frac{\partial n}{\partial x}\right]_{x = 0}$$

This is the usual boundary condition for molecular diffusion to surfaces in gases and liquids. Hence the results of experiment and theory for molecular diffusion can often be directly applied to particle diffusion.

Problems

1. Show that the solution (2·6) satisfies the one-dimensional diffusion equation and that the total number of particles in the gas-per-unit cross section parallel to the plane of release is conserved.

2. The first approximation to the coefficient of diffusion for a binary gas mixture of molecules that act as rigid elastic spheres is given by

$$D=\frac{3kT}{2p(d_1+d_2)^2}\left\{\frac{kT(m_1+m_2)}{2\pi m_1 m_2}\right\}^{1/2}$$

where 1 and 2 refer to the components of the mixture, d is the molecular diameter, and m is the mass. Under what circumstances does this expression reduce to kT/f with f the friction coefficient in the free molecule range ($d_p \ll l$)?

3. It is proposed to bubble an aerosol through water to remove the suspended particles. Estimate the efficiency of particle removal for the range $0.5 > d_p > 0.05$ μm for 1 mm bubbles rising through a water column 1 ft high. The temperature is 20°C. Assume that the aerosol in the bubble is initially mixed, that the bubbles behave like rigid spheres and that the bubbles come instantly to their constant rise velocity. Does this appear to be a potentially useful method for gas cleaning?

4. Derive an expression relating applied electric field potential, E, and particle diameter for which the charge acquired by field and diffusion charging are equal. Plot particle diameter as a function of E over the range 1 to 10 kiloV/cm for $Nt = 10^7$ ion sec/cm^3 at $T = 20$°C and 1 atm. The ion mobility is 2.2 cm^2/V sec and the dielectric constant of the particles is 8.0.

5. An aerosol containing 1 μm particles with a density of 2 g/cm^3 and a thermal conductivity of 3.5×10^{-4} cal/cm sec °K flows over a surface. Calculate the minimum temperature gradient at the surface necessary to prevent particle deposition by sedimentation. Neglect diffusion and assume the air flow is parallel to the surface, which is maintained at 20°C.

References

Carslaw, H. S., and Jaeger, J. C. (1959) *Conduction of Heat in Solids*, Oxford University Press, Oxford, 2nd ed. This work includes a compilation of solutions to the equation of unsteady heat conduction in the absence of flow for many different geometries, initial and boundary conditions. The basic equation is of the same form as the diffusion equation with the thermal diffusivity, $\kappa/\rho C$, in place of the diffusion coefficient. (Here κ, ρ, and C are the thermal conductivity, density, and specific heat of the continuous fluid.) Like D, the thermal diffusivity has c.g.s. dimensions of cm^2/sec.

Chandrasekhar, S. (1943) *Stochastic Problems in Physics and Astronomy*, *Reviews of Modern Physics*, **15**, 1 (Reprinted in N. Wax. (Ed.) (1954) *Noise and Stochastic Processes*, Dover, New York).

Chu, B. (1967) *Molecular Forces*: Based on the Baker Lectures of Peter J. W. Debye, Wiley-Interscience, New York. This small, concisely written book reviews the origins of the London–van der Waals forces and derives expressions for the interaction energy between bodies under the influence of these forces.

Davies, C. N. (1945) *Proc. Phys. Soc.*, **57**, 259.

Einstein, A. (1905) Ann. D Physik **17**, 549. This and related articles by Einstein have been translated and reprinted in R. Fürth, (Ed.) (1956) *Investigations on the Theory of the Brownian Movement*, Dover, New York.

Epstein, P. S. (1924) *Phys. Rev.*, **23**, 710.

Green, H. S. (1952) *The Molecular Theory of Fluids*, Interscience, New York.

Hidy, G. M., and Brock, J. R. (1970) *The Dynamics of Aerocolloidal Systems*, Pergamon, Oxford. An extensive review of the fundamentals of particle transport over a wide range of Knudsen numbers is given in this reference. Values for the accomodation coefficients collected from various sources are included.

Kennard, E. H. (1938) *Kinetic Theory of Gases*, McGraw-Hill, New York. A good discussion of the creep velocity at a particle surface resulting from a temperature gradient in the carrier gas is given on pp. 327 ff.

Perrin, F. (1936) *J. Phys. Radium*, Series 7, **7**, 1.

Perrin, J. (1910) *Brownian Movement and Molecular Reality*, Taylor and Francis, London.

Smith, W. B. and McDonald, J. R. (1975) J. Air Poll. Control Assoc., **25**, 168.

Waldmann, L., and Schmitt, K. H. (1966) Thermophoresis and Diffusiophoresis of Aerosols, Chap. VI in Davies, C. N. (Ed.) *Aerosol Science*, Academic, New York.

Wax, N. (Ed.) (1954) *Noise and Stochastic Processes*, Dover, New York. A good review of the foundations of the theory of the Brownian movement including applications to communication theory is given in this collection of papers.

Whitby, K. T., and Liu, Bi Y. H. (1966) The Electrical Behavior of Aerosols, Chap. III in Davies, C. N. (Ed.) *Aerosol Science*, Academic, New York.

White, H. J. (1963) *Industrial Electrostatic Precipitation*, Addison-Wesley, Reading, Mass. This highly readable monograph contains much practical operating data as well as discussions of fundamentals.

CHAPTER 3

Deposition by Convective Diffusion

Diffusional transport in flowing fluids is called *convective diffusion*. Particle deposition on surfaces by this mechanism is of fundamental importance to the functioning of gas-cleaning equipment, such as scrubbers and filters, as well as measurement instruments, such as the diffusion battery and certain types of filters. Convective diffusion contributes to the scavenging of small atmospheric particles by raindrops, and removal by vegetation and other surfaces, and is a significant mechanism of deposition in the lung. The particle size at which convective diffusion is effective depends on velocity and external force fields, but is usually in the range $d_p < 1$ μm.

The intensity of the Brownian motion increases as particle size decreases (Chap. 2). As a result, the efficiency of collection by diffusion for particles smaller than about 0.5 μm actually increases with decreasing particle size; as shown in this chapter, certain gas-cleaning devices are most efficient for very small particles.

In what follows, the equation of diffusion derived in Chap. 2 is generalized to take into account the effect of flow. There is a difference from the usual theory of convective diffusion because of the special boundary condition: The concentration vanishes at a distance of one particle radius from the surface. This causes considerable difficulty in the mathematical theory. For point particles ($d_p = 0$), rates of convective diffusion can often be predicted from theory or from experiment with aqueous solutions because the Schmidt numbers are of the same order of magnitude.

Expressions are derived in this chapter for rates of deposition for certain geometries and flow regimes. Most of these have been experimentally verified and can be used for design calculations.

3·1 Equation of Convective Diffusion

It is sometimes possible to predict rates of deposition by diffusion from flowing fluids by analysis of the equation of convective diffusion. This equation is derived by making a material balance on an elemental volume fixed in space with respect to laboratory coordinates (Fig. 2·1). Through this volume flows a gas carrying small particles in Brownian motion.

The rate at which particles are carried by the flow into the volume element across the face $ABCD$ is given by

$$\delta y\, \delta z \left[nu - \frac{\delta x}{2} \frac{\partial (nu)}{\partial x} \right]$$

where n is the particle concentration (number per unit volume) and u is the velocity in the x direction. The rate at which particles leave the volume across the opposite face is given by

$$\delta y\, \delta z \left[nu + \frac{\delta x}{2} \frac{\partial (nu)}{\partial x} \right]$$

The net rate of particle accumulation for the flow in the x direction is given by subtracting the rate leaving from the rate entering

$$-\delta x\, \delta y\, \delta z \frac{\partial nu}{\partial x}$$

Analogous expressions are obtained for the other four faces and summing up for all three pairs, the result for the net accumulation of particles in the volume element is given by

$$-\delta x\, \delta y\, \delta z \left[\frac{\partial nu}{\partial x} + \frac{\partial nv}{\partial y} + \frac{\partial nw}{\partial z} \right] = -\delta x\, \delta y\, \delta z\, \nabla \cdot n\mathbf{v}$$

The rate of particle accumulation in the volume $\delta x\, \delta y\, \delta z$ taking into account the flow, diffusion, and external force fields (Chap. 2) is obtained by summing the three effects:

$$\frac{\partial n\, \delta x\, \delta y\, \delta z}{\partial t} = -\delta x\, \delta y\, \delta z\, \nabla \cdot n\mathbf{v} + \delta x\, \delta y\, \delta z\, \nabla \cdot (D \nabla n - \mathbf{c}n)$$

where D is the coefficient of diffusion and $\mathbf{c}$ is the particle migration velocity resulting from the external force field. Dividing both sides by the volume $\delta x\, \delta y\, \delta z$ and noting that $\nabla \cdot \mathbf{v} = 0$ for an incompressible fluid, the equation becomes

$$\frac{\partial n}{\partial t} + \mathbf{v} \cdot \nabla n = D \nabla^2 n - \nabla \cdot \mathbf{c}n \qquad (3 \cdot 1)$$

when the diffusion coefficient is constant. As in Chap. 2, this result holds both for monodisperse and polydisperse aerosols. In the polydisperse case, n is the size distribution function, and both D and $\mathbf{c}$ depend on particle size. Coagulation and growth or evaporation are not taken into account; these are discussed in later chapters.

Values of D and $\mathbf{c}$ are determined by the factors discussed in Chap. 2. The new quantity entering (3·1) is the gas velocity distribution, $\mathbf{v}$, which is determined by the fluid mechanical regime. In some cases, $\mathbf{v}$ is obtained by solving the equations of fluid motion (Navier–Stokes equations) for which an extensive literature is available (Landau and Lifshitz, 1959; Rosenhead, 1963; Schlichting, 1968). In many cases, such as atmospheric transport and in complex gas-cleaning devices, experimental data are necessary for the gas velocity. In this chapter, velocity distributions are introduced without derivation but with reference to the literature. In all cases, it is assumed that particle concentration has no effect on the velocity distribution. This is true for the low aerosol concentrations usually considered.

There is an extensive literature on solutions to (3·1) for various geometries and flow regimes. Many results are given by Levich (1962) and by Schlichting (1968) for boundary layer flows. Results for heat transfer, such as those discussed by Schlichting, are applicable to mass transfer or diffusion if the diffusion coefficient, D, is substituted for the coefficient of thermal diffusivity, $\kappa/\rho C_p$, where κ is the thermal conductivity, ρ is the density, and C_p is the heat capacity of the fluid. We consider a limited number of cases with direct applications to problems of practical interest for aerosols.

3·2 Similitude Considerations

Let us consider the flow of an incompressible fluid, infinite in extent, over a body of a given shape placed at a given orientation to the flow. This is called an *external* flow. Bodies of a given shape are said to be geometrically similar when they can be obtained from one another by changing the linear dimensions in the same ratio. Hence, it suffices to fix one characteristic length, L, to specify the dimensions of the body. This would most conveniently be the diameter for a cylinder or sphere, but any dimension will do for a body of arbitrary shape. Similar considerations apply for *internal* flows through pipes or ducts.

For an external flow, it will be assumed that the fluid has a uniform velocity, U, except in the region disturbed by the body. If the concentration in the mainstream of the fluid is n_∞, a dimensionless concentration can be defined as follows:

$$n_1 = \frac{n}{n_\infty}$$

Limiting consideration to the steady state, the equation of convective diffusion in the absence of an external force field can be expressed in dimensionless form as follows

$$\mathbf{v}_1 \cdot \nabla_1 n_1 = \frac{1}{Pe} \nabla_1^2 n_1 \tag{3·2}$$

where $\mathbf{v}_1 = \mathbf{v}/U$ and $\nabla_1 = L\nabla$. The dimensionless group LU/D is known as the Péclet (Pe) number for mass transfer.

In many cases, the velocity field can be assumed to be independent of the diffusional field. The steady isothermal flow of a viscous fluid, such as air, in a system of given geometry depends only on the Reynolds number when the velocity is low compared with the speed of sound in the fluid.

The boundary condition for particle diffusion differs from the condition for molecular diffusion because of the finite diameter of the particle. For certain classes of problems, such as flows around cylinders and spheres, the surface condition can be written

$$n = 0 \quad \text{at} \quad \frac{\alpha}{L} = \frac{a_p}{L} = R$$

where α is a coordinate measured from the surface of the body. The dimensionless ratio $R = a_p/L$ is known as the interception parameter; particles within a distance a_p of the surface would be intercepted even if diffusional effects were absent.

Hence the dimensionless concentration distribution can be expressed in the following way:

$$n_1 = f\left(\frac{\mathbf{r}}{L}, Re, Pe, R\right) \tag{3·3}$$

Two convective diffusion regimes are similar if the Reynolds, Péclet, and Interception numbers are the same.

The local rate of particle transfer by diffusion to the surface of the body is given by

$$J = -D\left(\frac{\partial n}{\partial \alpha}\right)_{\alpha = a_p} = -\frac{Dn_\infty}{L}\left(\frac{\partial n_1}{\partial \alpha_1}\right)_{\alpha_1 = R}$$

Setting the local mass transfer coefficient $k = J/n_\infty$ and rearranging, the result is

$$\frac{kL}{D} = f(Re, Pe, R) \tag{3·4}$$

The dimensionless group kL/D, the Sherwood number, is a function of the Reynolds, Péclet, and Interception numbers. Rates of particle deposition measured in one fluid over a range of values of Pe, Re, and R can be used to predict deposition rates from another fluid at the same values of the dimensionless groups. In some cases, it is convenient to work with the Schmidt number $Sc = \nu/D = Pe/Re$ in place of Pe as one of the three groups, since Sc depends only on the nature of the fluid and the suspended particles.

For $R \to 0$ ("point" particles), theories of particle and molecular diffusion are equivalent. Schmidt numbers for particle diffusion are much larger than one (Table 2·1), often of the same order of magnitude as for molecular diffusion in liquids. The principle of dimensional similitude tells us that the results of diffusion experiments with liquids can be used to predict rates of diffusion of point particles in gases, at the same Reynolds number.

For certain flow regimes, it is possible to reduce the number of dimensionless groups necessary to characterize a system by properly combining them. This further simplifies data collection and interpretation in several cases of considerable practical importance.

3·3 Concentration Boundary Layer

Flow normal to a right circular cylinder is the basic model for the theory of aerosol filtration by fibrous and cloth filters, and of particle collection by pipes and rods in a flow. The aerosol concentration at large distances from the surface is uniform; at one particle radius from the surface, the concentration vanishes.

Referring to the nondimensional equation of convective diffusion (3·2), it is of interest to examine the conditions under which the diffusion term, on the one hand, or convection, on the other, is the controlling mode of transport. The Péclet number, dU/D, for flow around a cylinder of diameter d is a measure of the relative importance of the two terms. For $Pe \ll 1$, transport by the flow can be neglected, and the deposition rate can be determine approximately by solving the equation of diffusion in a nonflowing fluid with appropriate boundary conditions.

When the Péclet number is large, the physical situation is quite different. The mainstream flow then carries most of the particles past the cylinder. In the immediate neighborhood of the cylinder, the diffusional process is important since the cylinder acts as a particle sink. Thus at high Pe, there are two different transport regions: Away from the immediate vicinity of the cylinder, convective transport by the bulk flow predominates. Near the surface, the concentration drops sharply from its value in the mainstream to zero (Fig. 3·1). This surface region is known as the *concentration (or diffusion) boundary layer*. It is in many ways analogous to the velocity boundary layer that forms around the cylinder at high Reynolds numbers, with the Péclet number serving as a criterion similar to the Reynolds number. The role of the concentration boundary layer is fundamental to understanding the transport of Brownian particles to surfaces. It is possible to determine the dependence of the thickness of the concentration boundary layer on the Péclet number as shown in the following sections. The usefulness of this concept is not limited to flows around cylinders. It applies to flows around other bodies such as spheres and wedges and to flows inside channels under certain conditions as well.

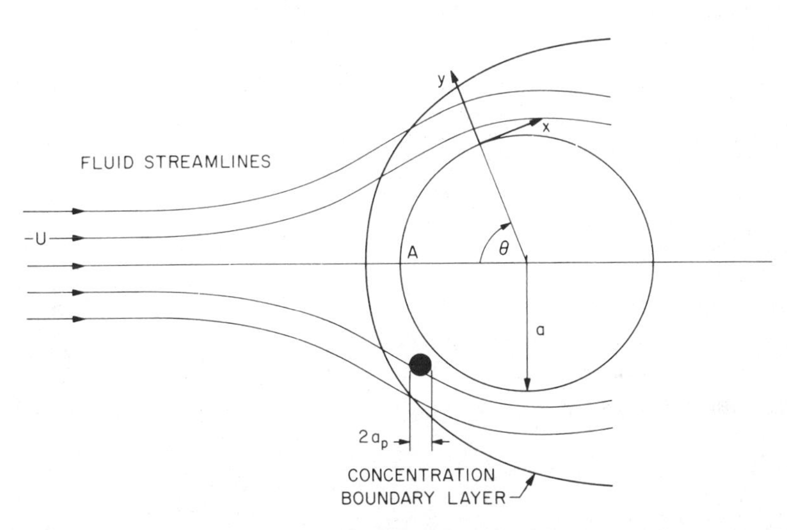

Fig. 3·1 Schematic diagram showing concentration boundary layer surrounding a cylinder (or sphere) placed in a flow carrying diffusing small particles. Curvilinear coordinate x, taken parallel to the surface, is measured from the forward stagnation point A. Particle concentration rises from zero at $r = a + a_p$ almost to the mainstream concentration (for example, to 99% of the mainstream value) at the edge of the boundary layer.

3·4 Diffusion to Cylinders at Low Reynolds Numbers: Concentration Boundary Layer Equation

Glass fiber and paper filters, often used in air cleaning, are highly porous mats of fine fibers, usually containing less than 10% solid material. The spacing between the individual fibers is much greater than the diameters of the particles filtered. As a result, in the absence of electrical effects, small particles are collected by diffusion to the fibers; larger particles are removed by inertial deposition (Chap. 4).

Filters are usually quite complex in structure with individual fibers arranged at various angles with respect to the flow. In studying their behavior, it is convenient to choose as an idealized model the single cylinder set normal to a uniform aerosol flow. The situation is illustrated in Fig. 3·1 and by the series of photographs shown in Fig. 3·2. When the fiber diameter is much greater than the mean free path of the air, continuum theory applies. The equation of convective diffusion for the steady state takes the following form in cylindrical coordinates:

$$v_\theta \frac{\partial n}{r\,\partial\theta} + v_r \frac{\partial n}{\partial r} = D\left[\frac{\partial^2 n}{\partial r^2} + \frac{1}{r}\frac{\partial n}{\partial r} + \frac{\partial^2 n}{r^2\,\partial\theta^2}\right] \tag{3·5}$$

where v_θ and v_r are the tangential and radial components of the velocity. For particles of radius a_p diffusing to a cylinder of radius, a, the boundary conditions are

$$\begin{aligned} \text{at} \quad r = a + a_p, &\quad n = 0 \\ r = \infty, &\quad n = n_\infty \end{aligned}$$

For fiber diameters smaller than 10 μm and air velocities less than 10 cm/sec, the Reynolds number is much less than unity. Near the cylinder, the stream function for the air flow can be approximated by the following expression

$$\psi = A\,Ua \sin\theta\left[\frac{r}{a}\left(2\ln\frac{r}{a} - 1\right) + \frac{a}{r}\right] \tag{3·6}$$

where $A = [2(2 - \ln Re)]^{-1}$ (Rosenhead, 1963).

Even though the Reynolds number is small, there are many practical situations in which $Pe = Re \cdot Sc$ is large because the Schmidt number, Sc, for aerosols is very large (Table 2·1). For $Pe \gg 1$, two important simplifications can be made in the equation of convective diffusion. First, diffusion in the tangential direction can be neglected in comparison with convective transport:

$$D\frac{\partial^2 n}{r^2\,\partial\theta^2} \ll \frac{v_\theta}{r}\frac{\partial n}{\partial\theta}$$

In addition, a concentration boundary layer develops over the surface of the cylinder with its thinnest portion near the forward stagnation

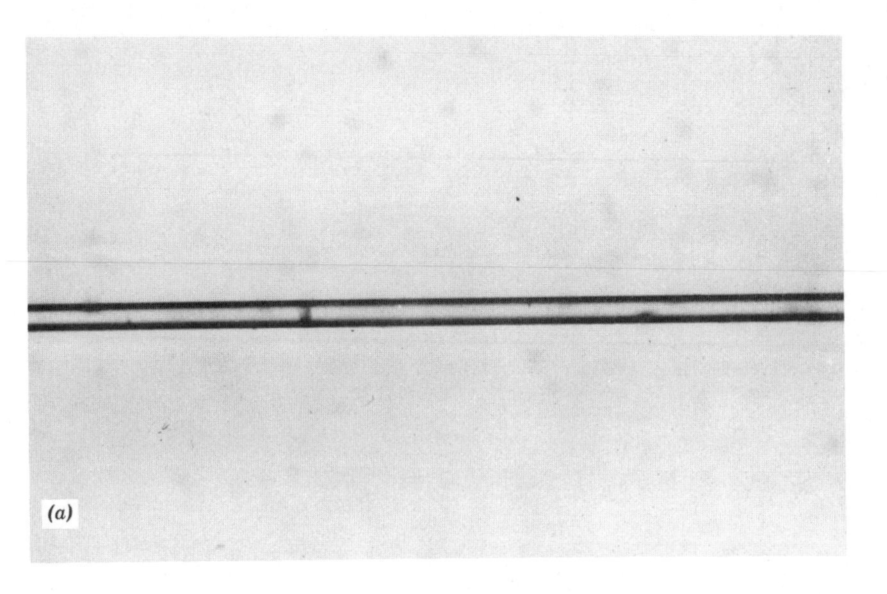

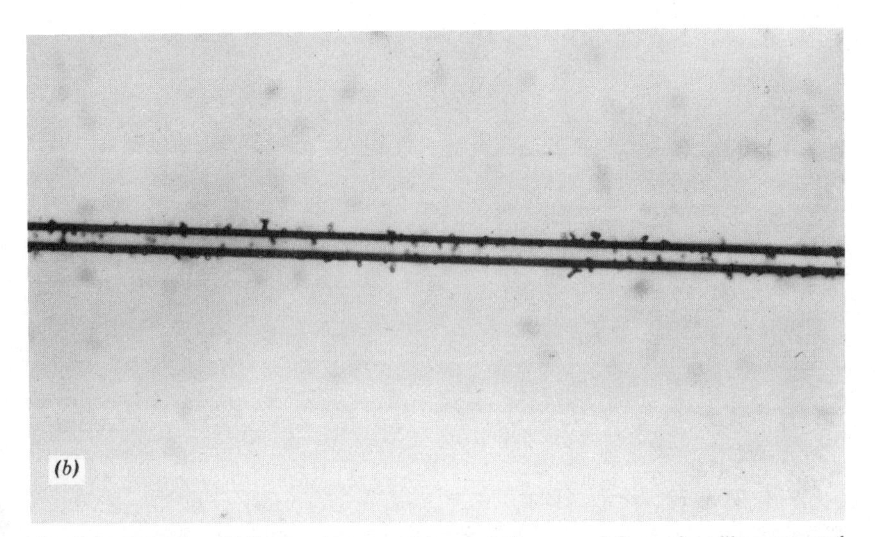

Fig. 3·2 Deposits of 1.3 μm polystyrene latex particles on an 8.7 μm glass fiber mounted normal to an aerosol flow and exposed for increasing periods of time. The air velocity was 13.8 cm/sec and the particle concentration was 1000 cm^{-3}. Photos by Dr. C. E. Billings. (A) 0 min; (B) 135 min; (C) 300 min; (D) 420 min.

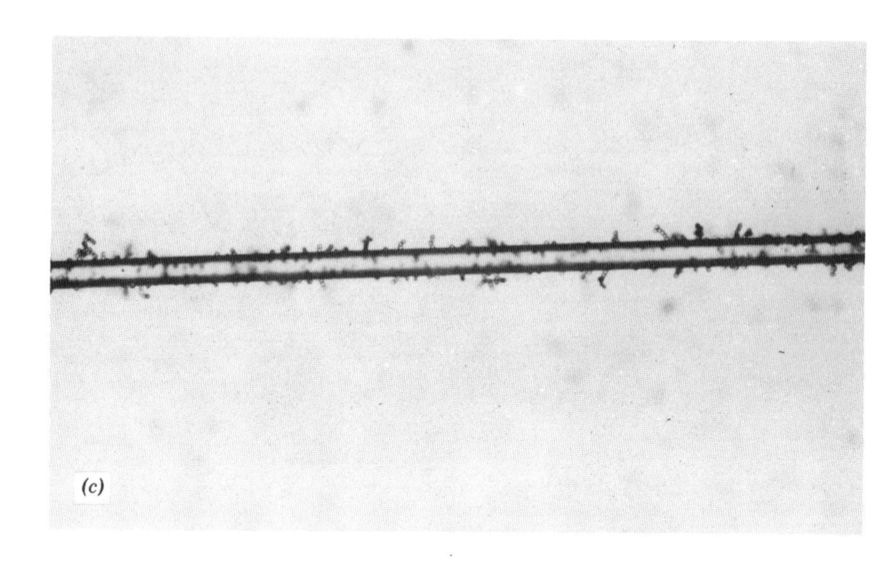

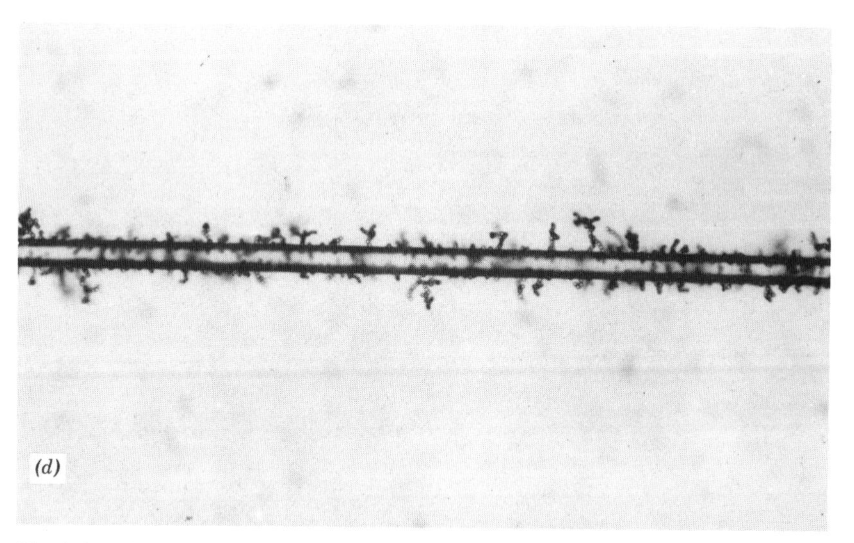

Fig. 3·2 (*Continued*)

point. When the thickness of the concentration boundary layer is much less than the radius of the cylinder, the equation of convective diffusion simplifies to the familiar form for Cartesian coordinates (Schlichting, 1968, Chap. XII):

$$u\frac{\partial n}{\partial x} + v\frac{\partial n}{\partial y} = D\frac{\partial^2 n}{\partial y^2} \tag{3·7}$$

where x and y are orthogonal curvilinear coordinates. The x-coordinate is taken parallel to the surface of the cylinder and measured from the forward stagnation point. The y-axis is perpendicular to x and measured from the surface. The velocity components u and v correspond to the coordinates x and y (Fig. 3·1).

3·5 Diffusion to Cylinders at Low Reynolds Numbers: Point Particles

First, we consider the case of point particles ($R \to 0$) for which the appropriate boundary condition is

$$\text{at} \quad y=0, \quad n=0$$
$$y=\infty, \quad n=n_\infty$$

The concentration boundary condition $n=n_\infty$ is set at $y=\infty$, even though the equation of convective diffusion (3·7) is valid only near the surface of the cylinder. This can be justified by noting that the concentration approaches, but does not quite reach, n_∞ at a distance very close to the surface for high *Pe*.

The components of the velocity are related to the stream function as follows:

$$u = \frac{\partial \psi}{\partial y}, \qquad v = -\frac{\partial \psi}{\partial x}$$

If x and ψ are taken as independent variables instead of x and y, (3·7) can be transformed into the following equation:

$$\left(\frac{\partial n}{\partial x}\right)_\psi = D\left[\frac{\partial}{\partial \psi}\left(u\frac{\partial n}{\partial \psi}\right)\right]_x \tag{3·8}$$

When the concentration boundary layer is thin, most of it falls within a region where the stream function (3·6) can be approximated by the first term in its expansion with respect to y:

$$\psi = 2AaUy_1^2 \sin x_1 \tag{3·9}$$

where $y_1 = y/a$ and $x_1 = x/a$. The x component of the velocity in this region is given by

$$u = \frac{\partial \psi}{\partial y} = \left(\frac{8AU}{a}\right)^{1/2} \sin^{1/2} x_1 \psi^{1/2}$$

Substitution in (3·8) gives

$$\frac{\partial n}{\partial \chi} = \frac{D}{aAU} \frac{\partial}{\partial \psi_1}\left(\psi_1{}^{1/2} \frac{\partial n}{\partial \psi_1}\right) \tag{3·10}$$

where

$$\chi = \int \sin^{1/2} x_1 \, dx_1$$

and

$$\psi_1 = \frac{\psi}{2AaU}$$

The boundary conditions in the transformed coordinates become

$$\begin{aligned} \text{at} \quad & \psi_1 = 0 \text{ (surface of cylinder)}, & n = 0 \\ & \psi_1 = \infty, & n = n_\infty \end{aligned}$$

By inspection of (3·10), we assume as a trial solution that n is a function only of the variable $\xi = \psi_1 / \chi^{2/3}$. This assumption must be checked by substitution of the expressions

$$\frac{\partial n}{\partial \chi} = -\frac{2}{3} \frac{\xi}{\chi} \frac{dn}{d\xi}$$

and

$$\frac{\partial n}{\partial \psi_1} = \frac{1}{\chi^{2/3}} \frac{dn}{d\xi}$$

in (3·10). The result of the substitution is an ordinary differential equation:

$$-\frac{APe}{3} \xi \frac{dn}{d\xi} = \frac{d}{d\xi}\left(\xi^{1/2} \frac{dn}{d\xi}\right) \tag{3·11}$$

with the boundary condition $n = 0$ at $\xi = 0$ and $n = n_\infty$ at $\xi \to \infty$. This supports the assumption that n is a function only of the variable ξ. Integration of (3·11) gives the following result for the concentration distribution:

$$n = \frac{n_\infty \int_0^{\xi^{1/2}} \exp\left(-\frac{2}{9} A Pe z^3\right) dz}{\int_0^\infty \exp\left(-\frac{2}{9} A Pe z^3\right) dz} \tag{3·12}$$

where $Pe = dU/D$ with d the diameter of the cylinder. The integral in the denominator can be expressed in terms of a gamma function, Γ, as

$$\left(\frac{9}{2}\right)^{1/3} \frac{1}{3} \Gamma\left(\frac{1}{3}\right) (A Pe)^{-1/3} = 1.45 (A Pe)^{-1/3}$$

The rate of diffusional deposition per unit length of cylinder is given by

$$2D \int_0^\pi \left(\frac{\partial n}{\partial y_1}\right)_{y_1=0} dx_1 = k_{av} \pi d n_\infty \tag{3·13}$$

The last expression defines the average mass transfer coefficient, k_{av}, for the cylinder. The concentration gradient at the surface is obtained by differentiating (3·12) with respect to y. The result is

$$\left(\frac{\partial n}{\partial y_1}\right)_{y_1=0} = \frac{(A Pe)^{1/3} n_\infty \sin^{1/2} x_1}{1.45 \chi^{1/3}} \tag{3·14}$$

Substituting in (3·13) and evaluating the integral with respect to x_1, the following result is obtained (Natanson, 1957):

$$\frac{k_{av} d}{D} = 1.17 (A Pe)^{1/3} \tag{3·15}$$

The concentration gradient at the surface can be expressed in terms of an effective boundary layer thickness, δ_c, as follows:

$$\left(\frac{\partial n}{\partial y}\right)_{y=0} = \frac{n_\infty}{\delta_c}$$

Substituting (3·14), the result is

$$\frac{\delta_c}{d} \sim (A Pe)^{-1/3}$$

with a proportionality constant of order unity near the forward stagnation point. Hence the thickness of the concentration boundary layer is inversely proportional to $Pe^{1/3}$; large Péclet numbers lead to thin concentration boundary layers as discussed in Section 3·4.

The theoretical expression (3·15) is in good agreement with data for diffusion in aqueous solutions over the high Pe range of interest in aerosol deposition. Recalling that $Pe = Sc \cdot Re$, (3·15) can be rearranged to give

$$\frac{k_{av.}d}{D} \Big/ \left(\frac{\nu}{D}\right)^{1/3} = 1.17(ARe)^{1/3}$$

for this low Reynolds number case. At higher Reynolds numbers, a different functional form is found for the Reynolds number dependence but the general relationship

$$\frac{k_{av}d}{D} \Big/ \left(\frac{\nu}{D}\right)^{1/3} = f(Re)$$

holds over a wide range of Reynolds numbers. The form of the function is shown in Fig. 3·3.

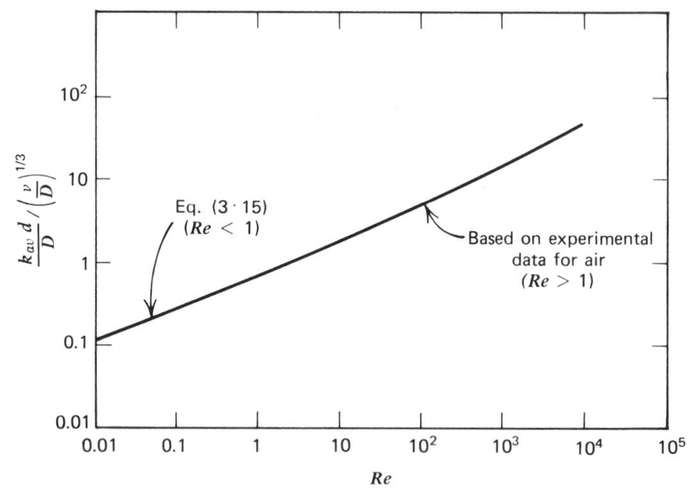

Fig. 3·3 Diffusion to single cylinders placed normal to an air flow. The theoretical curve for low Reynolds numbers is in good agreement with experimental data for diffusion in aqueous solution (Dobry and Finn, 1956). The curve for high Reynolds numbers is based on data for heat transfer to air (data of Hilpert reported by Schlichting, 1968) corrected by dividing the Nusselt number by $(\nu/D)^{1/3}$. This is equivalent to assuming laminar boundary layer theory (Section 3.9) is applicable. This curve applies to the diffusion of point particles ($R \rightarrow 0$).

The efficiency of removal, η_R, is defined as the fraction of the particles collected from the fluid volume swept by the cylinder:

$$\eta_R = \frac{k_{av}\pi d n_\infty}{n_\infty U d} = 3.68 A^{1/3} Pe^{-2/3} \tag{3·16}$$

Since A is a relatively slowly varying function of Reynolds number, the efficiency varies approximately as $d^{-2/3}$, which means that fine fibers are more efficient aerosol collectors than coarse ones. Since $Pe = dU/D$, $\eta_R \sim d_p^{-2/3}$ or $d_p^{-4/3}$ for the continuum and free molecule ranges, respectively. Hence small particles are more efficiently removed by diffusion than larger particles in the range $d_p < 0.5$ μm. The use of single fiber collection efficiencies in estimating the efficiency of fibrous filters is discussed in a later section.

3·6 Diffusion at Low Reynolds Numbers: Similitude Law for Particles of Finite Diameter

For particles of finite diameter, a useful similitude law can be derived as follows (Friedlander, 1967): It is assumed that the concentration boundary layer is thin and falls within the region where the first term in the expansion of the stream function (3·9) is applicable. The corresponding velocity distribution is given by

$$u = 4AU\left(\frac{y}{a}\right)\sin\left(\frac{x}{a}\right) \tag{3·17}$$

$$v = -2AU\left(\frac{y}{a}\right)^2\cos\left(\frac{x}{a}\right) \tag{3·18}$$

Substitution in (3·7) gives the following equation for convective diffusion:

$$4\frac{y}{a}\sin\left(\frac{x}{a}\right)\frac{\partial n}{\partial x} - 2\left(\frac{y}{a}\right)^2\cos\left(\frac{x}{a}\right)\frac{\partial n}{\partial y} = \frac{D}{AU}\frac{\partial^2 n}{\partial y^2} \tag{3·19}$$

We now introduce the following dimensionless variables:

$$n_1 = \frac{n}{n_\infty}; \qquad y_1 = \frac{y}{a_p}; \qquad x_1 = \frac{x}{a} \tag{3·20}$$

Then (3·19) becomes

$$4y_1 \sin x_1 \frac{\partial n_1}{\partial x_1} - 2y_1^2 \cos x_1 \frac{\partial n_1}{\partial y_1} = \left(\frac{Da^2}{AUa_p^3}\right)\frac{\partial^2 n_1}{\partial y_1^2} \tag{3·21}$$

with boundary conditions

$$\text{at} \quad y_1 = 1, \quad n_1 = 0$$
$$y_1 = \infty, \quad n_1 = 1$$

Only one dimensionless group $(Da^2/AUa_p^3) \sim (R^3PeA)^{-1}$ appears in (3·21) and the boundary conditions are pure numbers. Hence the concentration distribution is given by

$$n_1 = f(x_1, y_1, R^3PeA) \tag{3·22}$$

where $R = a_p/a$. In the boundary layer approximation, the particle deposition rate per unit length of cylinder is given by

$$2D\left(\frac{a}{a_p}\right)n_\infty \int_0^\pi \left(\frac{\partial n_1}{\partial y_1}\right)_{y_1=1} dx_1 = \eta_R n_\infty dU \tag{3·23}$$

Introducing (3·22) in (3·23) leads to the following functional relationship:

$$\eta_R RPe = \frac{\pi k_{av} d_p}{D} = f(RPe^{1/3}A^{1/3}) \tag{3·24}$$

This is the similitude law for the diffusion of particles of finite diameter but with $R \ll 1$. For fixed Re, the group $\eta_R RPe$ should be a single-valued function of $RPe^{1/3}$ over the range in which the theory is applicable ($Pe \gg 1, Re < 1, R \ll 1$). Experimental data collected for different particle and cylinder diameters and gas velocities and viscosities should all fall on the same curve when plotted in the form of (3·24).

In the limiting case, $R \to 0$, η_R is independent of the interception parameter R. By inspection of (3·24), this result is obtained if the function f is linear in its argument such that

$$\eta_R = C_1 \pi A^{1/3} Pe^{-2/3} \tag{3·25}$$

The constant $C_1\pi = 3.68$ according to (3·16). In the limiting case $Pe \to \infty$, particles follow the fluid and deposit when a streamline passes within one radius of the surface. This effect is sometimes called *direct interception*. The efficiency is obtained by integrating (3·18) for the normal velocity component over the front half of the cylinder surface:

$$\eta_R = -\frac{\int_0^{\pi/2} v_{y=a_p} dx}{Ua} = 2AR^2 \tag{3·26}$$

A result of this form can be obtained from (3·24) by noting that for $Pe \rightarrow \infty$, η_R is independent of Pe. Then the function f must be proportional to the cube of its argument.

Equations (3·25) and (3·26) are the limiting laws for the ranges in which diffusion and direct interception control, respectively. They show that for fixed velocity and fiber diameter, the efficiency at first decreases as d_p increases because of the decrease in the diffusion coefficient (3·25); further increases in d_p lead to an increase in η_R as $R = d_p/d$ increases (3·26).

3·7 Deposition Near the Forward Stagnation Point

An analytical solution to (3·21) does not seem possible, but a solution can be obtained for the region around the forward stagnation point (Fig. 3·1). Near the streamline in the plane of symmetry which leads to the stagnation point, $\sin x_1$ vanishes and $\cos x_1$ approaches unity so (3·21) becomes

$$-AR^3 Pe y_1^2 \frac{\partial n_1}{\partial y_1} = \frac{\partial^2 n_1}{\partial y_1^2} \qquad (x_1 \rightarrow 0) \tag{3·27}$$

with the boundary conditions

$$\text{at} \quad y_1 = 1, \quad n_1 = 0$$
$$y_1 = \infty, \quad n_1 = 1$$

There is a solution to (3·27) with these boundary conditions for which n_1 is a function only of y_1; the following expression is obtained for the coefficient of mass transfer at the forward stagnation point:

$$k_0 = \frac{-D(\partial n/\partial y)_{y=a_p;\, x=0}}{n_\infty}$$

$$= \frac{(D/a_p)e^{-AR^3Pe/3}}{\int_1^\infty \exp(-AR^3Pez^3/3)\,dz} \tag{3·28}$$

Although this result applies only at $x_1 = 0$, the deposition rate is greatest at this point and illustrates the general functional dependence on Pe and Re over the entire cylinder.

3.8 Filter Efficiency: Comparison with Experiment

The overall efficiency of a fibrous filter is directly related to the single fiber efficiency. In a regular array of fibers with uniform diameter, d, and a fraction solids, α, let the average concentration of particles of size, d_p, at a distance, z, from the filter entrance be N (Fig. 3·4). For a single fiber, the removal efficiency is defined as

$$\eta_R = \frac{b}{d} \tag{3·29}$$

where b is the width that corresponds to a region of flow completely cleared of all particles by the cylinder. In a differential distance, dz, in the flow direction, there are $\alpha\, dz/(\pi d^2/4)$ fibers per unit width; the removal over this distance by each fiber is

$$-dN \Big/ \frac{\alpha\, dz}{\pi d^2/4} = bN = (\eta_R d)N$$

Rearranging and integrating from $z=0$ to $z=L$, the thickness of the filter,

$$\eta_R = \frac{\pi d}{4\alpha L} \ln \frac{N_1}{N_2} \tag{3·30}$$

Since the fiber diameters are usually not all equal and the fibers are usually arranged in a more or less random fashion, η_R should be interpreted as an effective efficiency defined by (3·25) and based on an average diameter, $\bar{d}$, usually the arithmetic average. In an experimental determination of η_R, the practice is to measure N_1 and N_2, the inlet and outlet concentration of a monodisperse aerosol passed through the filter. The average fiber diameter, $\bar{d}$, can be determined by microscopic examination. As material accumulates in the filter, the character of the

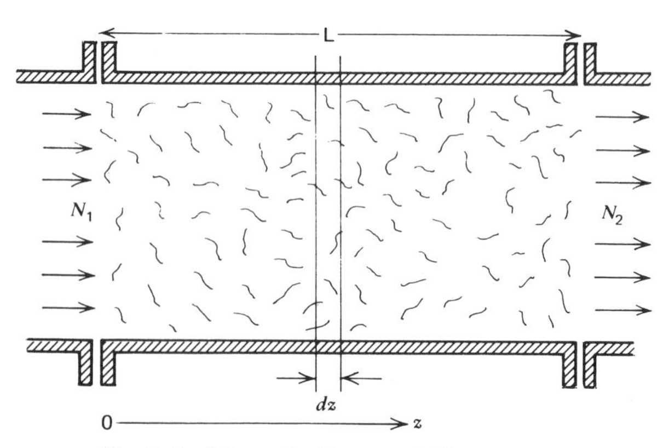

Fig. 3·4 Schematic diagram of fibrous filter.

bed changes. Hence η_R changes during the course of the filtration process.

Chen (1955) and Wong, Ranz, and Johnstone (1956) reported experimental values for single fiber efficiencies obtained in experiments with fiber mats and monodisperse liquid aerosols produced by a LaMer generator. The filter mats used by both sets of investigators were composed of glass fibers, distributed in size. The data extrapolated to zero fraction solids have been recalculated and plotted in Fig. 3·5 in the form resulting from the similitude analysis (3·24). The calculations are based on the average fiber diameter. The data of Chen covered the ranges $62 < Pe < 2.8\times10^4$, $0.06 < R < 0.29$, $1.4\times10^{-3} < Re < 7.7\times10^{-2}$, and $5.2\times10^{-4} < St < 0.37$. The Stokes number (Chap. 4), $St = mU/af$, where m is the particle mass, is a measure of the strength of the inertial effects and must be small for the diffusion theory to apply.

Most of the data fell in the range $10^{-3} < Re < 10^{-1}$, and theoretical curves for the forward stagnation point (3·28) are shown for the limiting values of the Reynolds number. Rough agreement between experiment and theory is evident. One would expect the experimental data, based on the average deposition over the fiber surface, to fall somewhat below the theoretical curves for the forward stagnation point. This is true over the whole range for Chen's data but not for those of Wong, Ranz, and Johnstone.

Based on the similitude analysis, the following semiempirical correlation was proposed to correlate experimental data:

$$\eta_R R Pe = 1.3 R Pe^{1/3} + 0.7 (R Pe^{1/3})^3 \tag{3·31}$$

For $R \to 0$, this expression reduces to the form of (3·25) and for $Pe \to \infty$, to (3·26). In both cases, however, an empirical coefficient independent of Reynolds number appears.

Equation (3·31) is based on data extrapolated to a filter with fraction solids, $\alpha = 0$. For real filters, there is always interaction among the fibers and the efficiency is a function of α. This effect was measured in the experiments discussed previously, and several approximate theoretical analyses appear in the literature (Davies, 1973).

As particles accumulate in the filter, both the efficiency of removal and the pressure drop tend to increase. Some data on this effect are available for which the literature should be consulted.

Example. Small particles are filtered through a mat composed of cylindrical fibers. Show that when diffusion and interception are operative there is a minimum in the efficiency with respect to particle size at

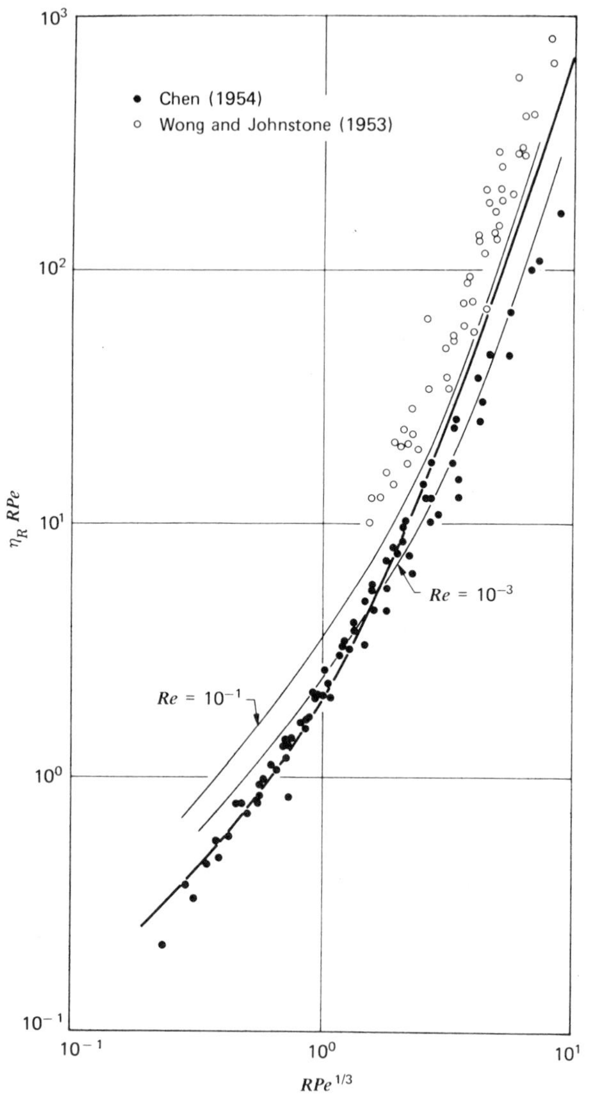

Fig. 3·5 Comparison of experimentally observed deposition rates on glass fiber mats for dioctyphthalate (Chen) and sulfuric acid (Wong) aerosols with theory for the forward stagnation point of single cylinders (Friedlander, 1967). The theoretical curves for $Re = 10^{-1}$ and 10^{-3} were calculated from (3.28). For all data points the Stokes number was less than 0.5. Agreement with the data of Chen is particularly good. Theory for the forward stagnation point should fall higher than the experimental transfer rates, which are averaged over the fiber surface. Equation (3·31), denoted by the heavy line, is recommended for design purposes. Values for η_R correspond to $\alpha \rightarrow 0$. For $\alpha < 0.1$, Wong, Ranz, and Johnstone (1956) found little variation with α, whereas Chen (1955) found a linear increase in η_R over this range.

constant gas velocity. Estimate the particle diameter corresponding to the minimum for a filter composed of 4 μm fibers and an air velocity of 150 cm/sec.

SOLUTION. For large particles, $Pe \to \infty$, interception controls and $\eta_R = 2AR^2$. For small particles, $R \to 0$, diffusion controls and $\eta_R = 3.68A^{1/3}Pe^{-2/3}$. Thus for large particles, the efficiency *increases* with particle diameter ($\sim d_p^2$) while for small particles, the efficiency increases as particle diameter decreases ($\sim d_p^{-4/3}$ for $d_p \ll l$).

To find the minimum, differentiate (3·31) with respect to particle diameter and set $d\eta_R/[d(d_p)] = 0$. The result is

$$-\left[\frac{d\ln D}{d\ln d_p}\right]_{\min} = 1.6\,Pe_{\min}^{2/3} R_{\min}^2$$

For $U = 13$ cm/sec and $d = 4$ μm, it is found that $d_p \approx 0.25$ μm based on values of D from Table 2·1. This result should be compared with Fig. 4·14.

3·9 Diffusion from Laminar Boundary Layer Flows

A problem of fundamental interest in fluid mechanics is the determination of the velocity distribution in the laminar boundary layer on a flat plate (Fig. 3·6). The fluid is of uniform velocity as it approaches the plate, but is retarded near the surface since the velocity vanishes at the wall. The retardation takes place in a thin region, the boundary layer, where the velocity rises from zero at the surface to approximately the free stream velocity. The boundary layer remains laminar, in the absence of disturbances, for Reynolds numbers $xU/\nu < 5 \times 10^5$, where x is the distance from the leading edge of the plate.

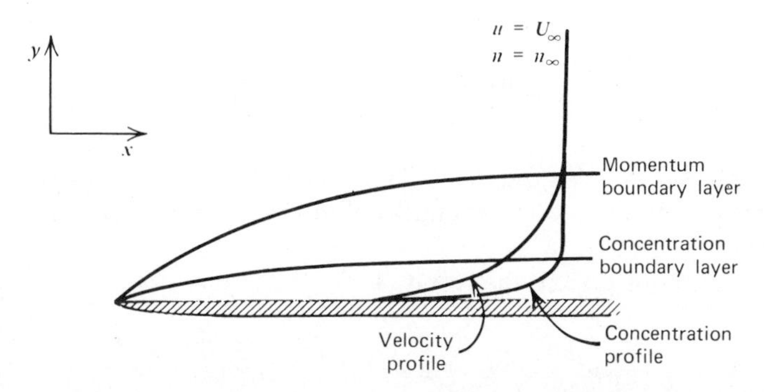

Fig. 3·6 Momentum and concentration boundary layers for laminar flow over a flat plate.

From aerodynamic theory, it can be shown that the thickness of the laminar boundary layer is given by

$$\delta_v = 1.72x\left(\frac{\nu}{xU}\right)^{1/2}$$

Suppose this equation is applied to the case of a tree leaf, parallel to the wind. At a point 2 cm from the leading edge of the leaf in a wind speed of 5 mi/hr, the boundary layer thickness would be about 0.05 cm.

Brownian particles present in the air boundary layer will diffuse to the surface and stick. This problem and its extensions have many important practical applications in particle deposition theory including lung deposition, transfer to leaves and other surfaces in the atmosphere, and deposition in the entrance region of a pipe with a well-faired entry.

For the steady state, the equation of convective diffusion takes the form

$$u\frac{\partial n}{\partial x} + v\frac{\partial n}{\partial y} = D\frac{\partial^2 n}{\partial y^2} \tag{3·32}$$

where diffusion in the direction of flow has been neglected in comparison with convection. The coordinate system is shown in Fig. 3·6. As the momentum boundary layer develops along the plate, a diffusion boundary layer is also established in which the concentration decreases from its value in the fluid mainstream to zero (for aerosols) on the wall. The Schmidt number, ν/D, represents a ratio of rates of diffusion of momentum to matter. For particle diffusion, $\nu/D \gg 1$, and the concentration boundary layer is thin compared with the velocity boundary layer. A simple analytical solution to the equation of convective diffusion can be obtained by using only the first terms in the expansions of the laminar boundary layer components u and v near the wall:

$$u = \alpha U \eta$$

$$(\eta \to 0)$$

$$v = \frac{\alpha U}{4}\left(\frac{\nu}{xU}\right)^{1/2}\eta^2$$

where $\eta = (U/\nu x)^{1/2}y$ and $\alpha = 0.332$. These expressions are obtained from series solutions to the Prandtl boundary layer equations (Schlichting, 1968).

Substituting in (3·32), the result is

$$\alpha\eta\frac{\partial n_1}{\partial x} + \frac{\alpha}{4}\left(\frac{\nu}{Ux}\right)^{1/2}\eta^2\frac{\partial n_1}{\partial y} = \frac{D}{U}\frac{\partial^2 n_1}{\partial y^2} \tag{3·33}$$

where $n_1 = n/n_\infty$ and n_∞ is the concentration at large distances from the surface of the plate.

The boundary conditions are

$$n_1 = 1 \quad \text{at} \quad y = \infty$$
$$n_1 = 0 \quad \text{at} \quad y = 0$$

Next it is assumed that n_1 is a function only of η, $n_1 = g(\eta)$, with the result that

$$\frac{\partial n_1}{\partial y} = \left(\frac{U}{\nu x}\right)^{1/2} \frac{dg}{d\eta}$$

$$\frac{\partial^2 n_1}{\partial y^2} = \left(\frac{U}{\nu x}\right) \frac{d^2 g}{d\eta^2}$$

$$\frac{\partial n_1}{\partial x} = -\frac{1}{2}\frac{\eta}{x}\frac{dg}{d\eta}$$

Substituting in (3·33), the result is

$$\frac{d^2 g}{d\eta^2} + \frac{\alpha}{4}\left(\frac{\nu}{D}\right)\eta^2 \frac{dg}{d\eta} = 0$$

This is an ordinary linear differential equation that can easily be integrated to give the following results (Schlichting, 1968) for the local diffusion flux:

$$|J| = \left| -D\left(\frac{\partial n}{\partial y}\right)_{y=0} \right| = 0.339\left(\frac{D}{x}\right)\left(\frac{xU}{\nu}\right)^{1/2}\left(\frac{\nu}{D}\right)^{1/3} n_\infty$$

and for the value averaged over the plate:

$$|J_{av}| = \left| \frac{1}{L}\int_0^L -D\left(\frac{\partial n}{\partial y}\right)_{y=0} dx \right| = 0.678\left(\frac{D}{L}\right)\left(\frac{LU}{\nu}\right)^{1/2}\left(\frac{\nu}{D}\right)^{1/3} n_\infty \quad (3\cdot34)$$

In the case of flow normal to a right circular cylinder or to a sphere, a laminar velocity boundary layer forms over the upstream surface at Reynolds numbers, based on diameter, greater than several hundred. The calculation of the local mass transfer coefficient near the stagnation point and over much of the upstream surface can be carried out numerically leading to an equation similar in form to (3·34). At an angle of somewhat less than 90° from the forward stagnation point the boundary layer separates, and such calculations are no longer valid. Average transfer coefficients over the entire surface have been *measured* for the cylinder and sphere. Results of such measurements for cylinders are shown in Fig. 3·3 along with the theory for low Reynolds number flows.

3·10 Diffusion from a Laminar Pipe Flow

When a gas enters a smooth pipe from a large reservoir through a well-faired entry, a laminar boundary layer forms along the walls. The velocity profile in the main body of the flow remains flat. The velocity boundary layer thickens with distance downstream from the entry until it eventually fills the pipe. If the Reynolds number based on pipe diameter is less than 2100, the pipe boundary layer remains laminar. The flow is said to be fully developed when the velocity profile no longer changes with distance in the direction of flow. The profile becomes approximately fully developed after a distance from the entry of about 0.04 $d(Re)$. For example, for $Re = 1000$, the entry length extends over 40 pipe diameters.

Small particles present in the gas stream diffuse to the walls as a result of their Brownian motion. Since the Schmidt number, ν/D, is much greater than unity, the diffusion boundary layer is thinner than the velocity boundary layer and the concentration profile tends to remain flat for much greater distances downstream from the entry than the velocity (Fig. 3·7). As a reasonable approximation for mathematical analysis, it can be assumed that at the pipe entry ($x = 0$), the concentration profile is flat while the velocity profile is already fully developed, that is, parabolic.

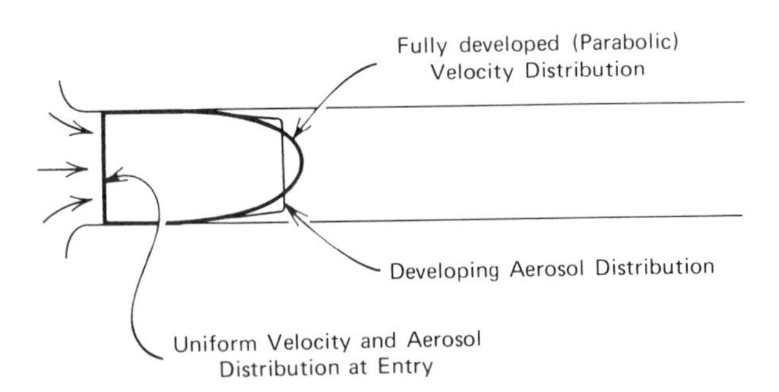

Fig. 3·7 Aerosol diffusion from a laminar flow. After a relatively short distance, the velocity distribution is fully developed. The slowly diffusing particles require a longer distance to reach a fully-developed concentration profile.

The problem of diffusion to the walls of a channel (pipe or duct) from a laminar flow is formally identical with the heat transfer (Graetz) problem when the particle size is small compared with the channel size ($R \to 0$). For a fully developed parabolic velocity profile, the steady-state equation of convective diffusion takes the following form in cylindrical coordinates:

$$u\frac{\partial n}{\partial x} = D\left[\frac{\partial(r(\partial n/\partial r))}{r\partial r} + \frac{\partial^2 n}{\partial x^2}\right] \tag{3·35}$$

where $u = 2U_{av}[1-(r/a)^2]$ and U_{av} is the average velocity. As boundary conditions, we require that the concentration be uniform at the tube inlet and vanish at the surface:

$$\begin{aligned} \text{at} \quad x=0, &\quad n=n_0 \text{ for } r<a \\ r=a, &\quad n=0 \end{aligned}$$

where a is the radius of the pipe. When $Pe > 100$, diffusion in the axial direction can be neglected. In nondimensional form, (3·35) can then be written as follows:

$$\frac{u_1}{2}\frac{\partial n_1}{\partial x_1} = \frac{\partial(r_1(\partial n_1/\partial r_1))}{r_1 \partial r_1} \tag{3·36}$$

where $u_1 = u/U_{av} = 2(1-r_1^2)$, $r_1 = r/a$, $n_1 = n/n_0$, and $x_1 = (x/a)/(dU_{av}/D) = (x/a)/Pe$ with the boundary conditions

$$\begin{aligned} n_1 = 1 &\quad \text{at} \quad x_1 = 0, \quad r_1 < 1 \\ n_1 = 0 &\quad \text{at} \quad r_1 = 1 \end{aligned}$$

The solution to (3·36) with these boundary conditions is obtained by separation of variables:

$$n_1 = R(r_1)X(x_1)$$

Substitution in (3·35) results in two ordinary differential equations

$$X' + \lambda^2 X = 0 \tag{3·37a}$$

$$R'' + \frac{1}{r_1}R' + \lambda^2 R(1-r_1^2) = 0 \tag{3·37b}$$

where $-\lambda^2$ is the separation constant and the primes refer to differentiation with respect to the independent variables.

The solution to (3·37*a*) is

$$X = C\exp(-\lambda^2 x_1)$$

where C is a constant of integration. The boundary conditions and symmetry require that $R'(0) = R(1) = 0$. Eq. (3·37*b*) can be solved for these conditions only for certain discrete values of $\lambda = \lambda_n$ called *eigenvalues*. Each value of λ_n corresponds to a new function $R_n(r_1)$. Since (3·36) is linear, a sum of solutions corresponding to each eigenvalue is also a solution:

$$n_1 = \sum_{n=0}^{\infty} C_n R_n(r_1)\exp(-\lambda_n^2 x_1) \tag{3·38}$$

The quantities of principal interest are the local mass transfer coefficient defined by the expression

$$J = -D\left(\frac{\partial n}{\partial r}\right)_{r=a} = k_x n_{av}$$

where n_{av} is the mixed mean concentration of the particles in the fluid:

$$n_{av} = \frac{2}{a^2 U}\int_0^a unr\,dr \tag{3·39}$$

and the mean transfer coefficient averaged over the surface of the tube up to x:

$$k_{av} = \frac{1}{x}\int_0^x k_x\,dx \tag{3·40}$$

The average concentration is given by (3·39) and (3·38):

$$\frac{n_{av}}{n_0} = 8\sum_{n=0}^{\infty} \frac{G_n}{\lambda_n^2}\exp(-\lambda_n^2 x_1) \tag{3·41}$$

Values of $G_n = (C_n/2)R_n'(1)$ and λ_n^2 are given in Table 3·1. Equation (3·41) can also be used to determine the size distribution of the particles remaining suspended if the initial distribution is known. The average mass transfer coefficient is most conveniently found by integration of a mass balance over a differential length of tube with (3·40). The result is

$$\frac{2k_{av}a}{D} = \frac{1}{2x_1}\ln\left(\frac{n_0}{n_{av}}\right) \tag{3·42}$$

TABLE 3·1

INFINITE-SERIES SOLUTION FUNCTIONS FOR THE CIRCULAR TUBE: CONSTANT SURFACE CONCENTRATION (SELLARS, TRIBUS, AND KLEIN, 1956)

n	λ_n^2	G_n
0	7.312	0.749
1	44.62	0.544
2	113.8	0.463
3	215.2	0.414
4	348.5	0.382

For $n>2$, $\lambda_n = 4n + \frac{8}{3}$; $G_n = 1.01276\lambda_n^{-1/3}$.

Mass transfer coefficients and the average concentration are shown in Fig. 3·8.

These results can be applied to the design of the diffusion battery, a device used to measure the particle size of submicron aerosols. The battery may consist of a bundle of capillary tubes, or of a set of closely spaced, parallel flat plates, through which the aerosol passes in laminar flow. The particle concentrations entering and leaving the diffusion battery are measured with a condensation nuclei counter (Chap. 6). From the measured value of the reduction in concentration, the value of $(x/a)/Pe$ can be determined from Fig. 3·8 or its equivalent for flat plates. The value of D, hence d_p, can be calculated since x, a, and U are

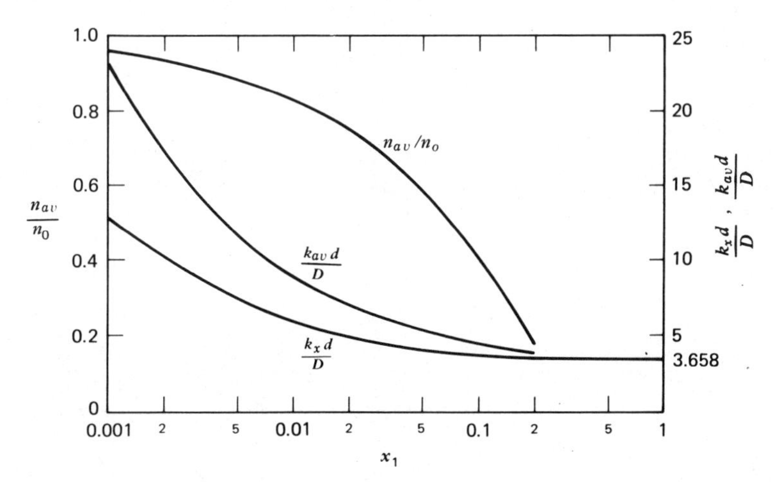

Fig. 3·8 Diffusion to the walls of a pipe from a fully developed laminar flow. Average concentration and local and average mass transfer coefficiencts are shown as a function of distance from the tube entrance.

known for the system. For polydisperse aerosols, the usual case, the method yields an average particle diameter that depends on the particle size distribution.

The theory also has application to efficiency calculations for certain classes of filters (Spurny et al., 1969) composed of a sheet of polymeric material penetrated by many small cylindrical pores. These are produced by exposing the sheet to α radiation and then washing in acid for a controlled time period to produce pores of uniform diameter in the micron and submicron range.

Example. In the region near the entrance to a pipe where $(x/a)/Pe$ is small, the series solutions (3·41) converge slowly. Derive an expression for the local mass transfer coefficient valid for the entry region. Assume point particles as before.

SOLUTION. The parabolic velocity distribution develops rapidly at the tube entrance, compared with the concentration distribution because $\nu/D \gg 1$. A thin concentration boundary layer forms in the region where $r \approx a$ and the velocity distribution can be represented by the first term in the expansion of the velocity with respect to the distance from the wall, $y = a - r$:

$$u = \frac{4U_{av}y}{a}$$

The equation of convective diffusion for the region near the wall takes the form

$$u\frac{\partial n}{\partial x} = D\frac{\partial^2 n}{\partial y^2} \tag{1}$$

where x is the distance along the tube axis. The boundary conditions are

$$n = n_\infty \quad \text{at} \quad y = \infty$$
$$n = 0 \quad \text{at} \quad y = 0$$

By inspection (experience in manipulation of the boundary layer equation helps), a solution can be obtained by introducing the dimensionless variable,

$$\eta = \left(\frac{U_{av}}{dD}\right)^{1/3} \frac{y}{x^{1/3}}$$

and assuming that n is a function only of η. Substitution in (1) leads to the following ordinary differential equation:

$$\frac{d^2n}{d\eta^2} + \frac{8}{3}\eta^2 \frac{dn}{d\eta} = 0$$

The solution that satisfies the boundary conditions is

$$n = \frac{n_\infty \int_0^\eta \exp(-\frac{8}{9}z^3)\, dz}{\int_0^\infty \exp(-\frac{8}{9}z^3)\, dz}$$

The mass transfer coefficient is given by

$$k_x = \frac{|J|}{n_\infty} = \frac{D}{n_\infty}\left(\frac{\partial n}{\partial y}\right)_{y=0}$$

$$= \frac{D}{x^{1/3}}\left(\frac{U_{av}}{dD}\right)^{1/3} \frac{1}{\int_0^\infty \exp(-\frac{8}{9}z^3)\, dz}$$

The integral in the denominator can be evaluated as a gamma function with the following result:

$$\frac{k_x d}{D} = 1.077\left(\frac{d^2 U_{av}}{Dx}\right)^{1/3}$$

This expression holds best over the region in which the concentration boundary layer thickness is smaller than the pipe radius. As an approximation, this corresponds to $x/d \ll Pe$.

3·11 Diffusion from a Turbulent Pipe Flow

When the pipe Reynolds number is greater than about 2100, the velocity boundary layer that forms in the entry region eventually turns turbulent. The velocity profile becomes fully developed at about 25 to 50 pipe diameters from the entry.

Small particles in such a flow are transported by turbulent and Brownian diffusion to the wall. An example of practical interest is the deposition of lead-containing particles from the exhaust gases in the tailpipe of an automobile. Thermophoresis probably plays a role in this case as well because the walls are cooler than the gas. In the sampling of atmospheric air through long pipes, wall losses result from turbulent diffusion.

Although the flow of turbulent fluids is much more complex than that of the laminar flows discussed in previous sections, semiempirical calculations of rates of diffusional transport are actually easier to make for fully developed turbulent duct flows.

In studying turbulent transport, it is often assumed that the pipe flow can be divided into three different zones (Fig. 3·9). The core of the pipe is a highly turbulent region in which molecular diffusion is negligible compared with transport by the turbulent eddies. Closer to the wall there is a transition region where both molecular and eddy diffusion are important.

Next to the wall itself, there is a thin sublayer in which the transfer of *momentum* is dominated by viscous forces, and the effect of weak turbulent fluctuations can be neglected. This applies also to heat and mass transfer for gases; the Schmidt and Prandtl number are near unity, which means that heat and mass are transported at about the same rates as momentum.

The situation is quite different for particle diffusion. In this case, $\nu/D \gg 1$ and even weak fluctuations in the viscous sublayer contribute significantly to transport. Consider a turbulent pipe flow. In the regions near the wall, the curvature can be neglected and the instantaneous

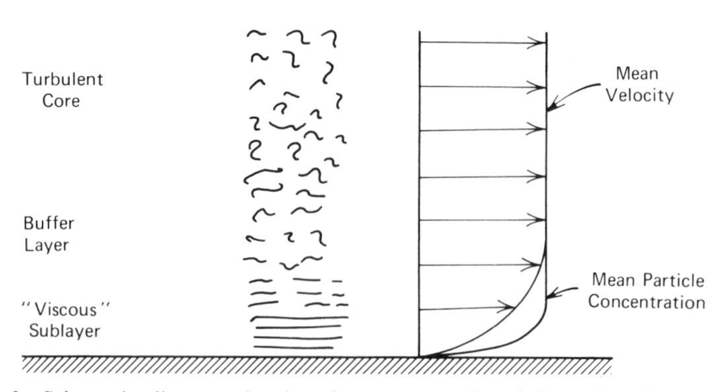

Fig. 3·9 Schematic diagram showing the structure of turbulent pipe flow. For convenience, the flow is divided into three regions. Most of the pipe is filled with the turbulent core, with the velocity rising rapidly over the viscous sublayer. The concentration drops more sharply than the velocity because $D \ll \nu$ and turbulent diffusion brings the particles close to the wall before Brownian diffusion can act effectively.

diffusion flux written as follows:

$$J = -D\frac{\partial n}{\partial y} + nv \qquad (3\cdot 43)$$

where y is the distance measured normal to the surface and v is the velocity in the y direction.

In analyzing turbulent pipe flows, it is assumed that the velocity and the concentration can be separated into mean and the fluctuating components:

$$v = v' \qquad (\text{since } \bar{v} = 0) \qquad (3\cdot 44a)$$

and

$$n = \bar{n} + n' \qquad (3\cdot 44b)$$

where the bar and prime refer to the mean and fluctuating quantities, respectively. Substituting in (3·43) and taking the time average, the result is

$$\bar{J} = -D\frac{\partial \bar{n}}{\partial y} + \overline{n'v'} \qquad (3\cdot 45)$$

We shall now examine the behavior near the wall of the quantities n' and v'. The continuity relation for an incompressible fluid takes the form:

$$\frac{\partial u}{\partial x} + \frac{\partial v}{\partial y} + \frac{\partial w}{\partial z} = 0 \qquad (3\cdot 46)$$

Taking the time average, the result is

$$\frac{\partial \bar{u}}{\partial x} + \frac{\partial \bar{v}}{\partial y} + \frac{\partial \bar{w}}{\partial z} = 0 \qquad (3\cdot 47)$$

Subtracting (3·47) from (3·46), the result is

$$\frac{\partial u'}{\partial x} + \frac{\partial v'}{\partial y} + \frac{\partial w'}{\partial z} = 0$$

At the wall, the velocity vanishes so that

$$\left(\frac{\partial u'}{\partial x}\right)_0 = \left(\frac{\partial w'}{\partial z}\right)_0 = 0$$

where the subscript 0 refers to the wall. Therefore,

$$\left(\frac{\partial v'}{\partial y}\right)_0 = 0$$

For fixed x, z, t, the fluctuating velocity can be expanded in a Taylor series as follows:

$$v' = v'_0 + \left(\frac{\partial v'}{\partial y}\right)_0 y + \left(\frac{\partial^2 v'}{\partial y^2}\right)_0 \frac{y^2}{2!} + 0(y^3)$$

and the fluctuating concentration:

$$n' = n'_0 + \left(\frac{\partial n'}{\partial y}\right)_0 y + O(y^2)$$

Since $v'_0 = (\partial v'/\partial y)_0 = n'_0 = 0$, the product $n'v'$ must be at least of order y^3.

Defining the eddy diffusion coefficient, ϵ, by the relationship

$$\overline{n'v'} = -\epsilon \frac{\partial \bar{n}}{\partial y} \tag{3·48}$$

we see that ϵ must be at least of order y^3 since $\partial \bar{n}/\partial y$ is finite at the wall. Based on experimental data for diffusion controlled electrochemical reactions in aqueous solution, the following expression was proposed by Lin, Moulton, and Putnam (1953) for the eddy diffusion coefficient in the viscous sublayer:

$$\epsilon = \nu \left(\frac{y^+}{14.5}\right)^3 \tag{3·49}$$

where $y^+ = [yU(f/2)^{1/2}]/\nu$ with U the average velocity, f the Fanning friction factor, and ν the kinematic viscosity. This expression for ϵ holds when $y^+ < 5$.

Substituting (3·48) in (3·45), the general expression for the diffusion flux is given by

$$\bar{J} = -(D + \epsilon) \frac{\partial \bar{n}}{\partial y} \tag{3·50}$$

For particle diffusion, $\nu/D \gg 1$. Compared with momentum transfer, particles penetrate closer to the wall by turbulent diffusion before

Brownian diffusion becomes important. (Viscous shear for momentum transfer is important at relatively large distances from the surface.) The particle concentration, which vanishes at the wall, rises rapidly practically reaching the mainstream concentration, $\bar{n}_\infty$, within the viscous sublayer in which ϵ is given by (3·49). The concentration distribution and deposition flux can be obtained by integrating (3·50) and assuming that $\bar{J}$ is a function only of x and not of y, the distance from the surface.

The boundary conditions are given by

$$\bar{n}=0 \quad \text{at} \quad y=0$$

$$\bar{n}=\bar{n}_\infty \quad \text{at} \quad y=\infty$$

The result of the integration is

$$\frac{kd}{D}=0.042\, Re f^{1/2} Sc^{1/3} \tag{3·51}$$

where the mass transfer coefficient is defined by the relationship $k=-(D/\bar{n}_\infty)(\partial\bar{n}/\partial y)_{y=0}$.

For $\epsilon \sim y^4$, the result is

$$\frac{kd}{D}=0.079\, Re f^{1/2} Sc^{1/4} \tag{3·52}$$

The value of the constant is that given by Deissler (1955). A theoretical justification for this form is given by Landau and Lifshitz (1959). Both expressions (3·51) and (3·52) can be used to correlate experimental data for transfer at high Schmidt numbers. For $Sc=10^3$, they disagree by about 5%, and for $Sc=10^4$, by 14%.

3·12 Convective Diffusion in an External Force Field: Electrical Precipitation

Electrical precipitation is the most widely used method for removing particles from power plant stack gases. In the most common type, the dusty gas flows between parallel plate electrodes that, however, may have quite complex geometries (Fig. 3·10). The particles are charged by ions generated in a corona discharge surrounding rods or wires suspended between the plates. Near the wire, the potential gradient is very high, and electron discharge and gas ionization take place. At some distance from the wire, the potential gradient drops below the value necessary to maintain the discharge.

The system is usually run with the discharge electrode negative, since this permits greater stability of operation and higher voltage before

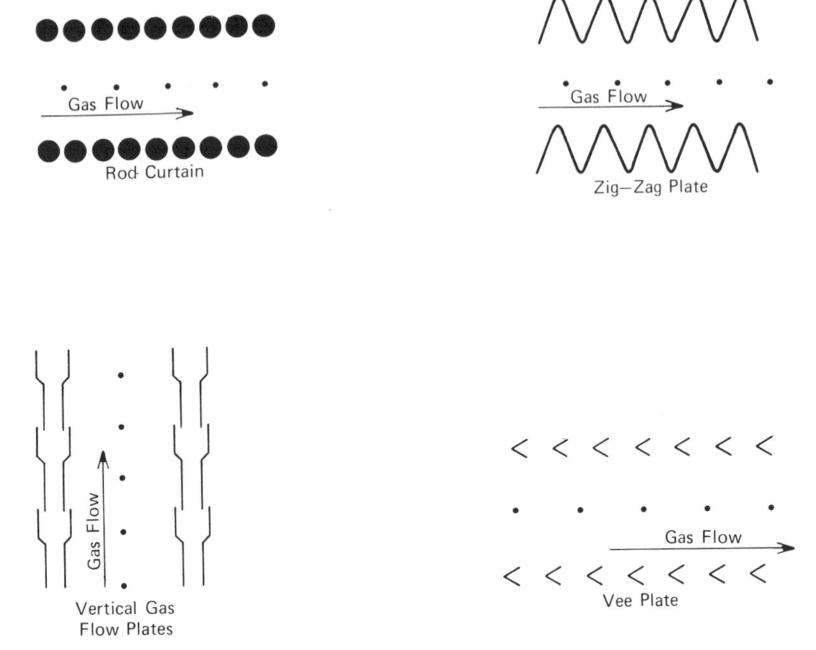

Fig. 3·10 Top views of two special collecting electrodes used in electrical precipitators. Many unusual designs have been developed to minimize dust particle reentrainment. The corresponding flow patterns are quite complex, making theoretical analysis of performance efficiency very difficult.

breakdown. The cloud of negative ions and electrons formed in the discharge moves toward the collecting electrodes. Particles are charged by field or diffusion charging depending on their size (Chap. 2).

With typical plate spacings of 6 to 15 in. and gas velocities of 3 to 10 ft/sec, corresponding to Reynolds numbers of 10^4 and greater, precipitator flows are turbulent. The particle flux in the direction normal to the collecting plate is given by

$$J = -D\frac{\partial n}{\partial y} + vn - c_e n \tag{3·53}$$

where c_e is a positive quantity for migration toward the plate. Substituting (3·44*a*) and (3·44*b*) for the velocity and concentration in the turbulent flow and taking the time average, the result is

$$\bar{J} = -(D+\epsilon)\frac{\partial \bar{n}}{\partial y} - c_e \bar{n} \tag{3·54}$$

The electrical migration velocity, assumed constant to simplify the analysis, actually varies because it depends on the field strength, which is a function of position, and on the charging time (Chap. 2). We next assume that $\bar{n}$ increases from zero on the surface to $\bar{n}_\infty$, the mainstream concentration, over a narrow region near the wall. For a given value of x in the direction of flow, the flux $\bar{J}$ can be assumed constant over the wall region, and (3·54) can be integrated to give

$$\bar{J}(x) = \frac{-c_e \bar{n}_\infty}{1 - \exp\left\{ -c_e \int_0^\infty dy/(D+\epsilon) \right\}} \tag{3·55}$$

The particle flux is negative for deposition on the surface. The integral $v_d = 1/[\int_0^\infty dy/(D+\epsilon)]$ represents a particle migration velocity resulting from combined Brownian and turbulent diffusion. For $v_d \gg c_e$, $|\bar{J}| = v_d \bar{n}_\infty$ and diffusion controls the transport process. In electrical precipitator design calculations, it is usually assumed that electrical migration is much faster than diffusional transport, that is, $c_e \gg v_d$. The exponential term in the denominator can then be neglected with the result $|\bar{J}| = c_e \bar{n}_\infty$, and the precipitator efficiency can be calculated from a material balance on a differential element of the precipitator (Fig. 3·11). The result is

$$\frac{(\bar{n}_{\infty 1} - \bar{n}_{\infty 2})}{\bar{n}_{\infty 1}} = 1 - \exp\left[\frac{-2c_e L}{Ub} \right] \tag{3·56}$$

where L is the length of the precipitator, b is the plate spacing, and U is the average gas velocity. This is the expression customarily used in precipitator design calculations.

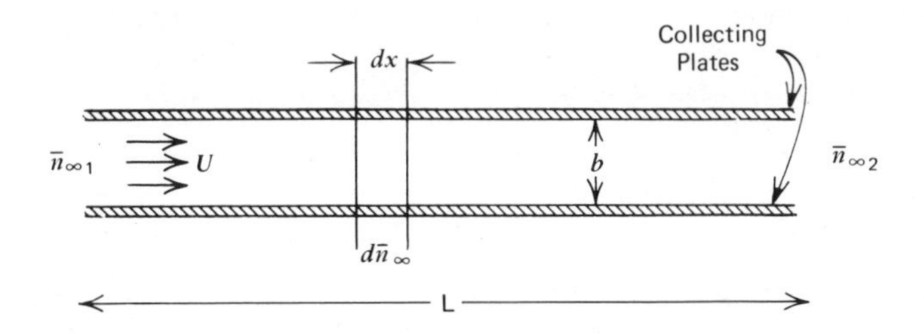

Fig. 3·11 Material balance on an element of precipitator

$$-Ubd\bar{n}_\infty = 2\bar{n}_\infty c_e dx$$

The migration velocity passes through a minimum corresponding to a particle size in the transition region between diffusion and field charging (Fig. 2·5). By differentiating (3·56), we see that the efficiency must also pass through a minimum at the same particle diameter. Indeed, a minimum has been observed experimentally in studies with a plant-scale precipitator (Fig. 3·12), in the particle size range corresponding to the minimum in the migration velocity.

In practice, it is not possible to calculate c_e from first principles because of the complexity of the interaction between the particles and the corona discharge. The mechanical and electrical behavior of the deposited dust layer are also difficult to characterize. Thus theory provides guidelines for precipitator design but in practice design is based to a great extent on empiricism.

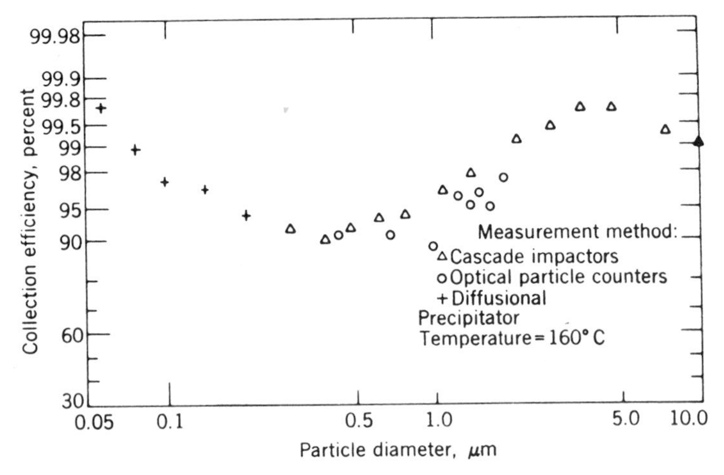

Fig. 3.12 Efficiency as a function of particle size for a pilot-scale electrical precipitator treating a side stream of flue gas from a utility boiler burning a low-sulfur coal (McCain, Gooch and Smith, 1975). The minimum near 0.5 μm probably results from the transition from diffusion to field charging. The decrease in efficiency for particles larger than 3 μm may result from reentrainment.

Problems

1. The velocity distribution for a low Reynolds number flow around a sphere is given by the following expressions due to Stokes:

$$u_\theta = -U\sin\theta\left(1 - \frac{3}{4}\frac{a}{r} - \frac{1}{4}\frac{a^3}{r^3}\right)$$

$$u_r = U\cos\theta\left(1 - \frac{3a}{2r} + \frac{a^3}{2r^3}\right)$$

Show that the mass transfer coefficient for particles of finite diameter at the forward stagnation point (in the concentration boundary layer approximation) is given by (Friedlander, 1967)

$$\frac{k_0 d}{D} = \frac{2e^{-\alpha}}{R\int_1^{\infty} e^{-\alpha z^3}\,dz}$$

where $\alpha = R^3 Pe/4$ and $Pe = dU/D$. This system represents a model for diffusional collection by small raindrops and fog droplets.

2. Estimate the collection efficiency of a filter mat ($\alpha = 0.0057$) composed of glass fibers 4 μm in diameter for 0.08 and 0.17 μm particles. The filter thickness is 0.8 cm, the air velocity is 13 cm/sec, and the temperature is 20°C. Compare your result with the experimentally measured values discussed in Chap. 4 (Fig. 4·14).

3. It is proposed to build a diffusion battery composed of a bundle of capillary tubes to determine the average particle size of an aerosol. For a tube diameter of 1 mm and a length of 20 cm, determine the velocity at which particles of $d_p = 0.05$ μm are 90% removed. Check to be sure the flow is laminar. The temperature is 20°C.

4. As a very crude model for air flow in the trachea, assume fully developed laminar pipe flow at a Reynolds number of 1000. The diameter is 2.0 cm.

(*a*) Estimate the relative rates of removal of SO_2 and sulfuric acid droplets per unit length of trachea. You may assume that the mucous layer is a perfect sink for SO_2 as well as particles. Express your answer in percent per centimeter as a function of particle size range $d_p < 0.5$ μm.

(*b*) Discuss the implication of your result in explaining the effects of sulfur dioxide and its oxidation products (sulfuric acid, for example) on the lung.

Make such assumptions as you deem necessary, but state all assumptions clearly.

5. Outside air is delivered to the instruments of an air monitoring station through a 2 in. duct at a velocity of 10 ft/sec. The duct is 12 ft long. Calculate the correction factor that must be applied for submicron particles as a result of diffusion to the walls of the duct. Express your answer in terms of the percentage by which the measured concentration must be multiplied to give the true concentration as a function of particle size.

6. For the particle size range of Fig. 2·5 and $E = 2$ kV/cm, determine the collection efficiency of a plate-type electrical precipitator, taking into account diffusional transport. The plate spacing is 12 in. and the plate length in the direction of flow is 5 ft. Make your calculations for velocities of 3 and 10 ft/sec, and plot efficiency as a function of particle diameter. The particles have a dielectric constant of 4, and the gas temperature is 160°C. Assume flat collecting plates and a fully developed turbulent flow.

7. A vertical plate is mounted in a large (effectively infinite) volume of air containing small particles (Fig. P7). The plate is heated, and a layer of warm air near the surface rises. The air velocity component parallel to the surface increases from zero at the surface to some maximum value and then falls to zero, the value far from the plate.

An important dimensionless parameter for this type of flow is the Grashof number

$$G_x = \frac{gx^3(T_0 - T_\infty)}{\nu^2 T_\infty}$$

where g is the gravitational constant, x is the length of the plate under consideration, and T_0 and T_∞ refer to the temperature at the wall and at large distances from the wall, respectively. For air over the range $10^4 < G < 10^8$, the flow is of the laminar boundary layer type. At higher values of G, the flow becomes turbulent, and for lower values, the layer becomes too thick for boundary layer theory to apply.

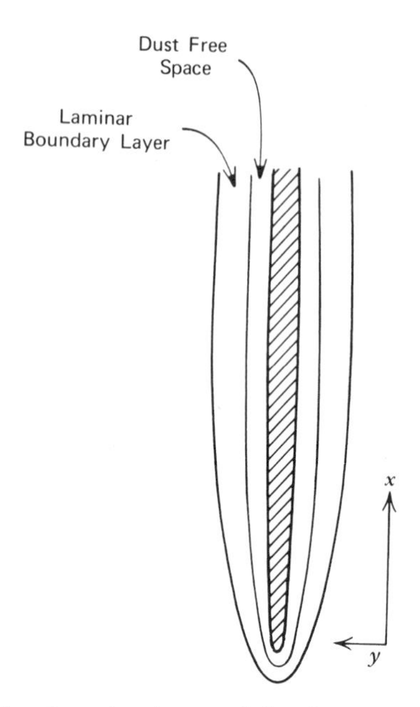

Fig. P7 Laminar boundary layer and dust-free space on a heated vertical plate.

A mathematical analysis has been carried out for the laminar boundary layer with gas properties independent of temperature and the results have been verified experimentally (Schlichting, 1968, pp. 300 ff.). It is found that the temperature gradient at the wall is given by

$$\left(\frac{\partial T}{\partial y}\right)_0 = -0.508(T_w - T_\infty)C_x{}^{-1/4}$$

where

$$C = \left[\frac{g(T_w - T_\infty)}{4\nu^2 T_0}\right]$$

The velocity component normal to the plate is directed toward the surface. Near the surface, this component can be represented by the first term in its expansion:

$$v_y = -0.675\nu C^3 y^2 x^{-3/4} + 0(y^3)$$

where the negative sign indicates that the flow is toward the plate.

Small particles in the boundary layer are exposed to two competing effects. The air velocity tends to carry the particles to the plate, but the temperature gradient at the wall tends to repel them. The result is a *dust-free space* near the surface (Watson, H. H. (1936) *Trans. Faraday Soc.*, **32**, 1073). Derive an expression for the thickness of the dust-free space assuming that it falls within a region where the velocity component normal to the surface can be represented by the first term in its series expansion. (Zernik, W. (1957) *Brit. J. Appl. Phys.*, **8**, 117).

Calculate the thickness of the dust-free space at a point 10 cm up from the leading edge of a plate for $d_p = 1\ \mu m$ and a particle thermal conductivity of 3.5×10^{-4} cal/cm sec °K. The air temperature is 25°C, and the plate temperature is 55°C. Compare your result with the thickness of the laminar boundary layer that can be shown to be of the order of $x/G_x^{1/4}$.

References

Chen, C. Y. (1955) *Chem. Rev.*, **55**, 595; also Filtration of Aerosols by Fibrous Media, Annual Report Eng. Exp. Station, Univ. of Illinois, January 30, 1954.

Davies, C. N. (1973) *Air Filtration*, Academic, London. This book includes an historical record of the subject as well as a fairly extensive bibliography. There is a review of the effects of fiber–fiber interaction on efficiency according to various models. Also discussed are effects of electrical forces. A few comparisons of theory and experiment are presented.

Deissler, R. G. (1955) NACA Report 1210.

Dobry, R., and Finn, R. K. (1956) *Ind. Eng. Chem.*, **48**, 1540.

Friedlander, S. K. (1967) *J. Colloid Interface Sci.*, **23**, 157.

Landau, L. D., and Lifshitz, E. M. (1959) *Fluid Mechanics*, Addison-Wesley, Reading, Mass.

Levich, V. G. (1962) *Physiochemical Hydrodynamics* (English translation), Prentice-Hall, Englewood Cliffs, N. J., pp. 80–85. This reference contains much information on diffusion in aqueous solution. Included are derivations of expressions for mass transfer coefficients for different flow regimes with thin concentration boundary layers.

Lin, C. S., Moulton, R. W., and Putnam, G. L. (1953) *Ind. Eng. Chem.*, **45**, 640.

McCain, J. D., Gooch, J. P., and Smith, W. B. (1975) *J. Air Poll. Control Assoc.*, **25**, 117.

Natanson, G. L. (1957) *Dokl Akad. Nauk SSSR*, **112**, 100.

Rosenhead, L. (Ed.) (1963) *Laminar Boundary Layers*, Oxford University Press, Oxford.

Schlichting, H. (1968) *Boundary Layer Theory*, McGraw-Hill, New York, 6th ed. This is the classical collection of velocity boundary layer solutions, including many different geometries and flow regimes. These velocity distributions can be used to calculate mass transfer coefficients using concentration boundary layer theory. Some solutions that are directly applicable to particle diffusion are given in a chapter on thermal boundary layers.

Sellars, J. R., Tribus, M., and Klein, J. S. (1956) *Trans. ASME*, **78**, 441.

Spurny, K. R., Lodge, J. P., Jr., Frank, E. R., and Sheesley, D. C. (1969) *Environ. Sci. Tech.*, **3**, 453.

Wong, J. B., Ranz, W. E., and Johnstone, H. F., (1956) *J. Appl. Phys.*, **27**, 161; also Tech. Report 11, Eng. Expt. Station, Univ. of Illinois, October 31, 1953.

C H A P T E R 4

Inertial Deposition

Suspended particles may not be able to follow the motion of an accelerating gas because of their inertia. This effect is most important for particles larger than 1 μm. It may lead to particle deposition on surfaces, a process known as *inertial deposition*. Inertial deposition is often the controlling mechanism for the removal of larger particles in gas-cleaning devices, such as filters, scrubbers, and cyclone separators. In filters, inertial deposition acts as the air stream passes around the individual fibers or grains composing the filter. The particles that cannot follow the air stream impact on the collecting element. In scrubbers, the collecting elements are water droplets, whereas in cyclone separators, it is the rotating gas stream that deposits the particles on the wall. In each case, it is the acceleration of the gas that leads to deposition. Inertial effects also play a role in atmospheric processes such as rain scavenging and deposition on vegetation and man-made structures.

Unlike diffusion, which is a stochastic process, particle motion in the inertial range is deterministic. The calculation of inertial deposition rates is usually based either on a force balance on a particle or on a direct analysis of the equations of fluid motion in the case of colliding spheres. Few simple, exact solutions of the fundamental equations are available, and it is usually necessary to resort to numerical computations. For a detailed review of many experimental and theoretical studies of the behavior of particles in the inertial range, the reader is referred to Fuchs (1964).

For particles smaller than about 1 μm, diffusion becomes increasingly important and the methods of calculation of Chap. 3 become applicable. A satisfactory theory for the transition from deposition by diffusion to deposition by inertial forces has not been developed.

4·1 Particle–Surface Interactions: Low Speeds

As two surfaces approach each other, the fluid between them must be displaced. First, we consider the case of two plane parallel circular disks of radius a approaching each other along their common axis (Fig. 4·1).

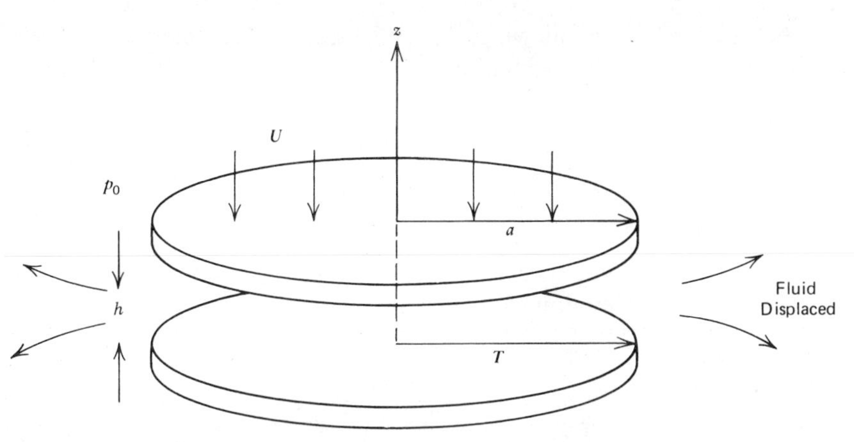

Fig. 4·1 Outward radial flow between two flat disks of radius a. Bottom disk is fixed, and top advances with velocity U, which can change slowly with time.

The disks are immersed in a fluid in which the pressure is p_0. Without loss of generality, it is possible to assume one of the disks is fixed and the other in relative motion. The motion is sufficiently slow to neglect the inertial and unsteady terms in the equations of fluid motion.

The flow is axisymmetric, and for low velocities, the pressure gradient across the gap in the z direction, $\partial p/\partial z$, can be neglected. Hence the pressure is a function only of r. The applicable equations are the r component of the equation of motion:

$$\mu\left(\frac{\partial^2 v_r}{\partial r^2}+\frac{1}{r}\frac{\partial v_r}{\partial r}-\frac{v_r}{r^2}+\frac{\partial^2 v_r}{\partial z^2}\right)=\frac{dp}{dr} \tag{4·1}$$

and the continuity relation:

$$\frac{1}{r}\frac{\partial r v_r}{\partial r}+\frac{\partial v_z}{\partial z}=0 \tag{4·2}$$

with the boundary conditions:

$$\text{at}\quad \begin{array}{ll} z=0 & v_r=v_z=0 \\ z=h & v_r=0,\quad v_z=-U \\ r=a & p=p_0 \end{array} \tag{4·3}$$

where h is the distance between the disks and p_0 is the pressure in the external region.

We now make the following "educated guesses" for the forms of the velocity and pressure distributions:

$$v_r = rZ(z) \tag{4·4}$$

$$\frac{dp}{dr} = Br \tag{4·5}$$

where B is a constant. These assumed forms are tested by substitution in the equation of motion (4·1):

$$\mu \frac{d^2Z}{dz^2} = B$$

Integrating with the boundary conditions (4·3) to obtain $Z(z)$, the following expression is found for the radial velocity:

$$v_r = \frac{1}{2\mu} \frac{dp}{dr} z(z-h) \tag{4·6}$$

If the continuity relation (4·2) is integrated with respect to z across the gap between the disks, making use of the boundary conditions on v_z, the result is

$$U = \frac{1}{r} \frac{d}{dr} \int_0^h r v_r \, dz = -\frac{h^3}{12\mu r} \frac{d}{dr}\left(r \frac{dp}{dr}\right)$$

Integrating twice and remembering that U is not a function of r, the result for the variation of the pressure with radial position is

$$p = p_0 + \frac{3\mu U}{h^3}(a^2 - r^2) \tag{4·7}$$

which is consistent with the assumption (4·5) for the pressure distribution. The drag force on the moving disk is calculated by integrating the pressure difference inside and outside the gap over the surface:

$$F = \int_0^a (p - p_0) 2\pi r \, dr$$

$$= \frac{3\pi\mu U a^4}{2h^3} \tag{4·8}$$

which is the result obtained by Reynolds in 1886. (See also Landau and Lifshitz, 1959, p. 70.) The resistance becomes infinitely great as the disks approach each other.

How, then, can two flat surfaces ever approach close enough to stick? In part, the explanation lies in the London–van der Waals forces which are attractive in nature. For the interaction between two flat plates, an approximate form can be derived for the attractive force per unit area:

$$F_{vdw} = -\frac{\pi Q}{6h^3} \tag{4·9}$$

where the constant Q is discussed in Section 2·8. Equating the resistance to the disk motion to the attractive force, the following result is obtained for the velocity:

$$U = \frac{Q}{9\mu a^2} \tag{4·10}$$

The resulting velocity is constant, and the two disks come in contact in a finite time period.

The drag on a sphere approaching a flat plate has been computed by solving the equations of fluid motion without inertia. The result of the calculation can be expressed in the form

$$F = 3\pi\mu d_p U G\left(\frac{h}{d_p}\right) \tag{4·11}$$

where the drag has dimensions of force and h is the minimum separation between sphere and plate. The dimensionless function $G(h/d_p)$, which acts as a correction factor for Stokes law is shown in Table 4·1.

TABLE 4·1

STOKES LAW CORRECTION FOR MOTION PERPENDICULAR TO A FLAT PLANE (HAPPEL AND BRENNER, 1965)

$(h+a_p)/a_p$	G
1	∞
1.128	9.25
1.543	3.04
2.35	1.837
3.76	1.413
6.13	1.221
10.07	1.125
∞	1

As the sphere approaches the surface, the drag increases to an infinite value on contact. To test the theory, experiments have been carried out with nylon spheres falling through silicone oil. Good agreement between theory and experiment was found up to distances close to the wall (Fig. 4·2).

For $h/d_p \ll 1$, the drag on a sphere approaches the form (Charles and Mason, 1960):

$$F = \frac{3}{2}\frac{\pi\mu d_p^2 U}{h} \tag{4·12}$$

In the case of a gas, when the particle arrives at a point of the order of a mean free path from the surface, the continuum theory on which the calculation of resistance is based no longer applies. van der Waals forces operate freely and lead to adhesion, as in the case of the disk, provided that the rebound effects discussed in the next section do not intervene.

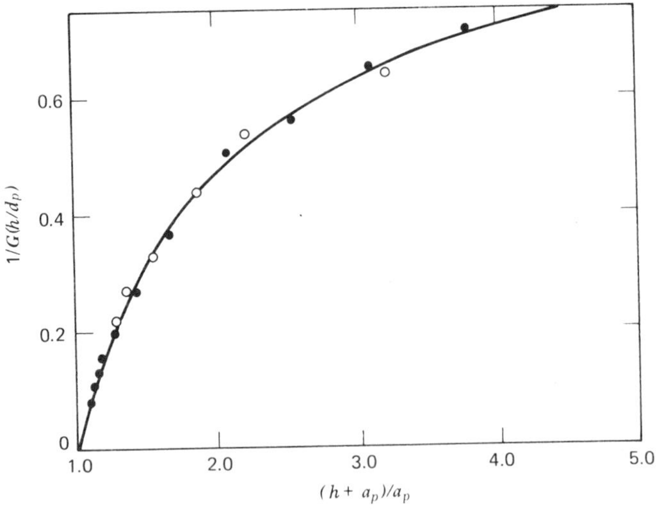

Fig. 4·2 Test of theoretical relation for the correction to Stokes law

$$\frac{1}{G(h/d_p)} = \frac{F}{3\pi\mu d_p}$$

for the approach of a sphere to a fixed plate as a function of the distance, h, from the plate. Data for nylon spheres of radii $a_p = 0.1588$ cm (open points) and 0.2769 cm (solid points) falling through silicone oil of viscosity 1040 poise. The line is calculated from the equations of fluid motion (Table 4.1) (MacKay, Suzuki, and Mason, 1963).

4·2 Particle–Surface Interactions: High Speeds

At low impact velocities, particles striking a surface adhere, but as the velocity increases, rebound occurs. For simplicity, we consider the case of a spherical particle moving normal to the surface of a semi-infinite target through a vacuum. In this way, fluid mechanical effects are eliminated. Bounce occurs when the kinetic energy of the rebounding particles is sufficient to escape attractive forces at the surface. Let $v_{1\infty}$ be the particle approach velocity at large distances from the surface (away from the immediate neighborhood of the attractive forces) and $\Phi_1(z)$ be the potential energy function for the particle surface interaction where z is the distance normal to the surface. The kinetic energy of the particle near the surface just before contact is given by $mv_{1\infty}{}^2/2+\Phi_{10}$, and the velocity of rebound v_{20} just after impact is given by

$$\frac{mv_{20}{}^2}{2}=e^2\left(\frac{mv_{1\infty}{}^2}{2}+\Phi_{10}\right) \tag{4·13}$$

where $\Phi_{10}=\Phi_1(0)$ and e is the coefficient of restitution. The velocity at large distances from the surface away from the interfacial force field, $v_{2\infty}$, is given by

$$\frac{mv_{2\infty}{}^2}{2}+\Phi_{20}=\frac{mv_{20}{}^2}{2} \tag{4·14}$$

Substituting, the result for $v_{2\infty}$ is

$$\frac{v_{2\infty}}{v_{1\infty}}=\left[e^2-\frac{\Phi_{20}-e^2\Phi_{10}}{mv_{1\infty}{}^2/2}\right]^{1/2} \tag{4·15}$$

When $v_{2\infty}=0$, a particle cannot escape the surface force field because all of the rebound energy is required to lift it out of the attractive field of the surface. The critical approach velocity corresponding to $v_{2\infty}=0$ is given by

$$v_{1\infty}^*=\left[\frac{2}{me^2}(\Phi_{20}-e^2\Phi_{10})\right]^{1/2} \tag{4·16}$$

Particles of higher velocity bounce, whereas those of lower velocity stick. For $\Phi_{10}=\Phi_{20}=\Phi_0$, this becomes

$$v_{1\infty}^*=\left[\frac{2\Phi_0}{m}\left(\frac{1-e^2}{e^2}\right)\right]^{1/2} \tag{4·17}$$

The coefficient of restitution depends on the mechanical properties of the particle and surface. For perfectly elastic collisions, $e=1$ and the particle energy is conserved after collision. Deviations from unity result from dissipative processes, including internal friction, that lead to the generation of heat and the radiation of compressive waves into the surface material.

Studies of particle–surface interactions in a vacuum can be carried out by means of the particle beam apparatus (Dahneke, 1975). Particle beams are set up by expanding an aerosol through a nozzle or capillary into a vacuum chamber. Because of their inertia, the particles tend to continue in their original direction through the chamber where the gas is pumped off. A well-defined particle beam can be extracted into a second low-pressure chamber through a small collimating hole. The beam can then be directed against a target to study the rebound process (Fig. 4·3).

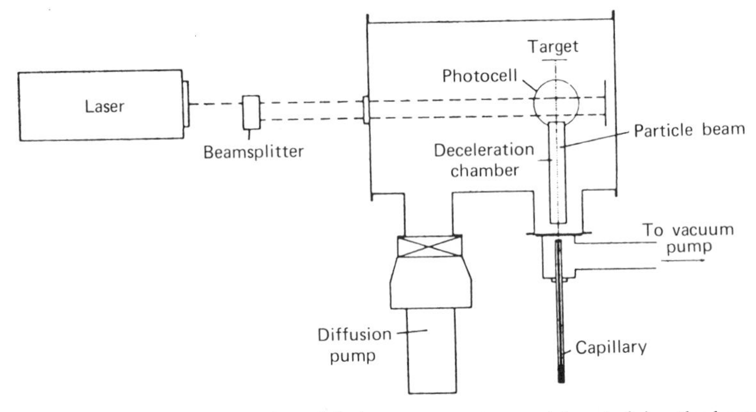

Fig. 4·3 Schematic diagram of particle beam apparatus used for studying the bouncing of small particles off surfaces (Dahneke, 1975).

In his experiments, Dahneke studied the bouncing of polystyrene latex spheres about 1 μm in diameter from polished quartz and other surfaces. Using a laser light source, he measured the velocities of the incident and reflected particles. The velocity of the incident particles was controlled by means of a deceleration chamber.

The coefficient of restitution was found from (4·15) by measuring $v_{2\infty}/v_{1\infty}$ for very large values of $v_{1\infty}$. Figure 4·4 shows the experimental data for 1.27 μm polystyrene latex particles and a polished quartz surface. Shown also is the curve representing (4·15) with $e=0.960$ and

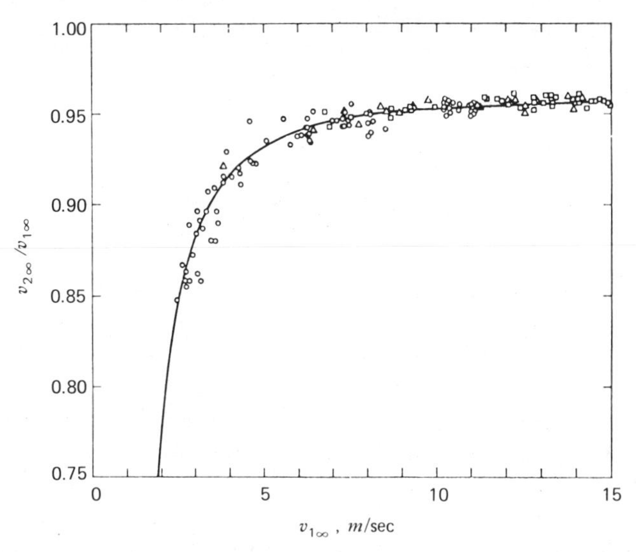

Fig. 4·4 Bouncing of 1.27 μm polystyrene latex particles from polished quartz surfaces. The curve is (4.15) with the value of the coefficient of restitution $e=0.960$ obtained by extrapolation to very large values of $v_{1\infty}$ (Dahneke, 1974).

$\Phi_{20}=\Phi_{10}$ chosen to give the best agreement between theory and experiment. Values of Φ_0 determined in this way were several hundred times higher than values calculated from van der Waals force theory. The value of the critical impact velocity, $v^*_{1\infty}$, was estimated to be 120 cm/sec by extrapolation to $v_{2\infty}/v_{1\infty}=0$.

A complete model for particle–surface interaction would include both fluid mechanical effects as the particle approaches the surface *and* elastic and surface forces. The fluid mechanical calculations would take into account free molecule effects as the particle comes to within one mean free path of the surface. The presence of thin films of liquids and surface irregularities further complicate the situation. In practice, the design of cascade impactors (Chap. 6) and other devices in which rebound may be important is carried out empirically, by experimenting with various particles, coatings, and collecting surfaces.

4·3 Particle Acceleration at Low Reynolds Numbers: Stokes Number

When a sphere moves at constant velocity through a fluid at rest, the drag force at low Reynolds numbers ($Re \ll 1$) is given by Stokes law. For an accelerating sphere moving in a straight line through a fluid at rest, the drag at low Reynolds numbers is given by

$$F = -3\pi\mu d_p u - \tfrac{1}{12}\pi\rho d_p^3 \frac{du}{dt} - \tfrac{3}{2}\pi\rho d_p^2 \left(\frac{\nu}{\pi}\right)^{1/2} \int_{-\infty}^{t} \frac{du}{dt'} \frac{dt'}{(t-t')^{1/2}} \qquad (4\cdot 18)$$

where ρ is the density of the fluid (Basset, 1910; Landau and Lifshitz, 1959, p. 97). The first term on the right-hand side is equivalent to the Stokes drag. The second and third terms arise from the particle acceleration, du/dt. The second term represents the resistance of an inviscid fluid to an accelerating sphere. It is equivalent to adding $\pi\rho d_p^3/12$ to the mass of the sphere, that is, one-half the mass of an equal volume of fluid. This term can be neglected for spheres of normal densities in air.

As a result of the integral term, the drag at any time depends on the past history of the sphere. This term, proportional to $\rho^{1/2}$, can also be neglected under normal circumstances for spheres of unit density in air (Fuchs, 1964). *Hence at low Reynolds numbers, the drag on accelerating spherical particles in a gas can usually be approximated by using Stokes law.* This assumption was made in the derivation of the Stokes–Einstein relation for the diffusion coefficient even though the accelerated motion was nonrectilinear (Chap. 2).

When a particle is projected into a stationary fluid with a velocity u_0, it will travel a finite distance before coming to rest as $t \rightarrow \infty$. The distance can be calculated to a close approximation by integration of the force balance on the particle:

$$m\frac{du}{dt} = -fu$$

to give

$$u = u_0 e^{-t/\tau}$$

and

$$x = u_0\tau(1 - e^{-t/\tau})$$

for the velocity and displacement, respectively. The characteristic time $\tau = \rho_p d_p^2/18\,\mu$. As $t/\tau \rightarrow \infty$, the distance that the particle penetrates, or stop-distance, is given by

$$s = \frac{\rho_p d_p^2 u_0}{18\,\mu} \qquad (4\cdot 19)$$

The dimensionless ratio of the stop distance to a characteristic length such as the diameter of a filter element (fiber or grain) or the viscous sublayer thickness is called the *Stokes number*. As shown in the sections that follow, the Stokes number plays an important role in the analysis of inertial deposition.

4·4 Similitude Law for Impaction: Stokesian Particles

When an aerosol flows over a collecting surface, particles will strike the body if their trajectories come within one particle radius of the surface (Fig. 4·5). Whether or not the particles adhere to the surface depends on the attractive and rebound energies, as discussed in previous sections.

At large distances from the collector, the fluid velocity is uniform, and the particle velocity is equal to that of the fluid. To determine the collection efficiency, it is, in general, necessary to solve the equations of fluid motion numerically. When the particles are much smaller than the collector, the flow fields for the particle and collector can be uncoupled. For the collector, the velocity distribution is determined by the Reynolds number based on collector diameter, independent of the presence of the particles. The particle is assumed to be located in a flow with a velocity at infinity equal to the *local* velocity for the undisturbed

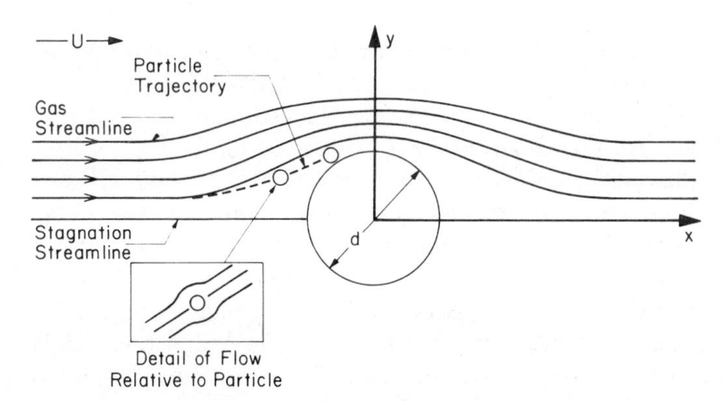

Fig. 4·5 Impaction of a small spherical particle on a cylinder placed normal to the flow. The particle is unable to follow the fluid streamline because of its inertia. The drag on the particle is calculated by assuming that it is located in a uniform flow with velocity at infinity equal to the local fluid velocity (see detail). Fore-and-aft symmetry of the streamline exists for low Reynolds number flows and for an inviscid flow, but the shapes of the streamlines differ.

gas flow around the collector; the drag on the particle is determined by the local relative velocity between particle and gas.

When the Reynolds number for the particle motion is small, a force balance can be written on the particle assuming Stokes law holds for the drag. For the x direction,

$$m\frac{du}{dt} = -f(u - u_f) \tag{4·20}$$

whereas for the vector velocity,

$$m\frac{d\mathbf{u}}{dt} = -f(\mathbf{u} - \mathbf{u}_f) \tag{4·21}$$

Here u is the particle velocity, u_f is the local fluid velocity, and f is the Stokes friction coefficient. We call particles that obey this equation of motion *Stokesian particles*. The use of (4·21) is equivalent to employing (4·18), neglecting the acceleration terms containing the gas density. Since (4·18) was derived for rectilinear motion, the extension to flows with velocity gradients and curved streamlines adds further uncertainty to this approximate method.

In dimensionless form, the equation of particle motion can be written:

$$St\frac{d\mathbf{u}_1}{d\theta} = -(\mathbf{u}_1 - \mathbf{u}_{f1}) \tag{4·22}$$

where $\mathbf{u}_1$ is the particle velocity normalized with respect to its velocity at large distances from the surface, U, and $\theta = tU/L$, where L is the characteristic length of the body. The dimensionless group $St = \rho_p d_p^2 U/18\,\mu L$, the Stokes number, was discussed in the last section.

For the mechanical behavior of two particle–fluid systems to be similar, it is necessary to have geometric, hydrodynamic, and particle trajectory similarity. Hydrodynamic similarity is achieved by fixing the Reynolds number for the flow around the collector. By (4·22) similarity of the particle trajectories depends on the Stokes number. Trajectory similarity also requires that the particle come within one radius of the surface at the same relative location. This means that the interception parameter, $R = d_p/L$, must also be preserved.

For Stokesian particles, two impaction regimes are similar when the Stokes, Interception, and Reynolds numbers are the same. The impaction efficiency, η_R, as in the case of diffusion, is defined as the ratio of the volume of gas cleared of particles by the collecting element to the total volume swept out by the collector. If all particles coming within one radius of the collector adhere,

$$\eta_R = f(St, R, Re) \tag{4·23}$$

As the particle approaches a surface, the use of Stokes law for the force acting on the particle becomes an increasingly poor approximation. Mean free path effects, van der Waals forces, and the hydrodynamic effects discussed in Section 4·1 complicate the situation. The importance of these effects is difficult to judge because reliable calculations and good experimental data are lacking.

4·5 Impaction on Cylinders: Stokesian Particles

Flow around single cylinders is the elementary model for the fibrous filter and is the geometry of interest for deposition on pipes, wires, and other such objects in an air flow (Chap. 3). The flow patterns at low and high Reynolds numbers differ significantly, and this affects impaction efficiencies.

For $Re \ll 1$, the stream function calculated from the Oseen approximation of the equations of fluid motion (3·9) is applicable in the region near the cylinder. This form is adequate for calculations of diffusional transport because the concentration boundary layer is confined to a thin region near the surface of the cylinder. For impaction, however, the behavior of particles at larger distances from the cylinder must be taken into account.

Numerical computations of the flow field around a circular cylinder have been made for several values of the Reynolds number in the range between 1 and 50. These are reviewed by Batchelor (1967) who cites the original references. Flow patterns for two different Reynolds number in this range are shown in Fig. 4·6.

For $Re > 100$, the velocity distribution can be determined from inviscid flow theory, outside the velocity boundary layer. The components of the velocity in the direction of the mainstream flow, x, and normal to the main flow, y, are

$$u_{f1} = 1 + \frac{y_1^2 - x_1^2}{(x_1^2 + y_1^2)^2} \tag{4·24a}$$

$$v_{f1} = -\frac{2x_1 y_1}{(x_1^2 + y_1^2)^2} \tag{4·24b}$$

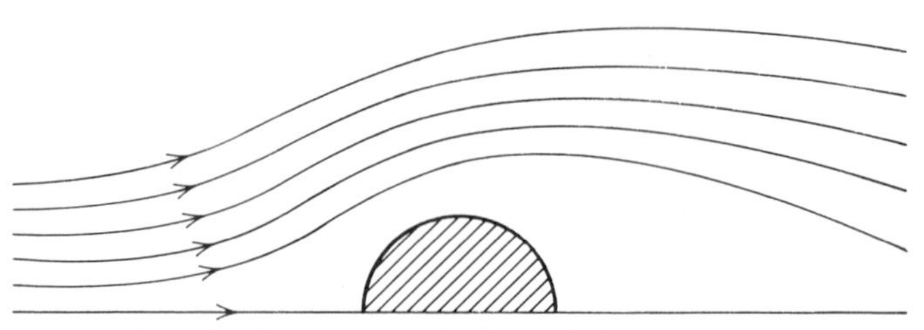

Fig. 4·6*a* Streamlines for flow past a circular cylinder at $Re=4$. There is marked fore-and-aft asymmetry of the flow pattern but no standing eddies.

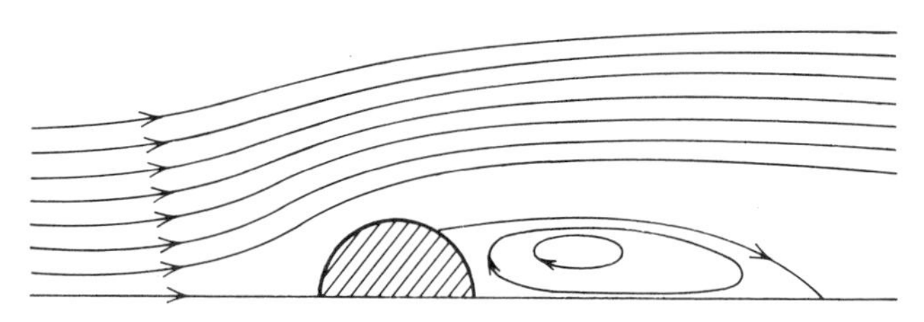

Fig. 4·6*b* Streamlines for flow past a circular cylinder at $Re=40$. Note the standing eddies behind the cylinder (after Batchelor, 1967).

where $x_1=x/a$ and $y_1=y/a$. Both x and y are measured from the axis of the cylinder.

Numerical calculations for the impaction of Stokesian particles have been carried out for the velocity distribution given by (4·24*a*) and (4·24*b*). In making the calculations, it is assumed that the particle velocity at large distances upstream from the cylinder is equal to the air velocity. A linearized form of the equations of particle motion is solved for the region between $x_1=-\infty$ and $x_1=-5$. A numerical solution is obtained for the region between $x_1=-5$ and the surface of the cylinder.

Measurements of the impaction efficiency for cylinders and spheres by Ranz and Wong (1952) represent the main body of data on which comparisons of theory and experiments are based. The experiments were carried out with a 3 mil platinum wire as the cylindrical collector and air velocities ranging from 428 to 5090 cm/sec, corresponding to Reynolds numbers between 13 and 330, based on the wire diameter. Monodisperse aerosols of sulfuric acid with diameters ranging from 0.56 to 1.30 μm were set up using a condensation aerosol generator. The amount of aerosol collected was determined by washing the wire with a known quantity of water and measuring the concentration of the wash water with a conductance cell.

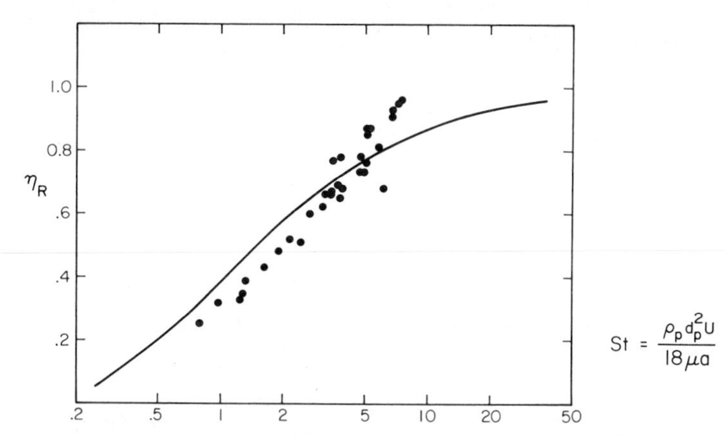

Fig. 4·7 Comparison of theory and experiment for impaction on single cylinders. Theoretical calculations are by Brun et al. (1955) for point particles in an inviscid flow, and the data are those of Wong and Johnstone (1953) for $100 < Re < 330$.

The results of the experiments are compared with inviscid flow theory in Fig. 4·7. Experimental data agree fairly well with the theory over the entire range of Stokes numbers. Deviations between theory and experiment arise for several reasons, aside from experimental error. The theory does not take into account the velocity boundary layer or the extra drag on a particle near the surface. These effects would tend to reduce the efficiency, as observed at the lower Stokes numbers. Flow separation and the formation of circulating eddies at the rear of the cylinder occur at these Reynolds numbers (Fig. 4·6*b*), tending to increase deposition efficiency. A quantitative estimate of this effect is difficult.

4·6 Impaction at High Reynolds Numbers: The Critical Stokes Number

By analyzing the motion of a small particle in the region near the stagnation point, it can be shown that for an inviscid flow, impaction does not occur until a critical Stokes number is reached. For an inviscid flow, the first term in an expansion of the velocity along the streamline in the plane of symmetry which leads to the stagnation point is given by

$$u_f = -bU(x_1 + 1) \qquad (4\cdot25)$$

where the dimensionless constant b depends on the shape of the body and the dimensionless coordinate $x_1 = x/a$ is negative (Fig. 4·5).

For a Stokesian particle, the equation of motion along the stagnation streamline takes the form

$$St\frac{d^2x_1'}{d\theta^2}+\frac{dx_1'}{d\theta}+bx_1'=0 \tag{4·26}$$

where $x_1' = x_1 + 1$ and $\theta = tU/a$. The solution to this equation is given by

$$x_1' = A_1e^{\lambda_1\theta} + A_2e^{\lambda_2\theta} \tag{4·27}$$

where A_1 and A_2 are integration constants and λ_1 and λ_2 are the roots of the characteristic equation

$$St\lambda^2+\lambda+b=0$$

and are given by

$$\lambda_{1,2} = -\frac{1}{2St}\left[1 \pm (1-4bSt)^{1/2}\right] \tag{4·28}$$

From (4·27), the velocity can be determined as a function of x_1 in the region near the stagnation point where the linear approximation (4·25) holds. For $St > St_{crit} = 1/4b$, the roots are complex conjugates and the u versus x diagram has a focal or spiral point at the stagnation point (Fig. 4·8). The first intersection of the spiral with the u-axis corresponds to

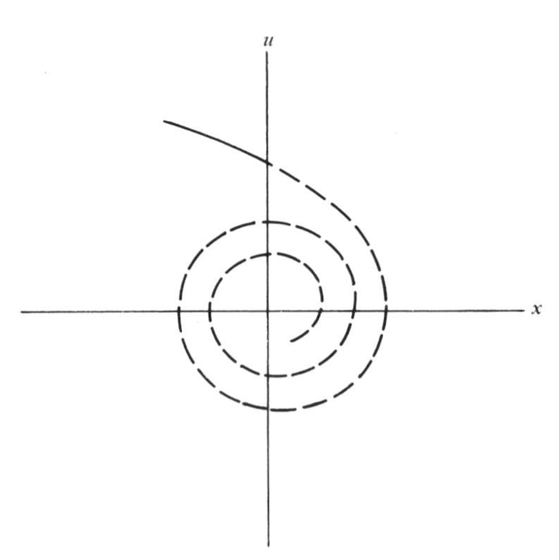

Fig. 4·8a Spiral point corresponding to $St > St_{crit}$. The particle velocity at the surface ($x = 0$) is positive, and the impaction efficiency is nonvanishing. Only the solid portion of the curve is physically meaningful.

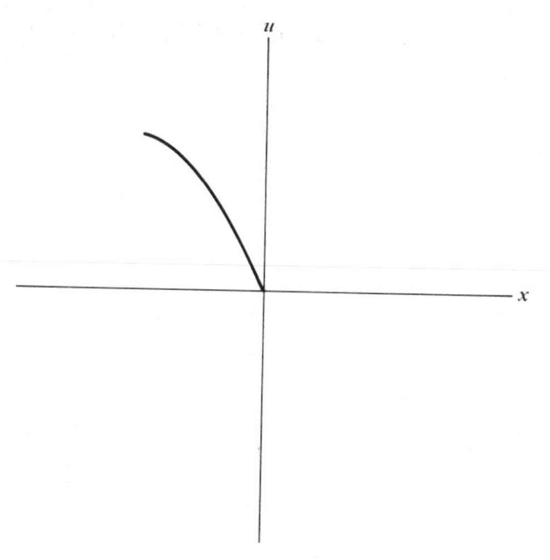

Fig. 4·8*b* Nodal point corresponding to $St < St_{crit}$. Particle velocity vanishes at the surface and the impaction efficiency is zero.

the finite velocity at the stagnation point. The rest of the spiral has no physical significance, since the particle cannot pass the surface of the collector.

For $St < 1/4b$, the roots are real and both are negative. The u versus x diagram has a nodal point that corresponds to zero particle velocity at the forward stagnation point and zero impaction efficiency. For $St = 1/4b$, the roots are equal and the system again has a node as its singularity. Thus $St = 1/4b$ represents a lower limit below which the impaction efficiency vanishes. For inviscid flow around cylinders, $b = 2$ and $St_{crit} = 1/8$, whereas for a sphere, $b = 3$ and $St_{crit} = 1/12$, based on the radius of the collector.

This analysis provides a lower anchor point for impaction efficiency diagrams, such as Fig. 4·7. It applies also to non-Stokesian particles (Section 4·7) because the point of vanishing efficiency corresponds to zero relative velocity between particle and gas. Hence Stokes law can be used for the particle motion near the stagnation point. This is one of the few impaction problems for which an analytical solution is possible. Impaction from a supersonic flow over a wedge is another (Serafini, 1954).

Example. In certain types of heat exchangers, a gas flows normal to a bank of tubes carrying fluid at a different temperature, and heat transfer occurs at the interface. Fouling of the outside surface of the tubes occurs as a result of particles depositing from the flow. If the tubes are 1 in. (outside diameter) and the gas velocity is 10 ft/sec, estimate the diameter of the largest particle ($\rho_p = 2$ g/cm^3) that can be permitted in the gas stream *without* deposition by impaction on the tubes. Assume the gas has the properties of air at 100°C ($\nu = 0.33$ cm^2/sec, $\mu = 2.18\ (10)^{-4}$ g/cm sec).

SOLUTION. As an approximation, take as a model particle deposition on a single cylinder placed normal to an aerosol flow. The Reynolds number for the flow, based on the cylinder diameter, is 2320, which is sufficiently large to use the potential flow approximation for the stagnation region. By Table 4·2, the critical Stokes number for the cylinder is

$$\frac{\rho_p d_p^2 U}{18\ \mu a} = \frac{1}{8}$$

Substituting and solving for d_p, the result is $d_p \approx 10\ \mu$m. Provided all particles are smaller than this, there will be no deposition by impaction on the tubes (to the extent that the model is applicable).

The analysis neglects boundary layer effects and is probably best applied when the particle diameter is larger than or of the order of the boundary layer thickness. The change in the drag law as the particle approaches the surface is also not taken into account. Hence the criterion provides only a rough estimate of the range in which the impaction efficiency becomes small.

For most real (viscous) flows, $u_f \sim (x_1 + 1)^2$ in the region near the stagnation point because of the no-slip boundary condition and the continuity relation. In this case, (4·25) does not apply, the equation of particle motion cannot be put into the form of (4·26), and the analysis developed previously is not valid.

4·7 Impaction of Non-Stokesian Particles

In previous sections, we have considered the case of small particles that follow Stokes law all along their trajectories. For high velocity flows around a collector, larger particles retain their velocities as they approach the surface, and the Reynolds number for their motion may be too large for the Stokes law approximation to hold. The approach taken previously must then be modified to account for the change in the form of the drag law.

A general expression for the drag on a fixed spherical particle in a gas of constant velocity, U, can be written as follows:

$$F = \frac{\rho \pi d_p^2}{8} C_D U^2 \tag{4·29}$$

This expression defines the drag coefficient, C_D, which by dimensional analysis is a function of the Reynolds number based on particle diameter and gas velocity. For $Re \ll 1$, Stokes law is applicable, and $C_D = 24/Re$. Rewriting (4·29) to include the Stokes form, we obtain

$$F = \frac{C_D Re}{24} (3\pi\mu d_p U) \tag{4·30}$$

In general, it is necessary to use the results of experiments and semi-empirical correlations to relate the drag coefficient to the Reynolds number. The expression

$$C_D = \frac{24}{Re}(1 + 0.158\ Re^{2/3}) \tag{4·31}$$

agrees well with experiment for $Re < 1000$ (Serafini, 1954).

We consider now the case of steady flow of an aerosol around a collector such as a cylinder or sphere. A force balance can be written on a particle by considering it to be fixed in a uniform flow of velocity $\mathbf{u} - \mathbf{u}_f$:

$$m \frac{d\mathbf{u}}{dt} = -\frac{C_D Re}{24} 3\pi\mu d_p (\mathbf{u} - \mathbf{u}_f) \tag{4·32}$$

where C_D is a function of the *local* Reynolds number of the particle:

$$Re = \frac{d_p |\mathbf{u} - \mathbf{u}_f|}{\nu}$$

$$= Re_p |\mathbf{u}_1 - \mathbf{u}_{f1}|$$

and

$$|\mathbf{u} - \mathbf{u}_f| = \left[(u - u_f)^2 + (v - v_f)^2 + (w - w_f)^2\right]^{1/2}$$

The particle Reynolds number is defined by

$$Re_p = \frac{d_p U}{\nu}$$

Also

$$\mathbf{u}_1 = \frac{\mathbf{u}}{U}$$

$$\mathbf{u}_{f1} = \frac{\mathbf{u}_f}{U}$$

where U is the gas velocity at large distances from the collector.

As in the case of Stokesian particles, the contribution of particle acceleration to the drag has been neglected. Clearly, (4·32) is on flimsy grounds from a theoretical point of view. Its application should be tested experimentally, but a rigorous validation has never been carried out. In nondimensional form, (4·32) can be written as follows:

$$\frac{d\mathbf{u}_1}{d\theta} = -\frac{C_D Re_p}{24}\frac{1}{St}|\mathbf{u}_1 - \mathbf{u}_{f1}|(\mathbf{u}_1 - \mathbf{u}_f) \tag{4·33}$$

where $\theta = tU/a$ and $St = \rho_p d_p^2 U/18\ \mu a$. The gas velocity, $\vec{u}_{f1}$, is a function of the Reynolds number based on the collector diameter. The drag coefficient is a function of $Re_p|\mathbf{u}_1 - \mathbf{u}_{f1}|$. Hence the trajectory of a non-Stokesian point particle is determined by Re, St, and Re_p; one more dimensionless group, Re_p, appears in the theory than in Stokesian particle theory.

Particle trajectories in inviscid flows around single elements of various shapes have been determined by solving (4.33) numerically; from the trajectories, impaction efficiencies were obtained. The results of such calculations have been reviewed by Golovin and Putnam (1962).

The case that has received most study is that of impaction on right circular cylinders placed normal to the air flow. Original applications were to the icing of airplane wings and to the measurement of droplet size in clouds. The results are also important for deposition on surfaces from the atmosphere and on pipes in heat exchangers.

Numerical calculations have been made by several investigators based on the velocity distribution for an inviscid flow. Limitations on the use of this theory have been discussed in Section 4·5. Results of calculations of impaction efficiency as a function of Stokes number are shown in Fig. 4·9. The curves were originally developed for use in the measurement of droplet sizes in supercooled clouds.

Since the drop size was not known, a new dimensionless group $P = Re_p^2/St = 18\rho^2 Ua/\mu\rho_p$, independent of drop size, was introduced. This new group is formed by combining the two groups on which the efficiency depends in the inviscid flow range. According to the rules of

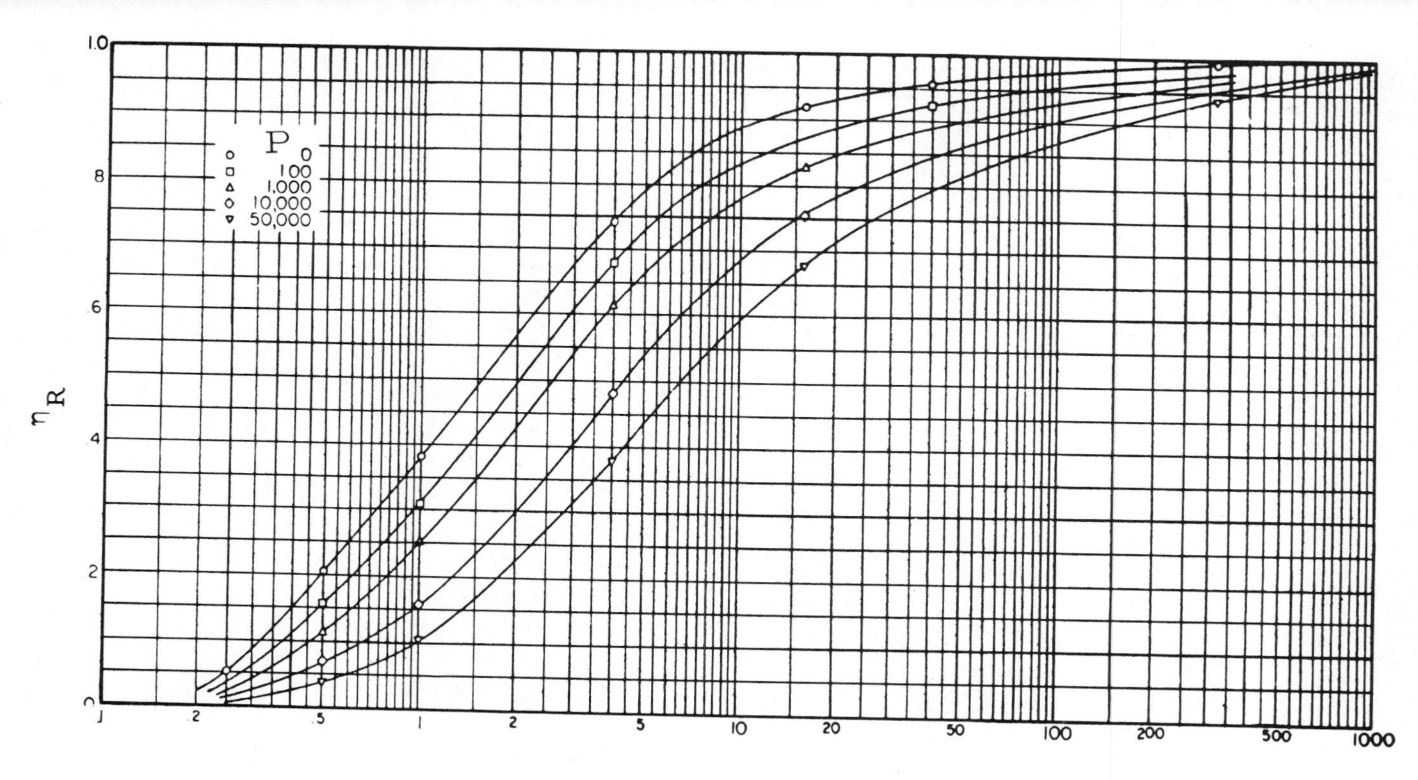

Fig. 4·9 Collection efficiency for cylinders in an inviscid flow with point particles (Brun et al., 1955). The rate of deposition, particles per unit time per unit length of cylinder, is $2\eta_R n_\infty Ua$ where n_∞ is the particle concentration in the mainstream.

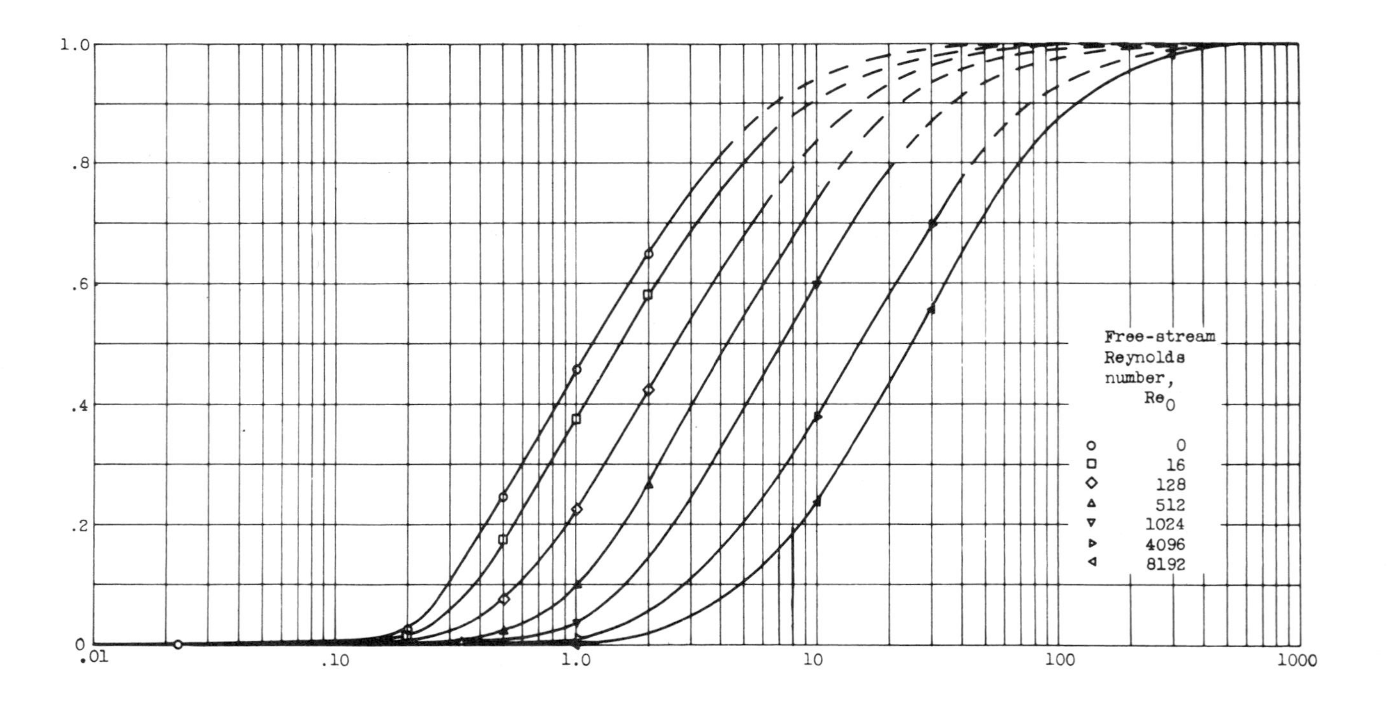

$$St = \rho_p d_p^{\,2} U / 18 \mu a$$

Fig. 4·10 Collection efficiency for spheres in an inviscid flow with point particles (Dorsch, Saper, and Kadow, 1955). The rate of deposition on the sphere, particles per unit time, is given by $\pi a^2 \eta_R U n_\infty$ where n_∞ is the particle concentration in the mainstream. The free stream Reynolds number, Re_0, is based on [illegible] dimensional analysis, this is permissible, but the efficiency is still determined by two groups, which for convenience are chosen to be *St* and *P*. If the collection efficiency is known, the particle size can be determined from Fig. 4·9. More often, the curves are used to determine collection efficiency at known values of *St* and *P*.

4·8 Impaction Efficiencies for Spheres

Of much interest in air pollution and cloud physics is the collision efficiency of drops moving through a cloud of small particles or droplets. Scrubbers used in gas cleaning depend on this process, and it produces a limited amount of atmospheric cleaning during rainfall. (In-cloud transport processes before precipitation occurs are believed to be most important in the atmospheric scavenging of particulate matter.)

For large spheres moving through a cloud of small particles ($R \rightarrow 0$), the efficiency can be calculated numerically from (4·33) as discussed in the foregoing sections on cylinders. For high Reynolds numbers, the velocity distribution outside the laminar boundary layer on the upstream side of the sphere is obtained from inviscid flow theory. Results of such calculations are shown in Fig. 4·10. They are subject to the inaccuracies resulting from neglect of boundary layer effects and of the separation of flow toward the rear of the sphere discussed in the case of the cylinder.

Numerical calculations of the deposition efficiency have also been made for flow around spheres at lower Reynolds numbers taking into account the effects of the wake and sedimentation (Beard, 1974). Fair agreement was found with the results of experiments with 0.4 μm particles and droplets falling under gravity at Reynolds numbers between 40 and 100 (about 0.5 mm drop diameter). The calculations indicated that deposition in the wake was important.

4·9 Deposition from a Rotating Flow: Cyclone Separator

Cyclone separators (Fig. 4·11) are frequently used for the removal of particles larger than a few micrometers from process gases. The dusty gas enters the annular space between the wall and the exit tube through a tangential inlet. The gas acquires a rotating motion, descends along the outer wall, and then rises, still rotating, to pass out the exit tube. Particles move to the outer wall as a result of centrifugal forces. They fall from the slowly moving wall-layer into a hopper at the bottom. Cyclones are inexpensive and can be constructed in local sheetmetal shops. They have no moving parts, and require little maintenance. They

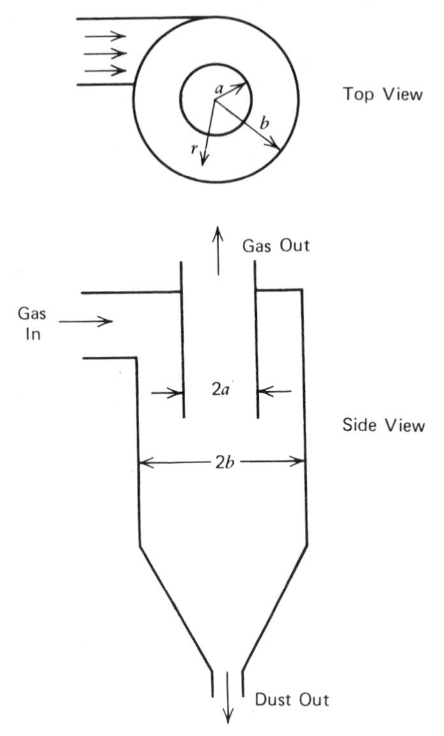

Fig. 4·11 Schematic diagram of cyclone separator. Dusty gas enters through the tangential inlet, spirals down the conical section, and moves up the exit section.

are often used for preliminary cleaning of the coarse fraction before the gases pass to more efficient devices, such as electrostatic precipitators. A common application is in the removal of flyash from gases from pulverized coal combustion.

An approximate analysis of the particle motion and cyclone performance can be carried out by setting up a force balance for Stokesian particles in the radial direction:

$$F_r = m\left[\frac{d^2r}{dt^2} - r\left(\frac{d\theta}{dt}\right)^2\right] = -3\pi\mu d_p (v_r - v_{rf})$$

where r and θ and the axis of the cyclone separator represent a set of cylindrical coordinates. For small particles, the acceleration term d^2r/dt^2 can be neglected; only the second term on the left-hand side—the "centrifugal force" term—is retained. Neglecting also the radial component of the gas velocity, v_{rf}, the result is

$$v_r = \frac{dr}{dt} = \frac{mr}{3\pi\mu d_p}\left(\frac{d\theta}{dt}\right)^2 \tag{4·34}$$

The particle trajectory is determined by rearranging (4·34):

$$\frac{dr}{d\theta} = \frac{mv_\theta}{3\pi\mu d_p} \tag{4·35}$$

For the motion in the θ direction, it is assumed that the particle and gas velocities are equal, $v_\theta = v_{\theta f}$. It is thus possible to solve (4·35) for the particle trajectories, if the gas velocity distribution, $v_{\theta f}$, is known.

Assume that the gas makes a well-defined set of turns in the annular space before exiting. Without specifying the nature of the flow field, the number of turns necessary for complete removal of particles of size d_p can be seen from Fig. 4·11 and (4·35) to be

$$N_t = \frac{\theta}{2\pi} = \frac{3\mu d_p}{2m}\int_a^b \frac{dr}{v_{\theta f}} \tag{4·36}$$

where the particle diameter is neglected by comparison with the dimensions of the channel. This result is independent of whether the flow is laminar or turbulent. Alternatively, the diameter of the smallest particle that can be removed from the gas in N_t turns is obtained by rearranging (4·36) to give

$$d_{p\,min} = \left[\frac{9\mu}{\pi\rho_p N_t}\right]^{1/2}\left[\int_a^b \frac{dr}{v_{\theta f}}\right]^{1/2} \tag{4·37}$$

Smaller particles are removed to an extent that depends on their distance from the wall at the entry.

The aerodynamic pattern is too complex for an exact analysis of the particle motion, and semiempirical expressions are used for $v_{\theta f}$ (Fuchs, 1964). An approximate result of the integration of (4·37) often used in design applications has the form

$$d_{p\,min} \sim \left[\frac{\mu(b-a)}{\rho_p N_t U}\right]^{1/2} \tag{4·38}$$

where U is the average velocity of the gas in the inlet tube. The number of turns, N_t, is usually determined empirically. For fixed gas velocity, performance improves as the diameter of the cyclone is reduced, because the distance the particle must move for collection decreases. It is likely the radial gas velocity components, associated with eddies, are also reduced in the smaller diameter cyclones.

4·10 Particle Eddy Diffusion Coefficient

Small particles in a turbulent gas diffuse from one point to another as a result of the eddy motion. The eddy diffusion coefficient of the particles will in general differ from that of the carrier gas. An expression for the particle eddy diffusivity can be derived for a Stokesian particle, neglecting the Brownian motion. In carrying out the analysis, it is assumed that the turbulence is homogeneous and that there is no *mean* gas velocity. The statistical properties of the system do not change with time. Essentially what we have is a stationary, uniform turbulence in a large box. This is an approximate representation of the core of a turbulent pipe flow, if we move with the mean velocity of the flow.

The analysis is similar to that used in Section 2·2 to derive the Stokes–Einstein relation for the diffusion coefficient. Again we consider only the one-dimensional problem. Particles originally present in the differential thickness around $x=0$ (Fig. 2·2) spread through the fluid as a result of the turbulent eddies. If the particles are much smaller than the size of the eddies, the equation of particle motion is given by (4·20):

$$m\frac{du}{dt} = -f(u-u_f)$$

where u_f is the local velocity of the gas.

Again we have used the Stokes form for the drag force, but based it on the *relative* velocity of the particle. The distance traveled by a particle in time t is obtained by integrating (4·20):

$$x+\frac{u-u_0}{\beta} = \int_0^t u_f(t')\,dt'$$

where $\beta=f/m$. Multiplying the left-hand side by $(1/\beta)(du/dt)+u$ and the right-hand side by u_f, which is permissible by (4·20),

$$\frac{x}{\beta}\frac{du}{dt}+xu+\frac{(u-u_0)}{\beta^2}\frac{du}{dt}+\frac{u^2-u_0u}{\beta} = \int_0^t u_f(t)u_f(t')\,dt' \tag{4·39}$$

We now average each term in (4·39) over all of the particles originally in the element around $x=0$. On a term-by-term basis,

$$\overline{x\frac{du}{dt}} = \frac{d\overline{xu}}{dt} - \overline{u^2} \qquad (4\cdot40)$$

The particle eddy diffusion coefficient is given by

$$\epsilon_p = \overline{xu} = \frac{1}{2}\frac{d\overline{x^2}}{dt} \qquad (4\cdot41)$$

and does not change with time after a sufficiently long interval from the start of the diffusion process. Hence

$$\frac{d\overline{xu}}{dt} = 0 \qquad (t\to\infty) \qquad (4\cdot42)$$

Since the statistical properties of the system do not change with time,

$$\overline{u\frac{du}{dt}} = \frac{1}{2}\frac{d\overline{u^2}}{dt} = 0 \qquad (4\cdot43)$$

After long periods of time, there is no correlation between the particle acceleration and its initial velocity:

$$\overline{u_0\frac{du}{dt}} = 0 \qquad (t\to\infty) \qquad (4\cdot44)$$

or between its velocity and its initial velocity:

$$\overline{u_0 u} = 0 \qquad (t\to\infty) \qquad (4\cdot45)$$

Averaging (4·39) over the particles and substituting (4·40) through (4·45) gives

$$\epsilon_p = \overline{u_f^2}\int_0^{t\to\infty} R(t')\,dt' \qquad (4\cdot46)$$

The coefficient of correlation, R, between gas velocities *in the neighborhood of the particle* at two different times, t_1 and t_2, is defined by the relation

$$\overline{u_f(t_1)u_f(t_2)} = \overline{u_f^2}\,R(t_2-t_1)$$

where $\overline{u_f^2}$ is the mean square gas velocity. The value of this coefficient is near unity when t_2 is near t_1 and then becomes very small as the interval $t_2 - t_1$ increases. The integral in (4·46), which has the dimensions of a time, is assumed to approach a limit rapidly.

By (4·46), the particle eddy diffusivity is proportional to the mean square fluid velocity multiplied by the time scale $\int_0^{t\to\infty} R(t')\,dt'$. This expression is of the same form as the Taylor eddy diffusion coefficient for the turbulent fluid (Goldstein, 1938, p. 217). However, the correlation coefficient in (4·46) applies to the gas velocities *over the path of the particle*. Heavy particles move slowly and cannot follow the fluid eddies that surge around them. Thus the time scale that should be employed in (4·46) ranges between the Lagrangian scale for small particles that follow the gas and the Eulerian time scale for heavy particles that remain almost fixed (Friedlander, 1957).

Some attempts have been made to determine these time scales. In one set of experiments with a homogeneous grid-generated turbulence, the ratio of the Lagrangian to the Eulerian scales was reported to be three (Snyder and Lumley, 1971). Thus the ratio of the eddy diffusivity of a particle that is too heavy to follow the gas motion to the eddy diffusivity of the gas would be 1/3. What is meant by a *heavy* particle in this context? One measure is the ratio of the characteristic particle time $\rho_p d_p^2/18\,\mu$ to the smallest time scale of the fluid motion $(\nu/\epsilon_d)^{1/2}$, where ν is the kinematic viscosity of the gas and ϵ_d is the turbulent energy dissipation (cm^2/sec^3). For a heavy particle, this ratio would be much greater than unity.

For a pipe flow, these conclusions would hold best in the turbulent core. Near the walls, the turbulence is highly nonuniform, and the analysis is no longer applicable.

4·11 Turbulent Deposition

When a turbulent gas carrying particles larger than 1 μm flows parallel to a surface, particles deposit because of the fluctuating velocity components normal to the surface. Particles are unable to follow the eddy motion and are projected to the wall through the relatively quiescent fluid near the surface. The net rate of deposition depends on the relative rates of transport and reentrainment. Experiments have been carried out with turbulent flows through vertical tubes whose walls are coated with a sticky substance to permit measurement of the deposition rate alone. The data are correlated by introducing a particle transfer coefficient $k = |\bar{J}|/\bar{n}_\infty$, where $\bar{J}$ is the particle flux at a given point on the pipe wall and $\bar{n}_\infty$ is the average particle concentration in the mainstream of the fluid at that cross section. The transfer coefficient has the dimensions of a velocity.

The deposition of 0.8 μm iron particles in an 0.58 in. tube with a well-faired entry is shown in Fig. 4·12. When the Reynolds number based on tube diameter is greater than 2100, the boundary layer becomes turbulent at some distance from the inlet. The transition usually occurs at a Reynolds number, based on distance from the entrance, Re_x, of between 10^5 and 10^6, depending on the roughness of the wall and the level of turbulence in the mainstream. As shown in Fig. 4·12, the deposition rate tends to follow the development of the turbulent boundary layer. No deposition occurs until Re_x is about 10^5; the rate of deposition then approaches a constant value at $Re_x = 2 \times 10^5$ in the region of fully developed turbulence.

A complete theory of turbulent deposition has not yet been developed. In the model most often used for predicting deposition rates (Friedlander and Johnstone, 1957), it is assumed that the particles migrate from the mainstream by turbulent diffusion with an eddy diffusion coefficient equal to that of the fluid. An expression for the eddy diffusion coefficient in the viscous sublayer near the wall is employed. However, instead of requiring the particles to reach the wall by diffusion, it is assumed that they need diffuse only to within one stop-distance (4·19) of the wall. Best agreement between experiment

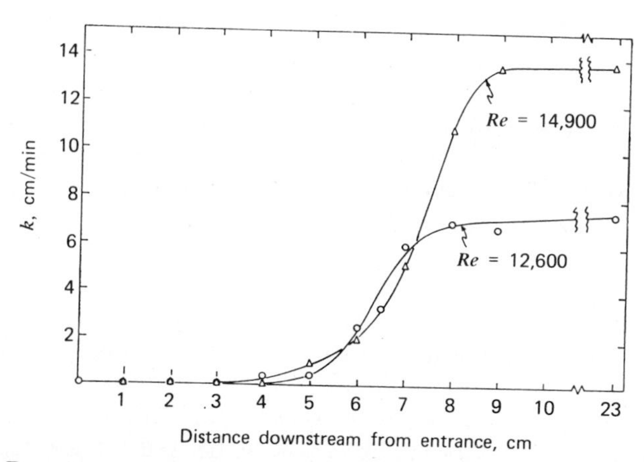

Fig. 4·12 Deposition of 0.8 μm iron particles near the entrance of an 0.58 cm tube. The Reynolds numbers shown are for the tube flow. Near the tube entrance, the flow is laminar, and no deposition takes place. In the transition region, the boundary layer turns turbulent, and deposition increases until it reaches the value corresponding to fully developed turbulent pipe flow (Friedlander and Johnstone, 1957).

and theory is obtained by basing the stop-distance on the mean radial fluctuating velocity in the fluid mainstream. Based on experiment, this velocity is given by

$$\left(\overline{v'^2}\right)^{1/2}=0.9U\left(\frac{f}{2}\right)^{1/2} \tag{4·47}$$

where U is the average gas velocity and f is the Fanning friction factor. It would seem more reasonable to base the stop-distance on values of the fluctuating velocity measured closer to the surface, but calculated deposition rates are then much too low compared with measured values.

In the steady state, the time average flux of particles to the wall is given by the relation

$$\bar{J}=-\epsilon_p\frac{d\bar{n}}{dy} \tag{4·48}$$

where ϵ_p is the particle eddy diffusivity, $\bar{n}$ is the time average particle concentration, and y the distance from the wall. Integrating from $y=s$, the stop-distance where $\bar{n}=0$ by assumption, to $y=\infty$ where $\bar{n}=\bar{n}_\infty$, the result is

$$k=\frac{|\bar{J}|}{\bar{n}_\infty}\left[\int_s^\infty\frac{dy}{\epsilon_p(y)}\right]^{-1} \tag{4·49}$$

The stop-distance is calculated from (4·19) with (4·47). When the stop-distance is smaller than the thickness of the viscous sublayer, no particles would deposit if the sublayer were truly laminar. Experimentally, however, deposition is observed in this case; it can be explained by the presence of fluctuations in the sublayer that contributes to particle transport. The eddy diffusivity for matter in the sublayer is given by (3·49). By using this expression with others for the eddy diffusivity in the buffer layer and turbulent core, the integration (4·49) can be performed. Measured and calculated values for one set of experimental data are shown in Fig. 4·13. Agreement between theory and experiment is fair considering the approximate nature of the model and the experimental uncertainties.

When the stop-distance is much smaller than the thickness of the viscous sublayer, the following expression is derived for the deposition velocity by substitution in (4·49) with ϵ_p given by (3·49):

$$\frac{k}{U}=\frac{\rho^2 d_p^4\rho_p^2U^4}{6.1(10)^5\mu^4}\left(\frac{f}{2}\right)^{5/2}\qquad\left(\frac{sU}{\nu}\left(\frac{f}{2}\right)^{1/2}<5\right)$$

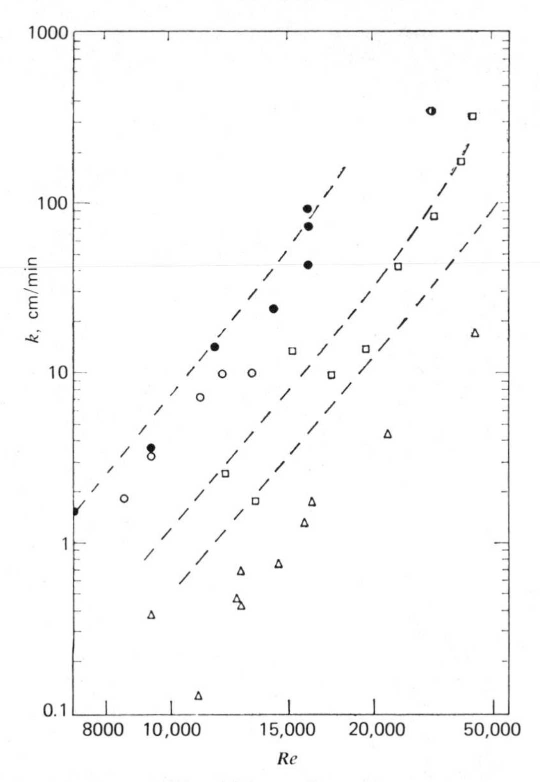

○. 1.57 μm iron particles, 1.30–cm glass tube
●. 1.57 μm iron particles, 1.30–cm glass tube, glycerol jelly adhesive
◑. 1.57 μm iron particles, 1.38–cm brass tube, pressure–sensitive tape
□. 1.81 μm aluminum particles, 1.38–cm brass tube
△. 0.8 μm iron particles, 1.30–cm glass tube
theory

Fig. 4·13 Deposition of iron and aluminum particles in glass and brass tubes of different sizes with coatings of various types. A goal of this set of experiments was to vary the electrical and surface characteristics of the system. The dashed curves are theoretical calculations based on (4.49) for 1.57 μm iron, 1.81 μm aluminum, and 0.8 μm iron particles (Friedlander and Johnstone, 1957).

Since f is a weak function of the Reynolds number, the transfer coefficient, k, is a strongly increasing function of both particle size and gas velocity, consistent with the inertial deposition mechanism.

When the stop-distance is much larger than the thickness of the buffer layer, the same type of model is applicable. By using (4·49) with an expression for the eddy diffusivity in the turbulent core, the following expression is obtained for the transfer coefficient:

$$\frac{k}{U(f/2)^{1/2}} = \frac{1-2s/d}{25\ln(d/s)-1.73} \qquad \left(\frac{s}{d}<0.5 \quad \text{and} \quad \frac{sU}{\nu}\left(\frac{f}{2}\right)^{1/2}>30\right)$$

Agreement between measured and calculated transfer coefficients is good (Forney and Spielman, 1974).

4·12 Transition from the Diffusion to Inertial Ranges

In the last two chapters, we have considered particle deposition from certain internal and external flows. At a fixed velocity, diffusion usually controls the deposition of the small particles, while for larger particles, inertial effects are important. In the submicron range, the collection efficiency increases with decreasing particle size as diffusional effects become more pronounced. For particles larger than a micrometer, the efficiency increases with increasing size because of interception and impaction. Often the result is a "window" in the efficiency curve for particles in the 0.1 to 1.0 μm size range.

Such a window has been observed in experimental studies of the performance of filters of various types. Whitby et al. (1961) generated monodisperse aerosols composed of methylene blue with particle sizes ranging from 0.08 to 14 μm. For the size range $d_p > 0.5$ μm, the monodisperse aerosols were produced with a spinning disk generator (Section 6·13). The aerosols were filtered through a mat composed of cylindrical glass fibers about 4 μm in diameter. The results of the experiments are shown in Fig. 4·14.

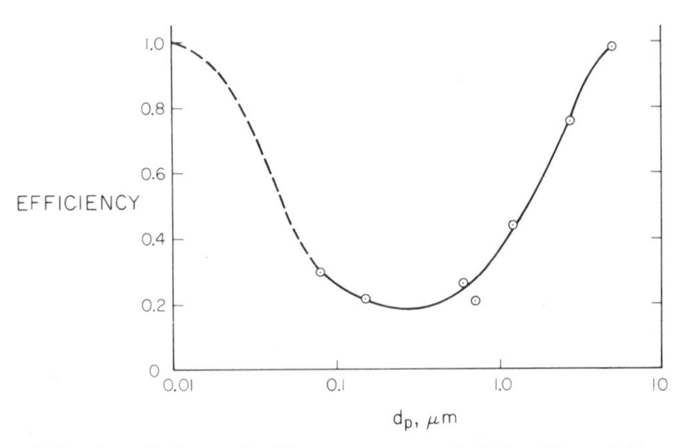

Fig. 4·14 Collection efficiency of a filter mat composed of glass fibers, 4.16 μm effective fiber diameter, filter thickness 0.8 cm, and fraction solids 0.0057. The approximately monodisperse aerosols were of methylene blue, and the air velocity was about 13 cm/sec. (Data of Whitby et al., 1961.) The dashed part of the curve is an extrapolation based on theory.

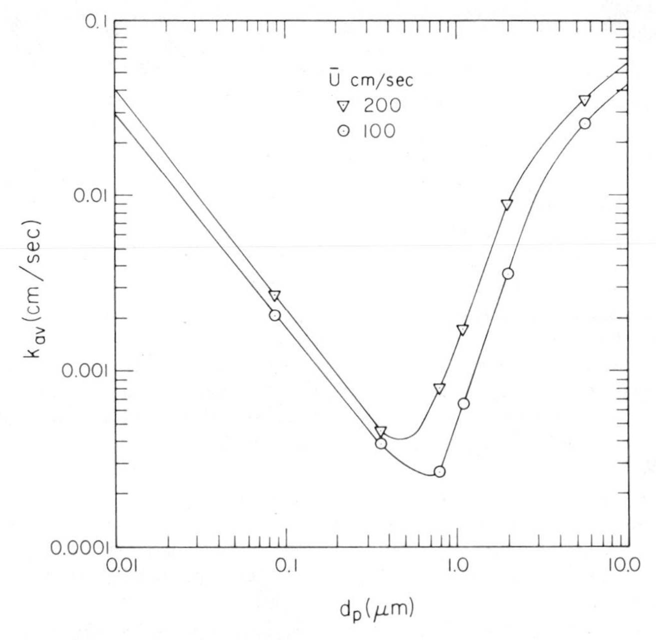

Fig. 4·15 Transfer coefficient averaged over a model bifurcation simulating the first branch of the human lung. Monodisperse aerosols were passed through the bifurcation in a pulse flow simulating inhalation at two average velocities, 100 and 200 cm/sec. Deposition of the smaller particles is controlled by diffusion and of the larger particles by impaction. There is a characteristic minimum in the 0.5 to 1.0 μm range (Bell, 1974).

Similar behavior has been observed in studies of aerosol deposition at a bifurcation under conditions simulating flow in the lung (Fig. 4·15). For turbulent pipe flow, a similar transition from the diffusion to turbulent deposition regimes would also be expected.

The result is that the often expressed intuitive belief that small particles are more difficult to remove from a gas than large ones is not usually correct. The particles most difficult to collect are those in the size range corresponding to the transition from diffusional to inertial deposition, usually between 0.1 and 1 μm.

A rigorous method of calculating the transition from the diffusion range, in which a stochastic model applies, to the inertial range, for which a deterministic model is used, has not been developed. The usual practice is to calculate the efficiencies separately for each effect and then assume additivity to produce a composite curve.

Problems

1. A particle is injected vertically with a velocity u_0 into a stationary gas. If the particle motion is initially upward, derive an expression for the maximum distance traveled by the particle against the gravitational field. Assume the motion can be described by Stokes law.

2. Derive an expression for the stop-distance of a non-Stokesian particle in terms of characteristic dimensionless parameters.

3. A duct 4 ft in diameter with a 90° bend has been designed to carry particles in the range $1 < d_p < 20$ μm, which adhere when they strike the wall. Before construction, it is proposed to carry out bench scale experiments to determine the particle deposition rate in the bend. The model is to be built to 1/10 scale, and the same aerosol will be used as in the full-scale system. Show that is is not possible to maintain both Stokes and Reynolds number similarity in the full-scale and model systems. If Stokes similarity is to be preserved, calculate the Reynolds number ratio for the model to full-scale systems. Why is it more important to preserve Stokes than Reynolds similarity in such experiments?

4. Estimate the efficiency of deposition by impaction of 0.1, 1, and 10 μm particles of density 1 g/cm^3 on the elements of a tree (trunk, limbs, twigs, and so on) for a wind velocity of 5 mi/hr. For this purpose, make some reasonable assumptions concerning tree element dimensions based on observation. Discuss your results.

5. An aerosol is to be filtered by passing it through a bed packed with a granular material 1 in. in diameter. Assuming that the packing elements are approximately spherical in shape, *estimate* the minimum size of the particles that can be collected by impaction for an air velocity of 2 ft/sec. The kinematic viscosity of the air is 0.15 cm^2/sec, and the viscosity is 1.8×10^{-4} g/cm sec. Particle density is 2 g/cm^3.

6. A bed packed with spherical elements 5 mm in diameter is to be designed to remove 5 μm particles of unit density from a gas stream at 20°C. The primary design criterion is to maximize the number of particles collected per unit of energy expended. Determine the optimal gas velocity basing your calculation on the performance of a single sphere.

7. Air at 20°C and 1 atm flows through a 6 in. duct oriented vertically at a velocity of 20 ft/sec. Plot the transfer coefficient (cm/sec), k, for deposition on the wall as a function of particle size over the range 10 μm $> d_p >$ 0.01 μm. Assume that the surface of the duct acts as a perfect sink for particles. Particle density is 2 g/cm^3.

References

Basset, A. B. (1910) *Quart. J. Math.*, **41**, 369.

Batchelor, G. K. (1967) *An Introduction to Fluid Dynamics*, Cambridge Univ. Press, Cambridge. Included are discussions of the changing patterns of flow around bluff bodies (such as cylinders) with increasing Reynolds number. References are given to original papers in which numerical calculations of the velocity distributions were made. Written for applied mathematicians, the book includes much useful expository material, and some comparisons with experimental data. Turbulent flows are not discussed.

Beard, K. V. (1974) *J. Atmos. Sci.*, **31**, 1595.

Bell, K. A. (1974) Aerosol Deposition in Models of a Human Lung Bifurcation, Ph.D. thesis in chemical engineering, California Institute of Technology.

Brun, R. J., Lewis, W., Perkins, P. J., and Serafini, J. S. (1955) NACA Report 1215.

Charles, G. E., and Mason, S. J. (1960) *J. Colloid Sci.*, **15**, 236.

Dahneke, B. (1975) *J. Colloid Interface Sci.*, **51**, 58.

Dorsch, R. G., Saper, P. G., and Kadow, C. F. (1955) NACA Technical Note 3587.

Forney, L. J. and Spielman, L. A. (1974) *Aerosol Sci.*, **5**, 257.
Friedlander, S. K. (1957) *AICh E J.*, **3**, 381.
Friedlander, S. K., and Johnstone, H. F. (1957), *Ind. Eng. Chem.*, **49**, 1151.
Fuchs, N. A. (1964) *Mechanics of Aerosols*, Pergamon, New York. This reference contains a thorough review of the literature on the behavior of aerosols in the inertial range through the 1950's. It lists 886 references! Both theory and experiment are covered.
Goldstein, S. (Ed.) (1938) *Modern Developments in Fluid Dynamics*, vol. 1, Oxford Univ. Press, Oxford.
Golovin, M. N., and Putnam, A. A. (1962) *Ind. Eng. Chem. Fund.*, **1**, 264.
Happel, J., and Brenner, H. (1965) *Low Reynolds Number Hydrodynamics with Special Applications to Particulate Media*, Prentice-Hall, Englewood Cliffs, N. J.
Landau, L. D., and Lifshitz, E. M. (1959) *Fluid Mechanics*, Addison-Wesley, Reading, Mass.
MacKay, G. D. M., Suzuki, M., and Mason, S. G. (1963) *J. Colloid Sci.*, **18**, 103.
Ranz, W. E., and Wong, J. B. (1952) *Ind. Eng. Chem.*, **44**, 1371.
Serafini, J. S. (1954) NACA Report 1159.
Snyder, W. H., and Lumley, J. L. (1971) *J. Fluid Mech.*, **48**, 47.
Whitby, K. T., Lundgren, D. A., McFarland, A. R., and Jordan, R. C. (1961) *J. Air Poll. Control Assoc.*, **11**, 503.
Wong, J. B., and Johnstone, H. F. (1953) Collection of Aerosols by Fiber Mats, Tech. Rept. 11, Eng. Expt. Station, Univ. of Illinois.

CHAPTER 5

Light Scattering and Visibility

Visibility degradation, one of the most obvious manifestations of air pollution, results primarily from light scattering by atmospheric particles. The effects of pollution on light scattering are particularly strong in urban and industrial air basins, but extend to regional and even global scales. Of great importance in air pollution control is the determination of the contribution of different emission sources to the total light scattering (Chap. 11). In laboratory research and atmospheric monitoring, light scattering methods are used to measure the sizes of individual particles or the integrated optical behavior of an aerosol cloud.

In broad outline, the problem of light scattering by clouds of small particles can be formulated as follows: Scattering by an individual particle depends on its size, refractive index and shape, and the wavelength of the incident light. The total light scattered from a collimated light beam is obtained by summing the scattering over particles of all sizes and refractive indices. In practice, light sources and sinks are disposed in a complex way; the radiation intensity at any point in space is determined by the arrangement of the sources and sinks, the spatial distribution of the aerosol, and its size distribution and composition.

In laboratory studies, it is possible to control these variables, and for certain relatively simple systems, good agreement can be obtained between theory and experiment. In field studies of light scattering by plumes and polluted atmospheres, the situation is much more complex. Certain model systems can be considered that can then be compared with the results of field experiments. In general, however, this branch of air pollution research is not well developed.

5·1 Interaction of Light with Suspended Particles: General Considerations

When aerosol particles interact with light, two different types of processes can occur. The energy received can be reradiated by the particle in the same wavelength. The reradiation may take place in all directions but usually different intensities in different directions. This process is called *scattering*. Alternatively, the radiant energy can be transformed into other forms of energy, such as heat, energy of chemical reaction, or radiation of a different wavelength. These processes are called *absorption*. In the visual range, light attenuation by absorption predominates for black smokes, whereas scattering controls for water droplets.

It is convenient to analyze the light attenuation process by considering a single particle of arbitrary size and shape, irradiated by a plane electromagnetic wave (Fig. 5·1). The effect of the disturbance produced by the particle is to diminish the amplitude of the plane wave. At a distance large compared with the particle diameter and the wavelength, the scattered energy appears as a spherical wave, centered on the particle and possessing a phase different from the incident beam. The total energy lost by the plane wave, the extinction energy, it equal to the scattered energy in the spherical wave plus the energy of absorption.

The most important characteristic of the scattered wave is its intensity, I, expressed in c.g.s. units as erg/cm^2 sec. At large distances from

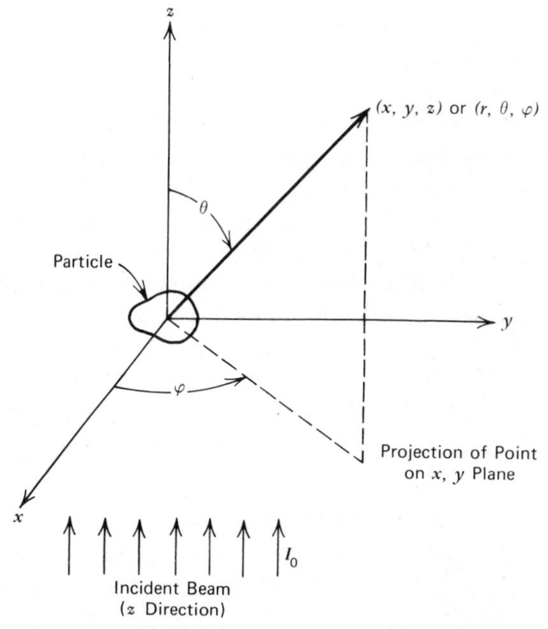

Fig. 5·1 The direction of scattering at any r is characterized by the scattering angle, θ, measured relative to the direction of the incident beam, and the azimuth angle, φ.

the origin, the energy flowing through a spherical surface element is $Ir^2 \sin\theta\, d\theta\, d\varphi$. This energy flows radially and depends on θ and φ but not on r. It is proportional to the intensity of the incident beam I_0 and can be expressed as follows:

$$Ir^2 \sin\theta\, d\theta\, d\varphi = I_0 \left(\frac{\lambda}{2\pi}\right)^2 F(\theta,\varphi,\lambda) \sin\theta\, d\theta\, d\varphi \tag{5·1a}$$

or

$$I = \frac{I_0 F(\theta,\varphi,\lambda)}{(2\pi r/\lambda)^2} \tag{5·1b}$$

The wavelength of the incident beam, λ, is introduced to make the scattering function, $F(\theta,\varphi,\lambda)$, dimensionless. In general, $F(\theta,\varphi,\lambda)$ depends on the wavelength of the incident beam and on the size, shape, and optical properties of the particles but not on r. For spherical particles, there is no φ dependence. The relative values of F can be plotted in a polar diagram as a function of θ for a plane in the direction of the incident beam. A plot of this type is called the *scattering diagram* for the particle.

The scattering function can be determined from theory for certain important special cases as discussed in the following sections. The performance of optical single particle counters (Chap. 6) depends on the variation of the scattering function with angular position, and much effort has been devoted to the design of such detectors.

The intensity function, by itself, is not sufficient to characterize the scattered light. Needed also are the polarization and phase of the scattered light, which are discussed in the standard references on the subject. For air pollution applications including visibility degradation and instrument design, the parameters of most interest are the intensity function and a derived quantity, the *scattering efficiency*.

The scattering efficiency, a dimensionless quantity, is defined as the energy scattered by the particle divided by the energy intercepted by the particle, based on its geometric cross section, s_g:

$$K_{scat} = \frac{\int F(\theta,\varphi) \sin\theta\, d\theta\, d\varphi}{(2\pi/\lambda)^2 s_g} \tag{5·2}$$

Similarly, the absorption efficiency is defined as the fraction of the incident beam absorbed per unit cross-sectional area of particle. The total energy removed from the incident beam, the extinction energy, is the sum of the energy scattered and absorbed. The corresponding extinction efficiency is given by

$$K_{ext} = K_{scat} + K_{abs} \tag{5·3}$$

In the next three sections, the dependence of the scattering efficiency on particle size is discussed; in the first two sections, very small and very large particles are considered. Both of these ranges can be treated from a relatively simple point of view. Unfortunately, most light scattering problems in air pollution fall into the difficult intermediate size range discussed in Section 3. For a detailed, readable monograph on light scattering by single particles, stressing the determination of $F(\theta,\varphi,\lambda)$, the reader is referred to van de Hulst (1957).

5·2 Scattering by Particles Small Compared to the Wavelength

Light, an electromagnetic wave, is characterized by electric and magnetic field vectors. For simplicity, we consider the case of a plane wave, linearly polarized, incident on a small spherical particle. The wavelength of light in the visible range is about 0.5 μm. For particles much smaller than the wavelength, say less than 0.05 μm, the local electric field produced by the wave is approximately uniform at any instant. This applied electric field induces a dipole in the particle. Since the electric field oscillates, the induced dipole oscillates, and according to classical theory, radiates in all directions. This type of scattering is called *Rayleigh scattering*.

The dipole moment, **p**, induced in the particle is proportional to the instantaneous electric field vector:

$$\mathbf{p} = \underline{\underline{\alpha}} \cdot \mathbf{E} \tag{5·4}$$

This expression defines the polarizability tensor, $\underline{\underline{\alpha}}$, which has the dimensions of a volume. In what follows, it will be assumed that the particles are isotropic, in which case α can be treated as a scalar. From the energy of the electric field produced by the oscillating dipole, an expression can be derived for the intensity of the scattered radiation:

$$I = \frac{(1+\cos^2\theta)k^4\alpha^2}{2r^2} I_0 \tag{5·5}$$

where the wave number $k = 2\pi/\lambda$. This expression applies to particles of arbitrary shape. The scattering is symmetrical with respect to the direction of the incident beam as shown in Fig. 5·2.

Since the intensity of the scattered light varies inversely with the fourth power of the wavelength, blue light (short wavelength) is scattered preferentially to red. This strong dependence leads to the blue of the sky in the absence of haze. It contributes to the red of the sunset when the red-enriched transmitted light is observed. In polluted atmospheres, however, molecular scattering is usually small compared with aerosol scattering. The principal contribution to scattering comes from a larger particle size range in which the Rayleigh theory does not apply. This is discussed in a later section.

For an isotropic spherical particle, it can be shown that

$$\alpha = \frac{3(m^2-1)}{4\pi(m^2+2)} v \tag{5·6}$$

where m is the refractive index of the particle and v is the volume, $\pi d_p{}^3/6$. When scattering without absorption takes place, the efficiency factor is obtained by substituting (5·6) in (5·5) and integrating:

$$K_{scat} = \frac{8}{3} x^4 \left\{ \frac{m^2-1}{m^2+2} \right\}^2 \tag{5·7}$$

where $x = \pi d_p/\lambda$ is the dimensionless optical particle size parameter.

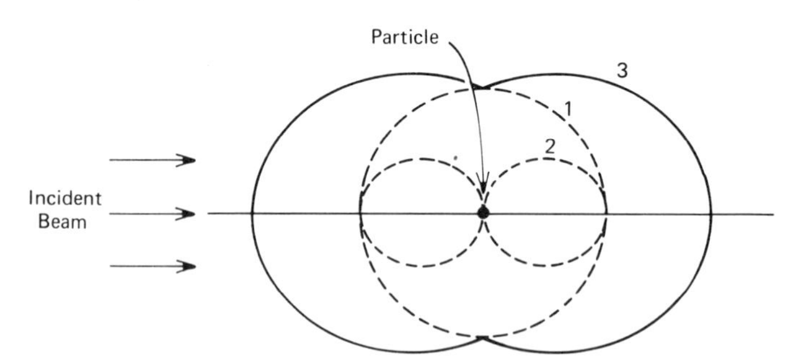

Fig. 5·2 Scattering diagram for very small isotropic particles with unpolarized incident radiation. This symmetric pattern holds for $d_p < 0.1\lambda$; for larger particles, strong asymmetries develop. 1, polarized with the electric vector perpendicular to the plane of page; 2, polarized with the electric vector in the plane of page; and 3 = 1 + 2 = Total. The intensity of scattered light is proportional to the radius vector to the curve.

Both scattering and absorption can be taken into account by writing the refractive index as the sum of a real and an imaginary component:

$$m = n - in' \tag{5·8}$$

where $n^2 + n'^2 = \epsilon$ and $nn' = \lambda\sigma/c$, where ϵ is the dielectric constant, σ is the conductivity, λ is the wavelength in vacuum, and c is the velocity of light. The imaginary term gives rise to absorption; it vanishes for nonconducting particles ($\sigma = 0$). Both ϵ and σ depend on λ, approaching their static values at low frequencies. For metals in the optical frequency range, both n and n' are of order unity. It can be shown (van de Hulst, 1957) that the scattering efficiency of small spherical absorbing particles is given by

$$K_{scat} = \frac{8}{3}x^4 \operatorname{Re}\left\{\frac{m^2-1}{m^2+2}\right\}^2 \tag{5·9}$$

where Re indicates that the real part of the expression is taken. The absorption efficiency can be shown to be given by

$$K_{abs} = -4x \operatorname{Im}\left\{\frac{m^2-1}{m^2+2}\right\} \tag{5·10}$$

where Im indicates that the imaginary part is taken.

5·3 Scattering by Large Particles: The Extinction Paradox

For particles much larger than the wavelength of the incident light ($x \gg 1$), the scattering efficiency approaches two. That is, a large particle removes from the beam *twice* the amount of light intercepted by its geometric cross-sectional area. What is the explanation for this paradox?

For light interacting with a large particle, the incident beam can be considered to consist of a set of separate light rays. Of those rays passing within an area defined by the geometric cross section of the sphere, some will be reflected at the particle surface and others refracted. The refracted rays may emerge again, possibly after several internal reflections and refractions. The reflected light and the refracted light that eventually emerges are part of the total scattering by the particle. Any of the incident beam that does not emerge is lost by absorption within the particle. Hence all of the energy incident on the particle surface is removed from the beam by scattering or absorption, accounting for an efficiency factor of unity.

There is, however, another source of scattering from the incident beam. The portion of the beam not intercepted by the sphere forms a plane wave front from which a region corresponding to the cross-sectional area of the sphere is missing. This is equivalent to the effect produced by a circular obstacle placed normal to the beam. The result, according to classical optical principles, is a diffraction pattern *within the shadow area* at large distances from the obstacle. The appearance of light within the shadow area is the reason why diffraction is sometimes likened to the bending of light rays around an obstacle.

The intensity distribution within the diffraction pattern depends on the shape of the perimeter and size of the particle relative to the wavelength of the light. It is independent of the composition, refractive index, or reflective nature of the surface. The total amount of energy that appears in the diffraction pattern is equal to the energy in the beam intercepted by the geometric cross section of the particle. Hence the total efficiency factor based on the cross-sectional is equal to two.

The use of the factor two for the efficiency requires that all scattered light be counted including that at small angles to the direction of the beam. In general, the observation must be made at a large distance from the particle. As van de Hulst points out, a flower pot in a window blocks only the sunlight falling on it, and not twice that amount, from entering a room; a meteorite of the same size in space between a star and a telescope on Earth will screen twice the amount falling on it.

5·4 Scattering in the Intermediate Size Range: Mie Theory

Rayleigh scattering for $x \ll 1$ and the large particle extinction law for $x \gg 1$ provide useful limiting relationships for the efficiency factor. Atmospheric visibility, however, is limited by particles whose size is of the same order as the wavelength of light in the optical range, from 0.1 to 1 μm in diameter. In this range, the theory of Rayleigh is no longer applicable since different parts of the particle interact with different portions of an incident wave. Such particles are still much too small for large particle scattering theory to be applicable. As a result, it is necessary to make use of a much more complicated theory due to Mie, which treats the general problem of scattering and absorption of a plane wave by a homogeneous sphere. Expressions for the scattering and extinction are obtained by solving Maxwell's equations for the regions inside and outside the sphere with suitable boundary conditions. It is found that the efficiency factors are functions of x and m alone. The calculations must be carried out numerically and the results have been tabulated for certain values of the refractive index (Penndorf, 1957*b*).

For water, $m = 1.33$, whereas for organic liquids it is often approximately 1.5. The scattering efficiency for these two values of m are shown in Fig. 5·3 as a function of the dimensionless particle diameter x. For $x \rightarrow 0$, the theory of Rayleigh is applicable. Typically, the curves show a sequence of maxima and minima, the maxima corresponding to the reinforcement of transmitted and diffracted light, and the minima to interference.

For absorbing spheres, the curve for K_{ext} is usually of simpler form, rising rapidly to reach a maximum at small values of x and then falling slowly to approach two at large values of x. Figure 5·4 shows the extinction efficiency for the case of carbon spheres. For such particles, nearly all of the scattering is due to diffraction, while almost all of the geometrically incident light is absorbed. The refractive index for absorbing spheres usually varies with wavelength, and this results in variation

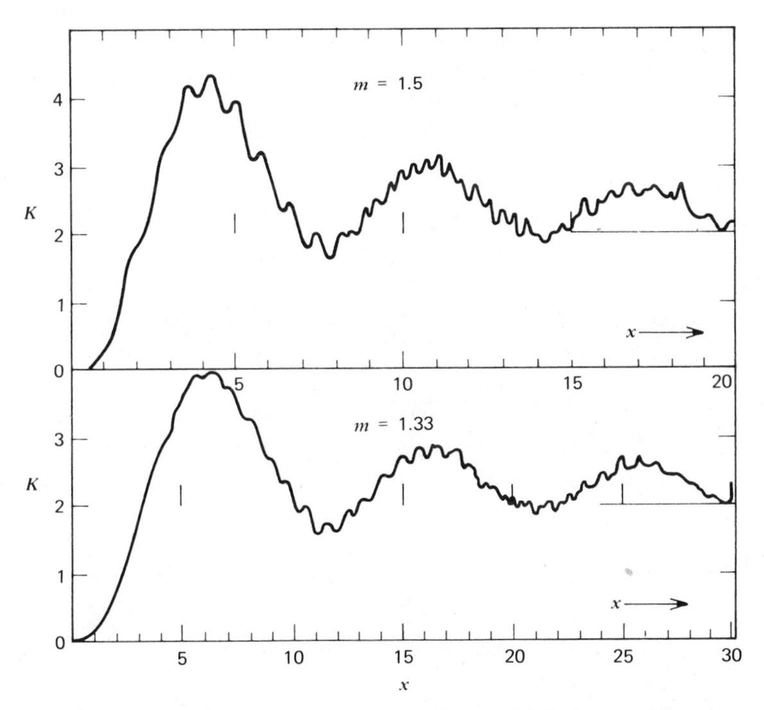

Fig. 5·3 Extinction curves calculated from the theory of Mie for $m = 1.5$ and $m = 1.33$ (van de Hulst, 1957). The curves show a sequence of maxima and minima of diminishing amputude, typical of nonabsorbing spheres with $2 > m > 1$. Indeed, by taking the abscissa of the curve for $m = 1.5$ to be $2x(m-1)$, all extinction curves for the range $2 > m > 1$ are reduced to approximately the same curve.

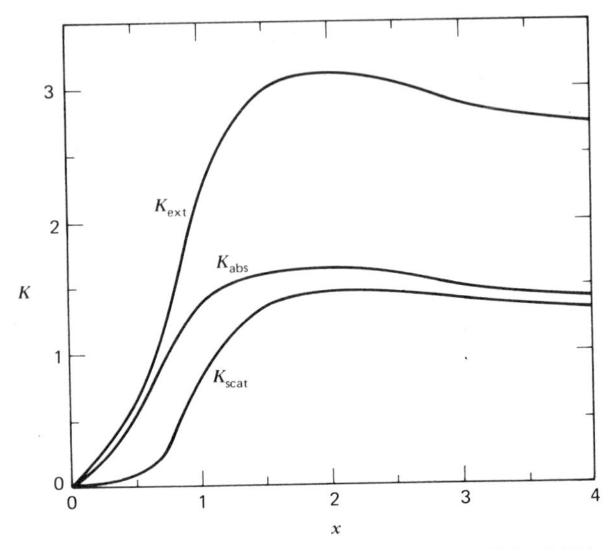

Fig. 5·4 Extinction efficiency for carbon particles with $m=2.00(1-0.33i)$, temperature not specified (McDonald, 1962). For small values of x, the extinction is due primarily to absorption, but for large x, scattering and absorption are of almost equal importance.

of K_{ext} as well. As shown in Table 5·1, however, the variation over the visible spectrum is not great.

TABLE 5·1
K_{ext} FOR CARBON SPHERES AT TWO DIFFERENT WAVELENGTHS
(McDonald, 1962)

	K_{ext}	
$x=\pi d_p/\lambda$	$\lambda=0.436\ \mu m$	$\lambda=0.623\ \mu m$
0.2	0.20	0.18
0.4	0.46	0.42
0.6	0.86	0.82
0.8	1.45	1.44
1.0	2.09	2.17
1.5	2.82	2.94
2.0	3.00	3.09
4.0	2.68	2.68
8.0	2.46	2.46

The angular dependence of the light scattering (Fig. 5·1) can be calculated from the Mie theory. The form of the angular dependence is of particular importance in the design of optical particle counters. By choosing a favorable geometry for the detector of the scattered light, the resolution of the instrument with respect to particle size can be optimized. This point is discussed further in Chap. 6 and an example given for a particular scattering geometry.

Scattering cross sections have been measured for liquid suspensions of transparent, irregular particles graded in size by sedimentation methods (Hodkinson, 1966). The shapes of the curves of the scattering cross sections were simpler than those of spherical particles, but theoretical predictions have not been made except for very small particles to which the Rayleigh theory is applicable.

5·5 Extinction by Particulate Clouds

We consider the case of an aerosol illuminated by a collimated light source of a given wavelength, that is, a monochromatic beam. The experimental arrangement is shown schematically in Fig. 5·5. A photometer of this type installed in a smoke stack or duct would be suitable for measuring the attenuation produced by the flowing aerosol. Long path instruments of this kind have also been used for measurements of extinction by the atmospheric aerosol.

At concentration levels of interest in air pollution, the particles are separated by distances large compared with their diameter and are distributed in space in a random fashion. Light scattered in a given direction from an incident beam by different particles will be composed of waves of different phases. The total energy of the scattered wave per unit area, that is, the intensity of the scattered wave in a given direction will be equal to the sum of the intensities of the individual particles in that direction. This type of behavior is referred to as *independent scattering*, and it simplifies calculation of the total scattering by particulate systems.

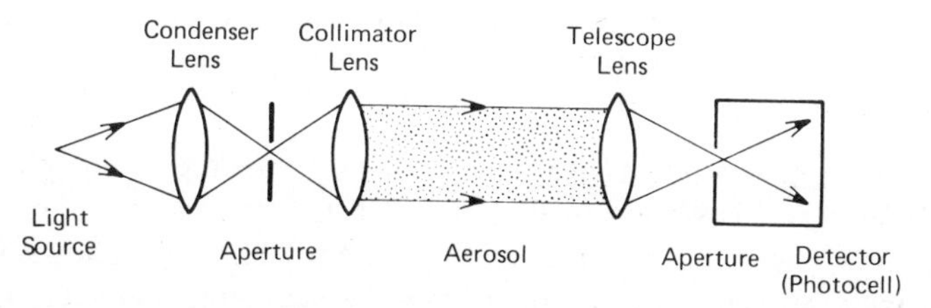

Fig. 5·5 Schematic diagram of an apparatus for the measurement of the extinction produced by a cloud of small particles. The goal is to measure only transmitted light and not light scattered by the particles. In practice, light of decreased intensity from the source is measured together with a certain amount of light scattered at small angles from the forward direction by the particles (after Hodkinson, 1966).

If there are dN particles in the size range d_p to $d_p + d(d_p)$ per unit volume of air, this corresponds to a total particle cross-sectional area of $(\pi d_p^2/4)\, dN dz$ over the light path length, dz, per unit area normal to the beam. The attenuation of light over this length is given by the relation

$$-dI = I\left[\int_0^\infty \frac{\pi d_p^2}{4} K_{ext}(x,m)\, n_d(d_p)\, d(d_p)\right] dz \tag{5·11}$$

where $dN = n_d(d_p)d(d_p)$. Hence the quantity

$$b = -\frac{dI}{Idz} = \int_0^\infty \frac{\pi d_p^2}{4} K_{ext}(x,m)\, n_d(d_p)\, d(d_p) \tag{5·12}$$

represents the fraction of the incident light scattered and absorbed by the particle cloud per unit length of path. It is called the *extinction coefficient* (sometimes the *attenuation coefficient* or turbidity), and it plays a central role in the optical behavior of particulate systems. In terms of the separate contributions for scattering and absorption (5·3),

$$b = b_{scat} + b_{abs} \tag{5·13}$$

where each term is understood to be a function of wavelength.

The contributions to $b(\lambda)$ from a given particle size range depend on the extinction cross section and on the particle size distribution function. The integral (5·12) can be rearranged as follows:

$$b = \int_{-\infty}^{\infty} \frac{db}{d\log d_p}\, d\log d_p \tag{5·14}$$

where

$$\frac{db}{d\log d_p} = \frac{3}{2}\frac{K_{ext}}{d_p}\frac{dV}{d\log d_p} \tag{5·15}$$

This function has been evaluated for a measured atmospheric size distribution and is shown in Fig. 5·6 as a function of particle size. The area under the curve is proportional to b. The figure shows that the principal contributions to b come from the size range between 0.1 and 3 μm.

If the aerosol is composed of a mixture of particles from separate sources with different refractive indices, the value of b must be summed over its contributions for the aerosol from each source (Chap. 11).

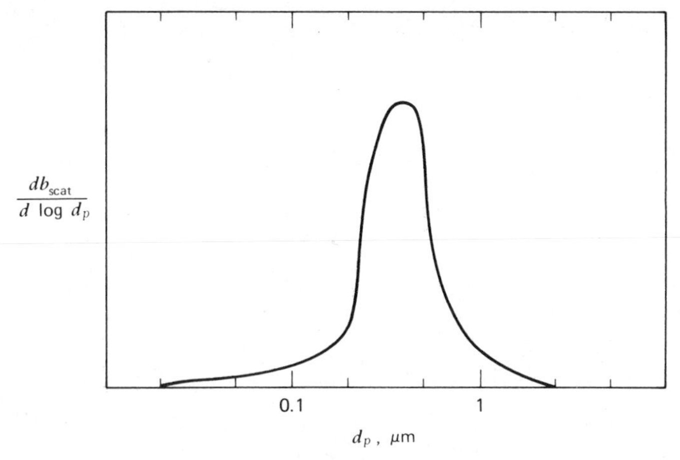

Fig. 5·6 Contributions to the scattering coefficient as function of particle size for the Pasadena, California, aerosol (August, 1969) based on the calculations of Ensor et al. (1972). The curve was calculated from the measured particle size distribution assuming $m = 1.5$. Largest contributions to light scattering came from the 0.2 to 0.5 μm size range for calculations made over the wavelength range 0.675 μm $> \lambda >$ 0.365 μm.

When the chemical properties of the aerosol are continuously distributed, and the refractive index depends on compostion, it would be proper to make use of the size–composition probability density function, but such calculations have not yet been made.

The reduction in the intensity of the light beam passing through the aerosol is obtained by integrating (5·12) between any two points, $z = L_1$ and $z = L_2$:

$$I_2 = I_1 e^{-\tau} \tag{5·16}$$

where the *optical thickness*, $\tau = \int_{L_1}^{L_2} b\, dz$, is a dimensionless quantity; b has been kept under the integral sign to show that it can vary with position, as a result of spatial variation in the aerosol concentration. Equation (5·16) is a form of *Lambert's law*.

In this calculation, only single scattering has been taken into account. In reality, each particle is also exposed to light scattered by other particles, but if the aerosol is sufficiently dilute and the path length sufficiently short, multiple scattering can be neglected (Section 5·9).

Example. Consider the following simplified model of coagulation: A monodisperse aerosol composed of very small particles coagulates but remains monodisperse. That is, the singlets combine simultaneously to form doublets, doublets to form quadruplets, and so on. (More realistic models are discussed in Chap. 7.) Assume that the particles are spherical and that the size in all cases is sufficiently small for the Rayleigh theory of light scattering to apply. Find the dependence of b_{scat} on particle diameter.

SOLUTION. Let N be the number of primary particles present in the cluster at any given time per unit volume of gas. By (5·6), the light scattered by a single particle is proportional to v^2. Summing over all particles in unit volume of gas, $b_{scat} \sim Nv^2$. Since the total volume of particulate matter per unit volume of gas is constant, $Nv = V$ is constant, and $b_{scat} \sim d_p^3$. Thus the scattering coefficient increases as the aerosol coagulates.

5·6 Scattering over the Visible Wavelength Range: Aerosol Contributions by Volume

In most cases of practical interest, the incident light, solar radiation for example, is distributed with respect to wavelength. The contribution to the integrated intensity I from the wavelength range λ to $\lambda + d\lambda$ is given by

$$dI = I_\lambda \, d\lambda$$

where I_λ is the intensity distribution function. The loss in intensity over the visible range, taking into account only single scattering, is determined by integrating (5·11) over the wavelength:

$$d\left(\int_{\lambda_1}^{\lambda_2} I_\lambda \, d\lambda\right) = -\left[\int_{\lambda_1}^{\lambda_2} b(\lambda) I_\lambda \, d\lambda\right] dz \qquad (5\cdot 17)$$

where λ_1 and λ_2 refer to the lower and upper ranges of the visible spectrum and b is now regarded as a function of λ. We wish to determine the intensity loss resulting from the particulate volume present in each size range of the size distribution function. For constant aerosol density, this is equivalent to the mass in each size range. Knowing the contributions of various sources to the mass in each size range, a quantitative link is provided between visibility degradation and source strengths. This relationship is explored further in Chap. 11.

Substituting (5·14) and (5·15) in (5·17), the result is

$$\bar{b} = \int_{\lambda_1}^{\lambda_2} b(\lambda) f(\lambda) \, d\lambda = \int_{-\infty}^{\infty} G(d_p) \frac{dV}{d\log d_p} \, d\log d_p \qquad (5\cdot 18)$$

where

$$G(d_p)=\frac{3}{2d_p}\int_{\lambda_1}^{\lambda_2}K_{ext}(x,m)f(\lambda)\,d\lambda$$

$f(\lambda)d\lambda$ is the fraction of the incident radiation in the range λ to $\lambda+d\lambda$, and $f(\lambda)$ has been normalized with respect to the total intensity in the range between λ_1 and λ_2. The quantity $G(d_p)$ represents the extinction over all wavelengths between λ_1 and λ_2 per unit volume of aerosol in the size range between d_p and $d_p+d(d_p)$. It is independent of the particle size distribution function. For a refractive index, $m=1.5$, $G(d_p)$ has been evaluated for the standard distribution of solar radiation at sea level, using Mie scattering functions. The result is shown in Fig. 5·7 as a function of particle size.

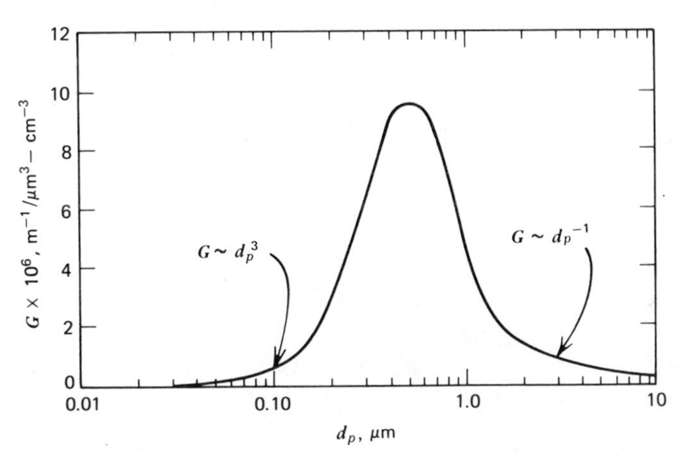

Fig. 5·7 Light scattering per unit volume of aerosol material as a function of particle size, integrated over all wavelengths for a refractive index, $m=1.5$. The incident radiation is assumed to have the standard distribution of solar radiation at sea level (Bolz, R. E. and Tuve, G. E. (Eds.) (1970), *Handbook of Tables for Applied Engineering Science*, Chemical Rubber Co., Cleveland, Ohio, p. 159). The limits of integration on wavelength were 0.36 to 0.680 μm. The limits of visible light are approximately 0.350 to 0.700 μm.

A number of interesting features are exhibited by this curve: The oscillations of the Mie functions (Fig. 5·3) are no longer present because of the integration over wavelength. For $d_p \rightarrow 0$ in the Rayleigh scattering range, $G(d_p) \sim d_p^{\,3}$. For large d_p, $G(d_p)$ vanishes since K_{scat} approaches a constant value (two) at all wavelengths; as a result, $G(d_p) \sim d_p^{-1}$ for $d_p \rightarrow \infty$. The most efficient size for light scattering on a mass basis corresponds to the peak in this function, which for $m = 1.5$ occurs in the size range between 0.5 and 0.6 μm. Particles of 0.1 μm diameter, on the one hand, and 3 μm on the other contribute only one-tenth the scattering on an equal mass basis. The volume distribution $dV/d\log d_p$ of the smog aerosol also often shows a peak in the 0.1 to 3 μm size range. This reinforces the importance of this range to total light scattering (Fig. 5·6).

5·7 Scattering in the Mie Range: Power Law Distributions

Often the size distribution of the atmospheric aerosol can be represented by a power law relationship in the light-scattering subrange, $3 > d_p > 0.1$ μm:

$$n_d(d_p) \sim d_p^{-\gamma} \tag{5·19}$$

This expression can be written without loss of generality in terms of the average particle diameter $\bar{d}_p = [6V/\pi N_\infty]^{1/3}$ as follows:

$$n_d = \frac{AN_\infty}{\bar{d}_p}\left(\frac{d_p}{\bar{d}_p}\right)^{-\gamma} \tag{5·20}$$

where A is a dimensionless factor that, with $\bar{d}_p$, may be a function of time and position. When A is constant, (5·20) represents a special form of the self-preserving spectrum (Chap. 7). Such forms may result from the interaction of various physical processes affecting the size distribution, but for the purposes of this discussion can be regarded as empirical.

Substituting (5·20) in the expression for the scattering coefficient, (5·12), the result is

$$b_{scat} = \frac{\lambda^{3-\gamma}AN_\infty}{4\pi^{2-\gamma}}\left[\frac{6V}{\pi N_\infty}\right]^{(\gamma-1)/3}\int_{x_1}^{x_2} K_{scat}(x,m)x^{2-\gamma}\,dx \tag{5·21}$$

where x_1 and x_2 correspond to the lower and upper limits, respectively, over which the power law holds. For the atmospheric aerosol, the lower limit of applicability of the power law is about 0.1 μm or somewhat less. This corresponds to $x_1 < 1$ and for this range K_{scat} is very small so that x_1 can be replaced by zero. The contribution to the integral for large values of x is small since γ is usually greater than 3 or 4 and K_{scat} approaches a constant, two. Hence x_2 can be set equal to infinity. The result is

$$b_{scat} = \frac{\lambda^{3-\gamma} A N_\infty}{4\pi^{2-\gamma}} \left[\frac{6V}{\pi N_\infty} \right]^{(\gamma-1)/3} \int_0^\infty K_{scat}(x,m) x^{2-\gamma} dx$$

$$= A A_1 \lambda^{3-\gamma} N_\infty^{(4-\gamma)/3} V^{(\gamma-1)/3} \tag{5·22}$$

where A_1 is a constant defined by this expression. Thus if the distribution obeys a power law (5·19) and (5·20), the order, γ, can be determined by measurement of the wavelength dependence of the extinction coefficient.

Experimentally, it is sometimes found that in a given geographical region

$$b_{scat} = A_2 V \tag{5·23}$$

where A_2 is a constant. This corresponds to $\gamma = 4$ and constant, A; by (5·20),

$$n_d = \frac{6AV}{\pi d_p^4} \tag{5·24}$$

a power law form that is often observed experimentally for the light-scattering subrange. However, (5·24) cannot extend to infinitely large particle diameters because the aerosol volumetric concentration becomes logarithmically infinite.

Equation (5·23) holds better when the value of V corresponds to the subrange $1 > d_p > 0.1$ μm rather than the total volumetric concentration. The constants, of course, differ. This indicates that the light-scattering subrange preserves its shape more consistently than the entire size spectrum.

It is also found experimentally that the dependence of b_{scat} on λ for the atmospheric aerosol can sometimes be represented by an equation of the form:

$$b_{scat} \sim \lambda^{-1.3} \tag{5·25}$$

corresponding to $\gamma = 4.3$, which is close to the value observed by direct measurement of the size distribution function. Equation (5·25) indicates more scattering in the blue (short wavelength) than in the red (long wavelength), with the result that the range of vision in hazy atmospheres is greater for red than for blue light.

5·8 Specific Intensity: Equation of Radiative Transfer

In the general case of aerosol/light interactions in the atmosphere or within a confined space, the light is neither unidirectional nor monochromatic; each volume element is penetrated in all directions by radiation. This requires a more careful definition of the intensity of radiation than used before. For the analysis of this case, an arbitrarily oriented small area, $d\sigma$, is chosen with a normal **n** (Fig. 5·8). At an angle θ to the normal we draw a line S, the axis of an elementary cone of solid angle $d\omega$. If through every point of the boundary of the area $d\sigma$ a line is drawn parallel to the nearest generator of the cone $d\omega$, the result is a truncated semi-infinite cone $d\Omega$, similar to the cone $d\omega$. Its cross-sectional area, perpendicular to S at the point P, will be $d\sigma \cos\theta$.

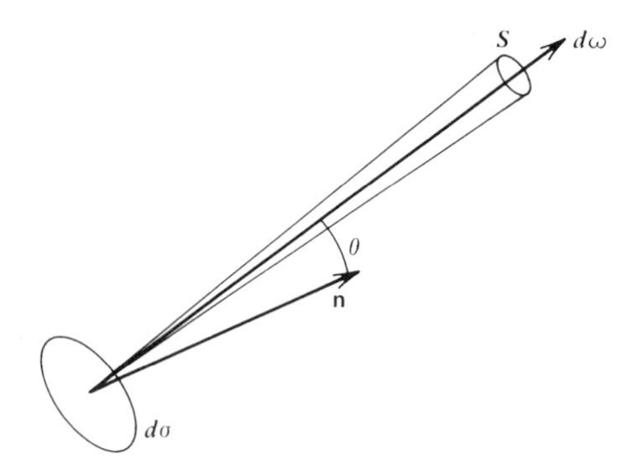

Fig. 5·8 Geometric factors determining specific intensity of radiation.

Let dE be the total quantity of energy passing in time dt through the area $d\sigma$ inside cone $d\Omega$ in the wavelength interval λ to $\lambda + d\lambda$. For small $d\sigma$ and $d\omega$, the energy passing through $d\sigma$ inside $d\Omega$ will be proportional to $d\sigma\, d\omega$.

The specific intensity of radiation or simply the intensity, I_λ, is defined by the relation

$$I_\lambda = \frac{dE}{d\sigma \cos\theta\, dt\, d\omega\, d\lambda} \tag{5·26}$$

The intensity is, in general, a function of the position in space of the point P, the direction **s**, time t, and wavelength λ:

$$I_\lambda = I_\lambda(P, \mathbf{s}, t, \lambda)$$

If I_λ is not a function of direction, the intensity field is said to be *isotropic*. If I_λ is not a function of position the field is said to be *homogeneous*. The total intensity of radiation is $I = \int_0^\infty I_\lambda \, d\lambda$. In the rest of this chapter, we suppress the suffix λ to simplify the notation

Now consider the radiant energy traversing the length, ds, along the direction in which the intensity is defined; a change in the intensity results from the combination of the effects of extinction (absorption and scattering) and emission:

$$dI\,(P, \mathbf{s}) = dI\,(\text{extinction}) + dI\,(\text{emission})$$

The loss by extinction can be written as before in terms of the extinction coefficient, b:

$$dI\,(\text{extinction}) = -\,bI\,ds$$

Emission by excited dissociated atoms and molecules in the air is usually small in the visible compared with solar radiation. Thermal radiation is important in the far infrared but not in the visible. Hence consistent with the assumptions adopted in this chapter, gaseous emissions can be neglected in the usual air pollution applications.

In an aerosol, however, a *virtual emission* exists because of rescattering in the $\mathbf{s}$ direction of radiation scattered from the surrounding volumes. The gain by emission is written in the form of a source term.

$$dI\,(\text{emission}) = bJ\,ds$$

This equation defines the source function, J.

Hence the energy balance over the path length, ds, takes the form

$$-\frac{dI}{b\,ds} = I - J \tag{5·27}$$

which is the *equation of radiative transfer*. This equation is useful, as it stands, in defining atmospheric visibility as discussed in a later section. Detailed applications require an expression for the source function, J, which can be derived in terms of the optical properties of the particles but this is beyond the scope of this book. For further discussion, the reader is referred to Chandrasekhar (1960) and Goody (1964).

5·9 Equation of Radiative Transfer: Formal Solution

The equation of radiative transfer is an energy balance; except for this concept, its physical content is slight. The physical problems of interest enter through the extinction coefficient and the source function. Many papers and monographs have been written on its solution for different boundary conditions and spatial variations of the optical path (Chandrasekhar, 1960; Goody, 1964). Some simple solutions are discussed in this and the next section. Most of the applications have been to planetary atmospheres and astrophysical problems rather than to smaller scale pollution problems.

The formal solution of the equation of transfer is obtained by integration along a given path from the point $s=0$ (Fig. 5·9):

$$I(s)=I(0)e^{-\tau(s,0)}+\int_0^s J(s')e^{-\tau(s,s')}b\,ds' \tag{5·28}$$

where $\tau(s,s')$ is the optical thickness of the medium between the points s and s':

$$\tau(s,s')=\int_{s'}^{s} b\,ds$$

The source function $J(s')$ over the interval 0 to s must be known to evaluate the integral in (5·28).

The interpretation of (5·28) is interesting: The intensity at s is equal to the sum of two terms. The first term on the right-hand side corresponds to Lambert's law (5·16), often used for the attenuation of a light beam by a scattering medium. The second term represents the contributions to the intensity at s from each intervening radiating element between 0 and s, attenuated according to the optical thickness correction factor. For aerosol scattering, this term can result with an originally unidirectional beam in the s direction from light scattered by the

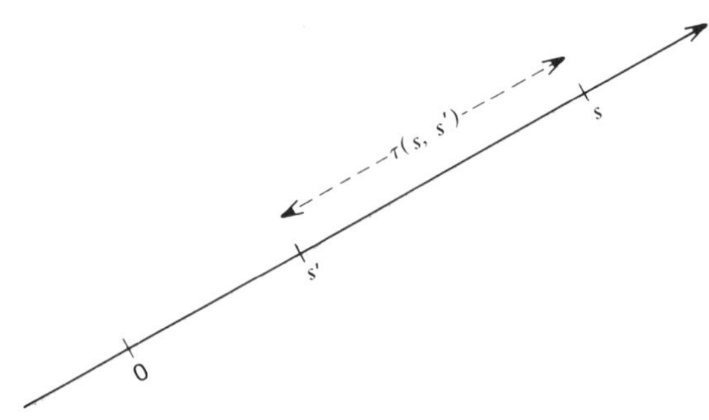

Fig. 5·9 Path of integration along the **s** vector. Light at point 0 reaches any point s attenuated according to Lambert's law. In addition, light is scattered toward s by particles between 0 and s such as those at the point s'.

surroundings into the particles along s and then rescattered by them in the s direction. This is the multiple scattering term neglected in the derivation of Lambert's law. The criterion for neglecting this term can be expressed in terms of the optical depth. For $\tau_{scat} < 0.1$, the assumption of single scattering is acceptable, while for $0.1 < \tau_{scat} < 0.3$, it may be necessary to correct for double scattering. For $\tau_{scat} > 0.3$, multiple scattering must be taken into account.

When the medium extends to $-\infty$ in the s direction, and there are no sources along s, it may be convenient not to stop the integration at the point 0 but to continue it indefinitely:

$$I(s) = \int_{-\infty}^{s} J(s') e^{-\tau(s,s')} b\, ds'$$

Thus the intensity observed at s is the result of scattering by all of the particles along the line of sight.

5·10 Visibility

Most of the information that we obtain through our sense of vision depends on our perception of differences in intensity or of color among the various parts of the field of view. An object is recognized because it has a different color or brightness than its surroundings, and also because of the variations of brightness or color over its surface. The shapes of objects are recognized by the observation of such variations.

Differences in intensity are particularly important and can be related to visibility as follows (Steffens, 1956): An isolated object on the ground is viewed from a distance (Fig. 5·10). The contrast between the test

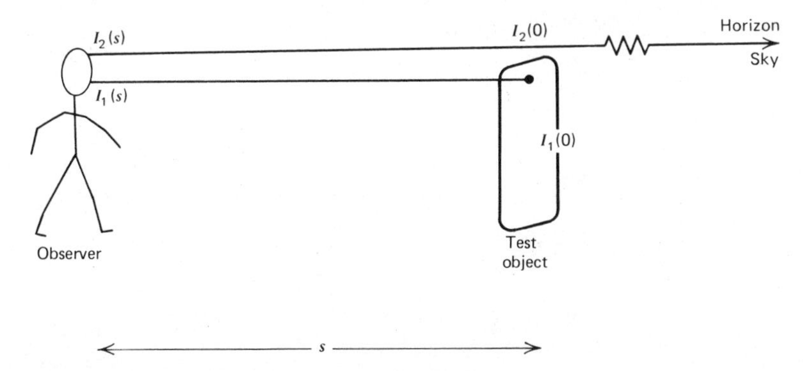

Fig. 5·10 Relative arrangements of observer, object, and horizon sky in definition of the visibility. The angle between the lines of sight corresponding to I_1 and I_2 is very small.

object and the adjacent horizon sky is defined by the expression

$$C = \frac{I_1 - I_2}{I_2} \tag{5·29}$$

where I_2 is the intensity of the background and I_1 is the intensity of the test object.

Expressions for the intensity can be obtained by integrating the equation of radiative transfer over the distance from the test object to the point of observation. If b and J are not functions of s, the integration gives

$$I(s) = I(0)e^{-bs} + J[1 - e^{-bs}] \tag{5·30}$$

where $s = 0$ corresponds to the location of the test object. Substituting in (5·29), the contrast as seen by an observer at a distance s from the object can be written

$$C = \frac{I_2(0)}{I_2(s)} \frac{[I_1(s) - I_2(s)]}{I_2(0)}$$

or with (5·30)

$$C = \frac{I_2(0)}{I_2(s)} \frac{[I_1(0) - I_2(0)]e^{-bs}}{I_2(0)} = \frac{I_2(0)}{I_2(s)} C(0)e^{-bs} \tag{5·31}$$

In viewing the horizon sky, the observer sees the virtual emission, J, resulting from the light from the sun and surroundings scattered in the direction of the observer by the atmosphere. This is sometimes referred to as the air light. By assumption, the air light is not a function of s. If 2 refers to the line of sight for the horizon sky, $I_2(s) = I_2(0) = J = \text{constant}$ and (5·31) becomes

$$C = C(0)e^{-bs}$$

If the test object is perfectly black, $I_1(0) = 0, C(0) = -1$, and

$$C = -e^{-bs}$$

The minus sign in this expression results because the test object is darker than the background.

The visual range or, more loosely, the visibility, is defined as the distance at which the test object is just distinguishable from background. Hence the minimum contrast that the eye can distinguish must now be introduced into the analysis. This contrast is denoted by C^* and the corresponding visual range $s = s^*$. For a black object at $s = s^*$,

$$C^* = -\exp(-bs^*)$$

or

$$s^* = -\frac{1}{b}\ln(-C^*)$$

Based on observation, the contrast threshold, C^*, is usually assumed to be -0.02:

$$s^* = -\frac{1}{b}\ln 0.02 = \frac{3.912}{b} \qquad (5\cdot 32)$$

Hence the visual range is inversely proportional to the extinction coefficient. Since b is a function of wavelength, the visual range defined in this way also depends on wavelength.

The total atmospheric extinction is the sum of contributions for the aerosol, molecular scattering and, perhaps, some gas absorption at certain wavelengths:

$$b = b_{aerosol} + b_{molecular}$$

Molecular scattering coefficients for air have been tabulated (Table 5·2). For $\lambda = 0.5\ \mu m$, the visibility calculated from (5·32) is about 220

TABLE 5·2
RAYLEIGH SCATTERING COEFFICIENT FOR AIR AT 0°C AND 1 ATM[a] (Penndorf, 1957*a*)

λ (μm)	$b_{scat} \times 10^8$ (cm^{-1})
0.2	954.2
0.25	338.2
0.3	152.5
0.35	79.29
0.4	45.40
0.45	27.89
0.5	18.10
0.55	12.26
0.6	8.604
0.65	6.217
0.7	4.605
0.75	3.484
0.8	2.684

[a] To correct for the temperature, $b_T = b_{T=0°C}(273/T°K)$ at 1 atm. This approximate formula does not take into account the variation of refractive index with temperature.

km or 130 mi. Hence the visibilities of a few miles or less, often observed in urban areas when the humidity is low, are due primarily to aerosol extinction. In some cases, however, there is probably a contribution to the extinction made by NO_2 absorption.

Problems

1. For a given mass of particles with the optical properties of carbon spheres, determine the particle size producing maximum extinction for $\lambda = 0.436$ μm.

2. Determine the concentration (μg/m^3) necessary to scatter an amount of light equal to that of air at 20°C and 1 atm. Assume a particle refractive index of 1.5 and a wavelength of 0.5 μm. Do the calculation for 0.1, 0.5, and 1 μm particles of unit density. Compare your result with the average concentration in the Los Angeles atmosphere, about 100 μg/m^3.

3. The California visibility standard requires that the visibility be greater than 10 mi on days when the relative humidity is less than 70%. Consider a day when the visibility controlling aerosol is composed of material with a refractive index of 1.5. Estimate the aerosol concentration in the atmosphere that would correspond to the visibility standard. Assume (1) the density of the spherical particles is 1 g/cm^3; (2) the aerosol is monodisperse with a particle size, $d_p = 0.5$ μm; and (3) the wavelength of interest is 0.5 μm. Express your answer in micrograms of aerosol per cubic meter of air.

4. The extinction of light by an aerosol composed of spherical particles depends on its optical properties and size distribution. Consider the distribution function $n_d(d_p) \sim d_p^{-4}$, often observed at least approximately. Suppose these particles are composed of an organic liquid ($m = 1.5$), on the one hand, or of carbon, on the other. This might correspond to a photochemical aerosol, such as experienced in Los Angeles ($m = 1.5$), and a soot aerosol generated by jet aircraft or other combustion processes. Calculate the *ratio* $b_{carbon}/b_{1.5}$ for fixed size distribution.

5. It is possible, in principle, to determine the size distribution of particles of known optical properties by measurement of the light scattered by a settling aerosol. In this method, the intensity of the light transmitted by the aerosol in a small cell, I, is recorded as a function of time. The aerosol is initially uniform spatially, and there is no convection.

The light scattered from a horizontal beam at a given level in the cell remains constant until the largest particles in the aerosol have had time to fall from the top of the cell through the beam. The scattering will then decrease as successively smaller particles are removed from the path of the beam.

Show that the size distribution function can be found from the relationship (Gumprecht and Sliepcevich, 1953):

$$-\frac{1}{I}\frac{dI}{dt} = LK_{scat}\frac{\pi d_p^{*2}}{4}n_d(d_p^*)\frac{d(d_p^*)}{dt}$$

where d_p^* is the maximum particle size in the beam at any time, t, and L is the length of the light path (cell thickness). Describe how d_p^* and $d(d_p^*)/dt$ can be determined.

In practice the system will tend to be disturbed by Brownian diffusion and convection, and this method is seldom used to determine $n_d(d_p)$.

References

Chandrasekhar, S. (1960) *Radiative Transfer*, Dover, New York. This is the classic reference on the equation of radiative transfer and its solutions.

Ensor, D. S., Charlson, R. J., Ahlquist, N. C., Whitby, K. T., Husar, R. B., and Liu, B. Y. H. (1972) *J. Colloid Interface Sci.*, **39**, 242; also in Hidy, G. M. (Ed.) (1972) *Aerosols and Atmospheric Chemistry*, Academic, New York.

Goody, R. M. (1964) *Atmospheric Radiation: I Theoretical Basis*, Oxford Univ. Press, Oxford. This monograph includes application of the theory of small particle scattering and the equation of radiative transfer to Earth's atmosphere.

Gumprecht, R. D., and Sliepcevich, C. M. (1953) *J. Phys. Chem.*, **57**, 95.

Hodkinson, J. R. (1966) The Optical Measurement of Aerosols in Davies, C. N. (Ed.) *Aerosol Science*, Academic, New York.

van de Hulst, H. C. (1957) *Light Scattering by Small Particles*, Wiley, New York. This is the classic reference on the subject. It is the first place to turn for detailed information on light scattering by single elements, including spheres and other bodies of simple shapes. It is primarily a theoretical work but includes several chapters on practical applications.

McDonald, J. E. (1962) *J. Appl. Meteor.*, **1**, 391.

Penndorf, R. B. (1957*a*) *J. Opt. Soc. Amer.*, **47**, 176.

Penndorf, R. B. (1957*b*) *J. Opt. Soc. Amer.*, **47**, 1010.

Steffens, C. (1956) Visibility and Air Pollution in Magill, P. L., Holden, F. R., and Ackley, C. (Eds.) *Air Pollution*, McGraw-Hill, New York.

CHAPTER 6

Experimental Methods

Most of the instruments used for measuring aerosol properties depend on particle transport or light scattering for size classification or collection; some examples—the thermal precipitator and diffusion battery—have already been mentioned in previous chapters on transport. Progress in aerosol instrumentation has been rapid, and designs change frequently. In this chapter, measurement methods are discussed in light of the principles on which they are based without going into the details of instrument design.

The principal applications of measurement systems are to the monitoring of air pollution; the testing of gas-cleaning equipment, such as filters and scrubbers; and the monitoring of process streams, such as stack gases.

Theory provides useful guidelines for instrument design, but it is rarely possible to predict performance from first principles. In almost all cases, it is necessary to calibrate the instruments using aerosols of known properties.

An ideal measurement instrument would automatically and continuously size and chemically analyze each particle individually, thereby permitting the determination of the size–composition probability density function, g (Chap. 1). From this function, most of the chemical and physical properties of aerosols can be obtained by integration. Such an instrument does not now exist. Existing devices measure integral moments of the size–composition p.d.f. (or of the size distribution function). Quantities often measured include the mass concentration, number concentration, limited ranges of the particle size distribution function, extinction coefficient, average chemical composition, and chemical composition over certain discrete size ranges. The most rapid developments in recent years have been in instruments for the measurement of size distribution. Least developed are methods for real-time determination of aerosol chemical composition.

The measurement instruments reviewed in this chapter are based on different physical principles. A limited number of these devices are discussed; the reader should refer to the references for more detailed information. Instruments with different sensing systems will, in general, have different response times, and the aerosol properties measured by each only partially overlap. A summary classification scheme in this chapter permits comparison of the various instruments according to the type of information they provide.

The instrument or group of instruments selected for a particular application depend on several factors. Most important, of course, is the type of information sought. Other factors include cost, portability, and reliability under the conditions of operation. Stack gas monitoring poses particularly difficult demands because of extreme conditions of temperature and humidity. In the case of measurement systems designed for routine monitoring, the maintenance required is an important factor.

Much effort and ingenuity have been devoted to the development of generators capable of producing monodisperse aerosols, and several are discussed at the end of this chapter. These are used for the testing of gas-cleaning equipment, the calibration of size measurement devices, and the investigation of aerosol behavior.

6·1 Sampling

Optical methods can, in some cases, be used to measure aerosol characteristics in the original gas stream without withdrawing a sample. In most cases, however, it is necessary to sample from a flowing gas through a tube into an instrument, such as those discussed in the following sections.

Care must be taken in the design of the sampling system to insure that a representative sample is obtained. The sampling stream intake should be designed to minimize preferential withdrawal of particles with respect to size. Deposition on the inside walls of the sampling tube and subsequent reentrainment must be minimized or taken into account. Precautions are also necessary to prevent condensation and other gas-to-particle conversion processes (Chap. 9). This problem is particularly acute in the sampling of hot, high-water content stack gases.

When there is a velocity difference between the gas stream and the gas entering the sampling probe, preferential withdrawal of particles with respect to size takes place. In analyzing the performance of a sampling probe oriented in the direction of the flow, it is assumed that the flow is uniform both in the mainstream of the gas and in the entrance to the probe. On dimensional grounds, for the particle size range $d_p > 1$ μm, the ratio of the concentration in the sample, n_s, to that in the mainstream, n_m, depends on the velocity ratio, U_m/U_s, and on the Stokes number:

$$\frac{n_s}{n_m} = f\left(\frac{U_m}{U_s}, St\right)$$

where s and m refer to the sampling and mainstreams, respectively. The goal of the sampling procedure is to insure that $n_s = n_m$.

The dependence of n_s/n_m on the velocity ratio found experimentally is shown in Fig. 6·1 for particles of varying size. To explain the shapes of these curves, we consider the case of fixed sampling velocity. For low mainstream velocities ($U_m/U_s \to 0$), the sample tends to be representative ($n_s \approx n_m$). The sampling orifice acts as a point sink and the streamlines of the flow are practically straight (Fig. 6·2a). As a result, inertial effects can be neglected. The n_s/n_m ratio initially decreases as mainstream velocity increases because inertial effects carry the particles around the sampling orifice (Fig. 6·2b). Further increase in the

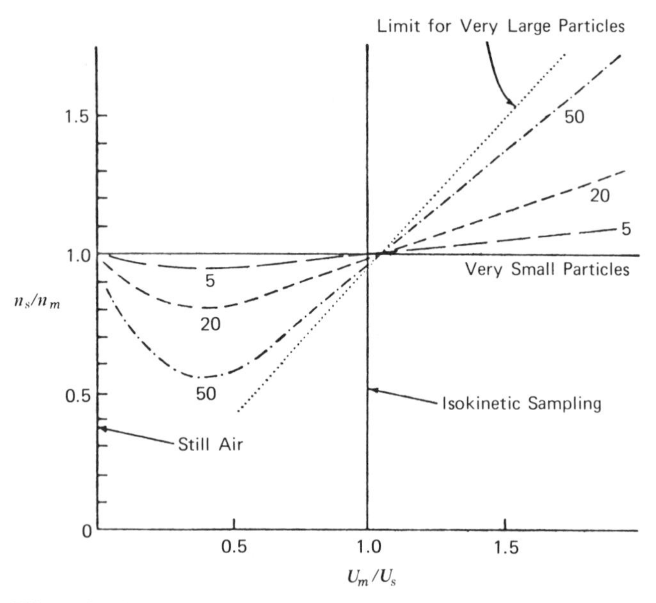

Fig. 6·1 Effect of velocity ratio on concentration ratio for a sampling tube oriented in the direction of the mainstream flow. The curves are approximate representations of the data of various experimenters for unit density particles of diameters (in μm) as indicated. The displacement of the point $n_s/n_m = 1$ from $U_m/U_s = 1$ results from the finite particle diameter. The curves apply to a nozzle of 1 to 2 cm diameter sampling at about 5 m/sec (May, 1967).

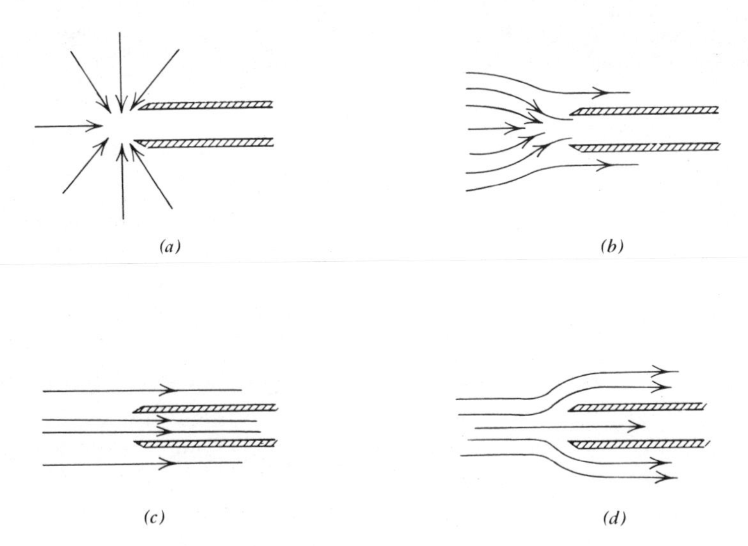

Fig. 6·2 Patterns of gas flow at the entrance to a sampling probe for different ratios of sampling to gas velocities.

mainstream velocity leads to an upturn in n_s/n_m, which approaches unity for $U_s = U_m$ (Fig. 6·2*c*). Sampling at the same velocity as that of the gas is known as *isokinetic sampling*. At higher mainstream velocities, particles are preferentially carried into the sampling tube (6·2*d*).

Since the effect is inertial, it increases for larger particles. For particles of unit density, orifice diameters about 1 cm, and velocities of 500 cm/sec, the effect of nonisokinetic sampling can be neglected for particles smaller than 5 μm.

Deposition on the walls of the sampling tube occurs as a result of diffusion in the small particle range and of turbulent deposition and sedimentation for larger particles. These effects have been discussed in previous chapters, and deposition rates can be estimated from the correlations for flow through tubes. Continued deposition on the walls of the tube eventually leads to reentrainment. The reentrained particles are likely to be agglomerates. Their presence modifies the size distribution, increasing the concentration of larger particles. The effect depends on the surface behavior of the deposited material and cannot be predicted from theory.

6·2 Microscopy

Small particles deposited on a surface can be observed by optical or electron microscopy, depending on their size. This is the primary measurement method upon which most aerosol sizing methods are ultimately based.

The magnification of a microscope depends on the focal lengths of the lenses making up the optical system. Any desired magnification is attainable in principle by proper selection of the focal lengths. There is, however, a magnification beyond which the image formed in the microscope does not gain *in detail* because of the effects of diffraction. Thus the image of a point object produced by an ideal lens (all aberrations corrected) is not a point but a diffraction pattern consisting of a circular disk surrounded by alternating dark and light rings of diminishing intensity.

Supposing we are looking at a particle under a microscope. The sizing of the particle depends on our being able to distinguish one edge of the particle from another on the opposite side. The ability of a microscope to size in this way is measured by its *resolving power*, the closest distance to which two objects under observation can approach and still be recognized as separate. By definition, the resolving power or limit of resolution of the instrument is the radius of the central disk of the diffraction pattern. From diffraction theory, this is given by

$$l_{res} = \frac{0.61\lambda}{m \sin\theta}$$

where λ is the wavelength of the light, m is the refractive index of the medium in which the object is located, and θ is the half-angle of the light rays coming from the object (Fig. 6·3). To maximize the resolution, λ should be small, and m and θ, large. The refractive index can be increased by immersing the object under observation in an oil ($m \approx 1.5$) instead of air ($m = 1$). The highest numerical aperture, $m \sin\theta$, attainable in this way is about 1.4. Thus taking $\lambda = 0.5$ μm, the best resolution achievable with the optical microscope is about 0.2 μm.

A significant improvement in resolving power over the oil immersion optical microscope is possible using the much shorter wavelengths

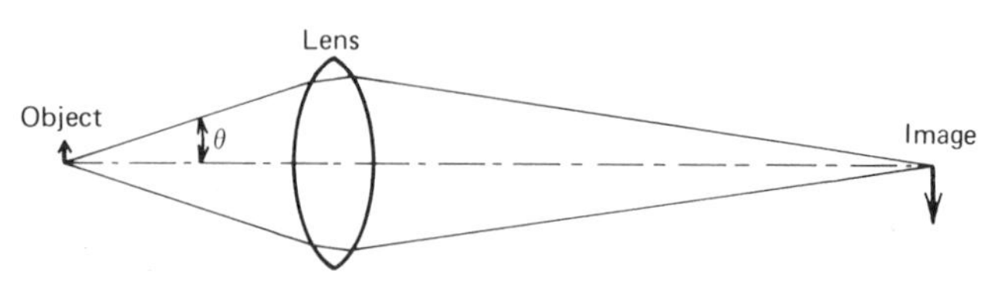

Fig. 6·3 Light from a single point on the object is intercepted over a half-angle (aperture) θ by the lens.

associated with high-speed electron beams. Both electric and magnetic fields can be used as lenses for electrons so the elements of a microscope are available. If electron lenses could be made as free from aberrations as optical lenses, a resolution of the order of 1 Å, smaller than atomic dimensions, would be possible (Zworykin et al., 1945), but this has not been achieved. However, practical units of resolution of 10 Å are easily attained (Table 6·1).

TABLE 6·1
APPLICATIONS OF MICROSCOPY (AFTER ZWORYKIN ET AL., 1945)

Type of Instrument	Resolving Power (μm)	Magnification	Type of Particle That Can Be Observed
Eye	200	1	Ordinary objects
Magnifying glass	25–100	2–8	Fog droplets
Low-power compound optical microscope	10–25	8–20	Pollen
Medium-power compound optical microscope	1–10	20–200	Airborne soil dust
High-power compound optical microscope; electron microscope	0.25	800	Cigarette smoke
Electron microscope; high-power compound optical microscope using ultraviolet light	0.10	2000	Surface area peak in urban aerosols
Electron microscope	0.001–0.05	4000–200,000	Aerosols formed by homogeneous nucleation

Since the electron beam must operate in a vacuum, volatile particles will evaporate leaving behind a residue that, however, can provide some information on the original particle. In addition, heating by the electron beam tends to modify the original particles.

Taking the limit of resolution of the average human eye as 0.2 mm, the maximum useful magnification can be defined as the ratio of the resolving power of the eye to that of the microscope:

$$\text{Magnification} = \frac{0.033\, m \sin\theta}{\lambda}$$

Values of the resolving power and magnification necessary for the observation of particles of various types are shown in Table 6·1.

Particles can be deposited on surfaces for optical or electron microscopy using a number of different devices including thermal and electrical precipitators, filters, and cascade impactors. The preparation of samples for microscopy including the use of these devices is reviewed by Silverman, Billings, and First (1971). For detailed information on the size and morphological characteristics of particle matter, there is no substitute for microscopy. However, for routine monitoring and for studies of aerosol dynamics, it is usually more convenient to use calibrated continuous, automatic counters of the types described in this chapter. A catalog of photomicrographs of particles of different origins, showing the remarkable variations in size and shape has been prepared (McCrone, Draftz, and Delly, 1973).

6·3 Mass Concentration: Filtration

The Federal ambient air quality standard for particulate pollution is expressed in terms of the mass loading ($\mu g/m^3$) (Chap. 11):

$$\rho = \int_0^\infty \rho_p n(v) v \, dv$$

where ρ_p, the density of the aerosol material, can be a function of particle size. Mass loadings are determined by weighing the filter before and after filtration under conditions of controlled humidity. Samples are also collected by filtration for chemical analysis.

Many different types of filters are available commercially. They can be broadly classified into two types with, however, some overlap. *Fibrous filters* are composed of mats of fibers that may be made of cellulose, glass, polymeric materials, metals, or asbestos. Mixtures of fibers such as cellulose and asbestos are sometimes used. The asbestos fibers are much finer than the cellulose fibers and have correspondingly higher collection efficiencies. The cellulose fibers serve to support the asbestos. Particles collected by fibrous filters are distributed internally through the filter.

Membrane filters are usually composed of thin films of polymeric materials sufficiently porous for air to flow through under pressure. Pore size is controlled in the manufacturing process. A significant fraction of the particles may be caught on the upstream surface of the filter, but some particles may also penetrate and be caught inside the pores of the medium as well.

The use of the term membrane is somewhat misleading since these polymeric films do not serve for gas transport by diffusion. Instead, gas passes through the pores of the film by a macroscopic flow process, driven by the pressure gradient.

The principal mechanisms of particle deposition for both fibrous and membrane filters are the diffusion and impaction of particles of finite diameter. Settling and electrostatic effects may contribute to removal.

The National Air Sampling Network, maintained by the U. S. for the measurement of atmospheric pollutants, makes use of a standard high-efficiency glass fiber filter for the sampling of aerosols. Glass fiber filters are less hygroscopic than paper filters and can withstand higher temperatures. They can be extracted with benzene, water, and acid for subsequent chemical analysis. Glass fiber filters are usually more expensive than paper filters and have poorer mechanical properties. A continuing problem has been the variability in their properties resulting from the difficulties in quality control during the manufacturing process.

The penetration of these filters as a function of air velocity by 0.3 μm particles is shown in Fig. 6·4. The maximum in the penetration curve corresponds to the transition from diffusion to impaction. Filter

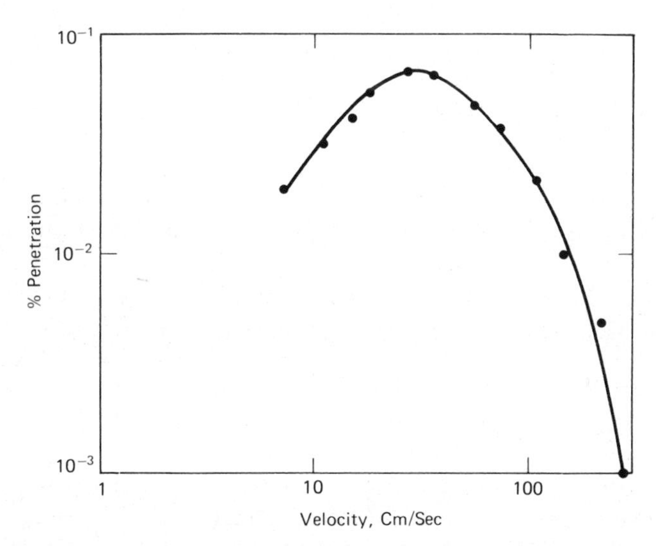

Fig. 6·4 Percentage of 0.3 μm dioctylphthalate particles passing a glass fiber filter (type MSA 1106B). This is the type used by the National Air Sampling Network of the Environmental Protection Agency. The penetration shows a maximum at a velocity of about 30 cm/sec, perhaps corresponding to the transition from collection by diffusion to collection by impaction. The test aerosol was produced by a condensation aerosol generator (Section 6·12) (Data of Lockhart, Patterson, and Anderson, 1964). This reference gives efficiency and pressure drop data for many other filter materials.

efficiencies are often reported for 0.3 μm particles, since these fall in the size range corresponding to the minimum in the efficiency curve. Hence particles of other sizes would, in general, be expected to be collected more efficiently.

6·4 Total Number Concentration: Condensation Nuclei Counter

A method was devised by Aitken in the nineteenth century to measure the total number concentration, N_{∞}. A modern version (Fig. 6·5) consists of a vertical tube about 60 cm long and 2 to 3 cm in diameter, lined with a wetted, porous ceramic material. The tube is sealed at the top and bottom by electrically heated glass plates. The tube is fitted with a light source at the top and a detector, a photomultiplier tube, at the bottom.

The tube is flushed several times with aerosol, usually room or atmospheric air, to replace the contents. A fixed quantity of filtered, particle-free air is then introduced into the tube under pressure.

A period of 50 sec is allowed for the air to become saturated and to equilibrate thermally. After reading the photo cell current, I_0, to obtain a measure of the initial extinction, the exit valve is opened and the pressure released. The gas expands adiabatically, and condensation takes place on the particles which grow into the light-scattering range. Since the growth process takes place in an essentially uniform gas mixture and the process is diffusion limited, the nuclei tend to grow into droplets all of which are of about the same size (Chap. 9). The new photo cell current, I, is read, and the droplet concentration, which is equal to the original particle concentration, is obtained from the ratio, $(I_0 - I)/I_0$.

The instrument can be calibrated by allowing the droplets to settle on a slide. The concentration is determined by counting the deposited droplets corresponding to each value of $(I_0 - I)/I_0$. Commerical instruments of this type are usually factory calibrated by comparison with a counter built and operating according to the original specifications of Pollak and his co-workers (Pollak and Metnieks, 1957).

A major uncertainty in the use of this type of instrument is the size of the smallest detectable particle. This probably varies depending on the chemical nature of the nuclei, and the saturation ratio at the end of the expansion, but is usually in the range between 10 and 100 Å.

An independent calibration of the instrument has been carried out by Liu et al. (1975). They compared the condensation nuclei counter with an electrical aerosol detector using monodisperse aerosols. Particle

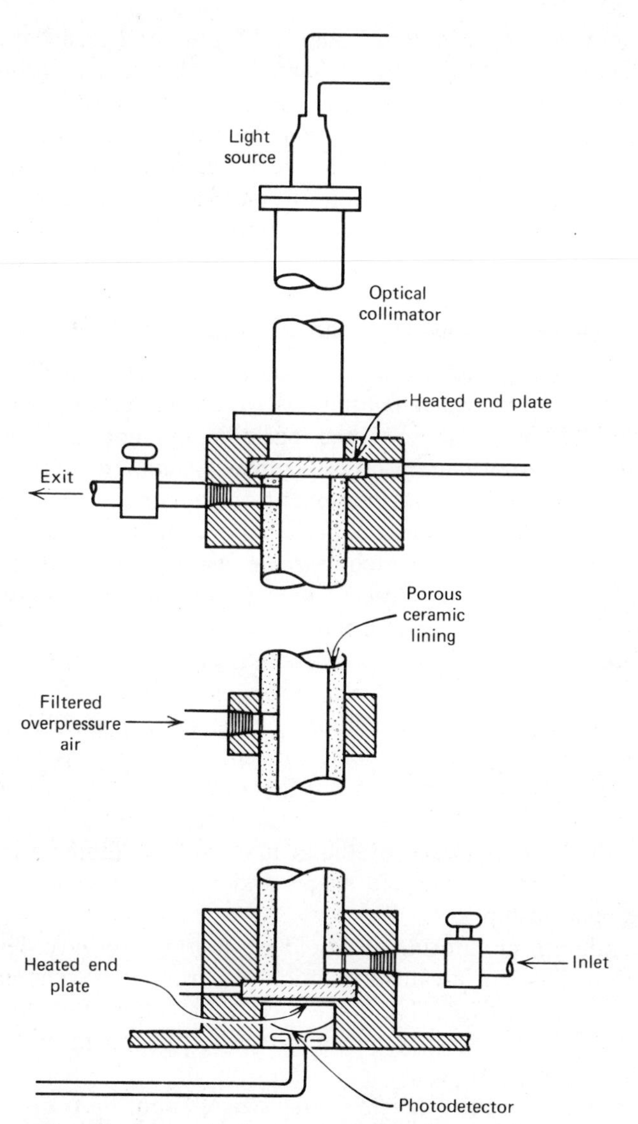

Fig. 6·5 Schematic diagram of condensation nuclei counter.

diameters ranged between 0.025 and 0.15 μm, and concentrations, between 500 and 2.5×10^5 particles/cm^3. Two different aerosols were used, one of sodium chloride and the other material volatilized from a heated nichrome wire. Very good agreement between the two methods was found indicating that over this size range, a properly calibrated nuclei counter does indeed detect all particles. This conclusion is probably not limited to the chemical compositions of the two aerosols studied.

6·5 Size Distribution Function: Single Particle Optical Counters

For particles larger than a few tenths of a micrometer, single particle optical counters can provide a continuous, on-line record of particle size distributions. Many versions of the basic instrument are marketed commercially, differing chiefly in the optical system. In most designs, the aerosol stream enters the instrument surrounded by a sheath of filtered air to prevent instrument contamination. Light from a source of illumination is scattered by each particle as it passes through the illuminated region, and the scattered light is collected and passed to a photomultiplier tube. The signal from the tube is classified as a function of pulse–height, which is related to particle size.

Of particular interest are the *size resolution* of the counter, or its ability to distinguish between neighboring particle sizes, and the *limit of detection*, or smallest size to which the counter responds. The size resolution depends on the relationship between pulse–height and particle size, the *response curve*. For particles of given optical properties, this relationship is determined by the geometry of the illumination and light collection systems. Particle shape and refractive index also influence the relationship.

For air pollution monitoring, it would be desirable to have detectors whose response is a single-valued function of particle size (volume) and not of shape or refractive index, since these parameters may vary from particle to particle. Practical systems fall far short of this ideal. In some commercial instruments, local minima occur in the response curve, which means that the pulse–height versus size relationship is not unique (Mercer, 1973). In other cases, the response is very sensitive to refractive index.

The characteristics of a modified commercial instrument with an ellipsoidal mirror light collector have been studied by Husar (1974); we describe this system as an example (Fig. 6·6). The light source is a filament lamp, and in the sensing volume, the light beam is 1.5 mm wide and 1 mm high. The sensing volume is located at one of the focal points

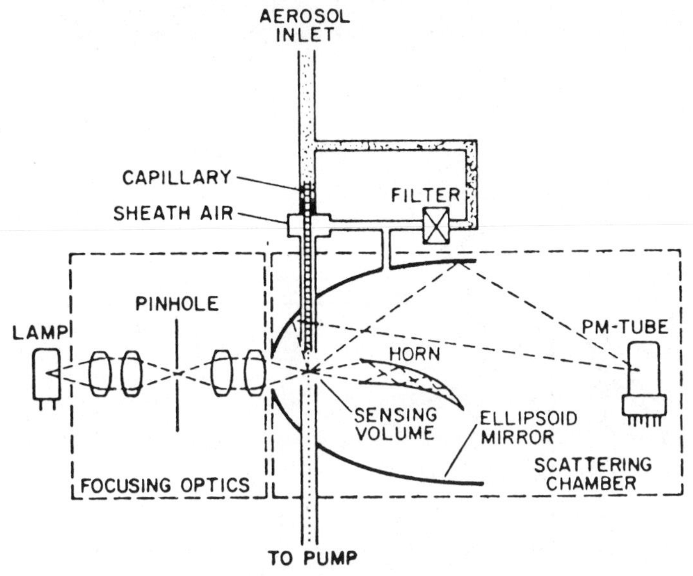

Fig. 6·6 Schematic diagram of the ellipsoidal mirror optical counter (Husar, 1974).

of the ellipsoidal mirror collector. Light scattered from the sensing volume in the range between 35 and 100° is reflected by the mirror into the other focal point at which the photomultiplier tube is located.

Response curves calculated from Mie theory are compared with an experimentally measured curve for polystyrene latex particles in Fig. 6·7. For both calculated and measured curves, pulse height increases in a fairly smooth way with particle diameter. No local minima or maxima, or size independent ranges are found in the curve. For nonabsorbing spheres over the range of refractive index studied, the error in the indicated size is always less than 30% if the polystyrene latex calibration is used.

The limit of detection of an optical particle counter depends on instrument noise, Rayleigh scattering by the air molecules, and stray light resulting from imperfect optics. In carefully designed laboratory instruments, the detection limit may be as low as 0.15 μm; for commercial counters, it is usually greater than 0.4 μm. Scattering by air molecules determines the theoretical limit of detection. This could, in principle, be reduced by pumping off the air as in the particle beam apparatus (Chap. 4).

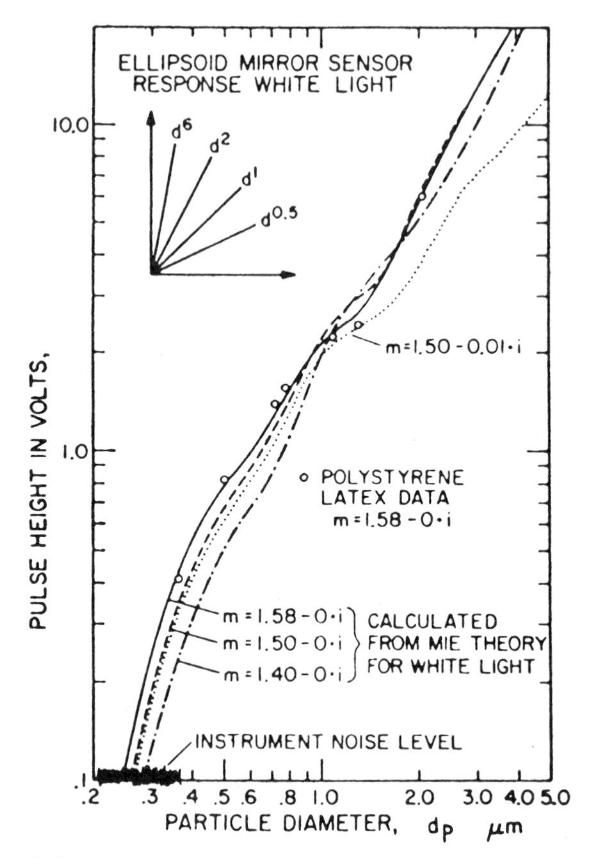

Fig. 6·7 The relative response curves of the ellipsoidal mirror optical counter. The lines were calculated from Mie theory and the points were measured experimentally (courtesy S. L. Heisler and R. B. Husar).

Difficulties are encountered in the use of the counters with the atmospheric aerosol because the refractive index varies from particle to particle in an unknown way. As a result, it is necessary either to assume a refractive index for the particles and calculate a modified calibration curve or to report data in terms of equivalent particle diameter for a standard aerosol.

6·6 Size Distribution Function: Electrical Mobility Analyzer

By charging the particles under controlled, reproducible conditions, a unique relationship can be established between particle size and electrical mobility over certain size ranges (Chap. 2). If the particles are then passed through a well-defined electric field, they can be classified according to size. A system of this kind, shown in Fig. 6·8, is described by Whitby et al. (1972). Its optimal range of operation is for particles between 0.01 and 0.3 μm. The particles are charged by mixing them in a chamber with negative ions generated by a corona discharge. The rate of charging is controlled by the diffusion of ions to the particles; for the

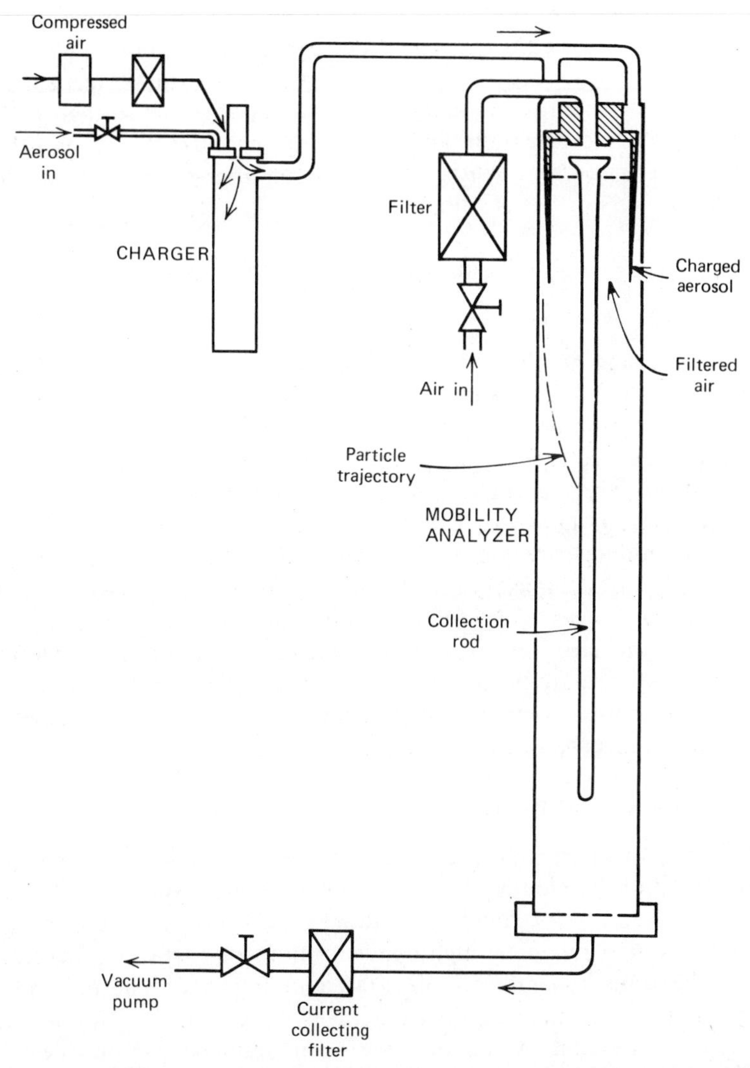

Fig. 6·8 Schematic diagram of an electrical mobility analyzer of a type available commercially (Whitby et al., 1972).

systems that have been studied, the charge acquired was independent of particle chemical composition.

From the charger, the aerosol enters the analyzer through an annular space where it joins a laminar core flow of filtered air. As the two streams flow through the mobility analyzer, the charged particles move in the electrical field through the clean air core to the positively charged inner collecting rod. Particles not deposited because their mobility is too small are collected by a glass fiber filter supported on a porous stainless steel disk. The current flow to the filter carried by the charged particles is measured by an electrometer connected to the stainless steel support.

The size distribution is determined by varying the voltage on the collecting rod incrementally and measuring the current coming off the filter. From the current–voltage curve, the number–size distribution can be calculated using the relation:

$$\Delta N = \frac{\Delta I}{e n_e Q}$$

where ΔN is the number concentration in a given size range, ΔI is the current increment corresponding to a given voltage step, n_e is the particle charge, and Q is the volumetric flow of aerosol through the analyzer. Each voltage setting corresponds to a maximum particle size collected; all larger particles have mobilities too small to deposit. This size and n_e are obtained by calibrating the instruments.

Several minutes are required to measure a size range divided approximately logarithmically into 10 or 15 channels. This means that the instrument averages over different time intervals in measuring each portion of the size spectrum. Simultaneous operation of a mobility analyzer, optical particle counter, and condensation nuclei counter has provided the most complete information yet available on particle size distributions in urban atmospheres.

6·7 Scattering and Extinction Coefficients

A variety of instruments have been designed for the measurement of light extinction by clouds of small particles. *Transmissometers* are available commercially for stack installation and for use with other process gases containing relatively high concentrations of particles. The principle of the transmissometer is illustrated in Fig. 5·5. The design and calibration of an in-stack instrument is described by Conner and Hodkinson (1967). Instruments have been designed for the measurement of atmospheric extinction usually over a long path (Middleton,

1952). When absorption by the particles can be neglected, the extinction results from scattering, and measurements of scattering can be used as a substitute for extinction. This assumption is probably acceptable for marine hazes and, perhaps, for photochemical aerosols such as those in Los Angeles where concentrations of absorbing particles are usually low. Aerosol absorption may be more important in the air of industrial regions of the midwest and east where carbon and soot concentrations are relatively high.

A schematic diagram of a compact instrument designed for the measurement of the scattering coefficient (Chap. 5) is shown in Fig. 6·9. This device is known as an *integrating nephelometer*, since it integrates the scattering over almost all angles along the axis of the detector to give the scattering coefficient.

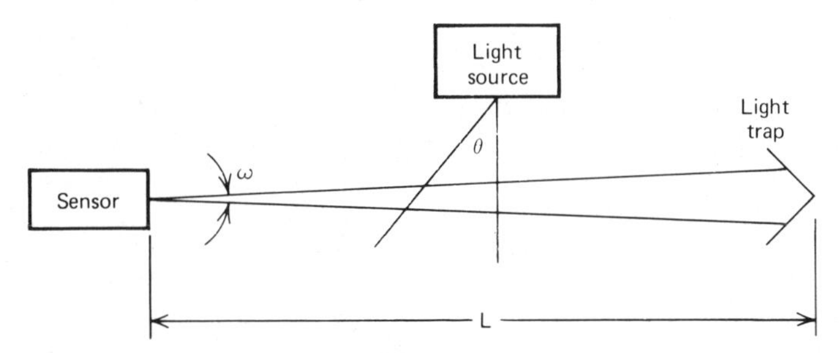

Fig. 6·9 Principle of the integrating nephelometer. The sensor detects the light scattered by the particles present in the region defined by the solid angle ω. The light source has the special property that its intensity in any direction θ is given by $I_0 \cos\theta$. When the sensor and light trap are sufficiently far apart, it can be shown (Middleton, 1952) that for this special light source the sensor signal is proportional to the integral of the scattered light. By proper calibration, the quantity b_{scat} can be obtained. Commercial instruments are about 6 ft long and 6 in. in diameter.

Important features of the instrument include a light source whose radiation intensity follows a cosine law, together with a specially designed light collection system for defining the shape of the scattering volume. The measurement volume is small compared to the scale over which b_{scat} changes in the atmosphere. Hence the instrument gives a local light-scattering coefficient that, for a uniform atmosphere, is inversely related to the visibility or visual range (Chap. 5). The local value of b_{scat} measured in this way can be related to the properties of the aerosol, including the size distribution and chemical composition, also measured locally. This is of great importance in relating emission source contributions to visibility degradation (Chap. 11). The use of the integrating nephelometer in atmospheric measurements is discussed by Butcher and Charlson (1972).

6·8 Mass and Chemical Species Distribution: The Cascade Impactor

The cascade impactor is the instrument most commonly used for the classification of aerosol particles according to size for subsequent chemical analysis. The device consists of a series of stages, each composed of an orifice through which the aerosol flows normal to a collecting surface (Fig. 6·10). The orifice may be rectangular or circular in shape. The air flows over the collecting surface and on to the next stage, whereas particles too large to follow the air motion deposit on the surface. The basic mechanism of collection is inertial impaction.

The orifice diameter (or width) is largest at the first stage, and the gas velocity is lowest here. The largest suspended particles deposit at the first stage; smaller ones pass on to succeeding stages of smaller orifice diameter and progressively increasing efficiency of removal.

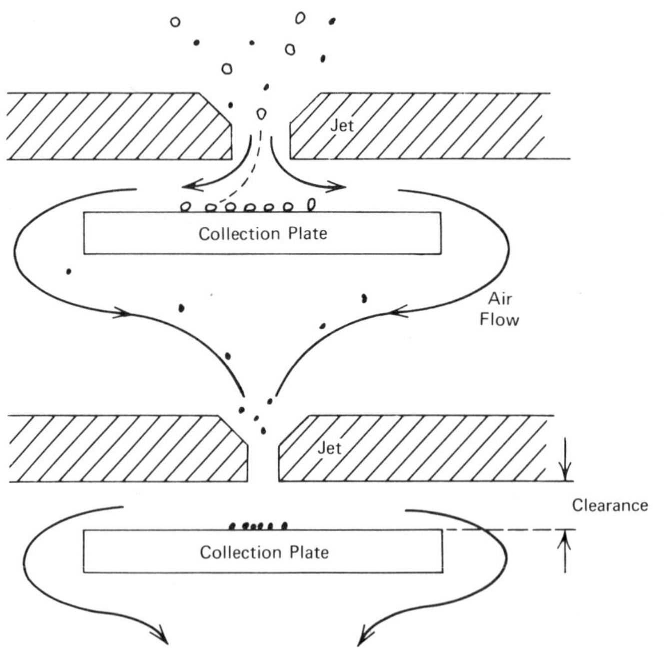

Fig. 6·10 Schematic diagram showing two stages of a cascade impactor. The last stage is often followed by a filter.

The efficiency of a stage for particles of a given size is defined as the fraction of the particles removed from the gas flowing through the stage. At a given flow rate, in the absence of particle reentrainment or rebound, the stage efficiency depends on the Stokes number,

$$St = \frac{C\rho_p U d_p^2}{18\ \mu d}$$

where U is the average velocity through the jet and d is the jet width or diameter. The slip correction factor, C, is given by (2·17). Ideally, the efficiency curve should be a step function corresponding to a given Stokes parameter. All larger particles would be caught at the stage, whereas all smaller particles would pass. In practice, the efficiency curve is "S" shaped as shown in Fig. 6·11. A stage is usually characterized by the diameter corresponding to 50% efficiency:

$$d_p^* = \left[\frac{18 \mu d\, St^*}{C\rho_p U} \right]^{1/2} \tag{6·1}$$

where the asterisk refers to the value at 50% efficiency. For round jets, $St^* \approx 0.2$ for values of the ratio of clearance to diameter (Fig. 6·10) over

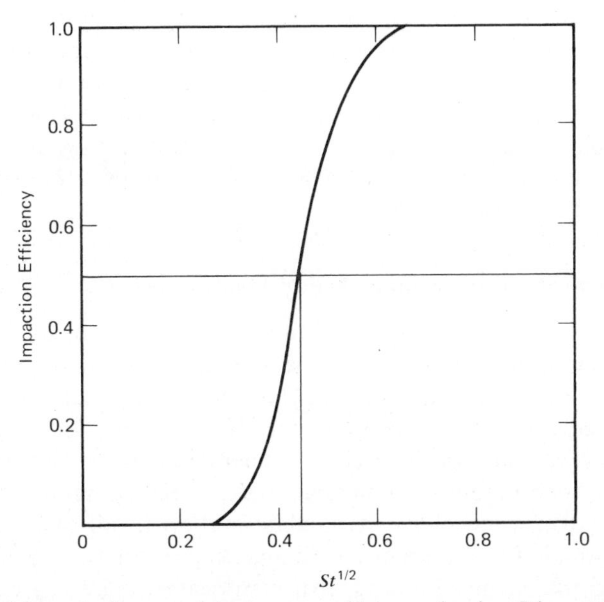

Fig. 6·11 Schematic diagram of jet impactor efficiency showing St^* corresponding to 50% impaction efficiency. For round jets, the lower tail of the efficiency curve may not exist (Marple, V. A. and Liu, B. Y. H. (1974), *Environ. Sci. Technol.*, **8**, 648).

the range between 1 and 10. For rectangular jets, $St^* \approx 0.66$ for clearance to diameter ratios of 1 to 5. Reynolds number similarity (Chap. 4) has not been investigated in detail. It is usually ignored, and this is supported by the data of Ranz and Wong (1952) over a limited range. It is good practice to calibrate each stage of an instrument, whether commercial or home-made, using monodisperse aerosols.

In the case of a complex aerosol, such as atmospheric particulate matter, size, shape, and density vary among particles that deposit on the same impactor stage. This is true even for a stage with perfect (step function) performance for spherical particles of constant density. How, then, are the results of measurements with the impactor to be interpreted for complex aerosols? Impactor data are often reported in terms of the *aerodynamic diameter*, defined as the diameter of a hypothetical sphere of unit density with the same Stokes number (or settling velocity) as the particle in question. Particles of different size, shape, and density may have the same aerodynamic diameter. The aerodynamic diameter is particularly useful for calculations of deposition by impaction.

Jet velocities usually range from about 3 to 4 m/sec on the first stage to between 100 and 200 m/sec on the last stage. Conventional cascade impactors are usually satisfactory for the classification of particles larger than 0.3 μm (aerodynamic diameter). According to (6·1), smaller particles can be separated by increasing the velocity, decreasing jet diameter or width, or increasing the slip correction factor, C. Increasing velocity may result in particle rebound, whereas decreasing jet dimensions may pose problems in machining small holes. The most satisfactory method of separating smaller particles by impaction has been to increase C by operating the impactor at lower pressures (Stern, Zeller, and Schekman, 1962).

6·9 Elemental Composition: X-Ray Fluorescence

A variety of microanalytical techniques can be used for the chemical analysis of atmospheric particulate matter collected by filters or on impactor slides. These include wet chemical methods, atomic absorption spectroscopy, neutron activation analysis, and x-ray fluorescence. Until now, however, no commercial instrument has been developed that is capable of continuous, on-line monitoring of *any* elemental or molecular component of atmospheric particulate matter. In all cases, it is necessary to sample for a more or less long period of time and then await the results of the chemical analysis. In many cases, there is a considerable delay. This has proved to be a serious handicap in studies of the origin and behavior of particulate pollution and in epidemiological studies as well.

It would not be feasible to review any substantial portion of the available microanalytical techniques. Instead, two methods with specific aerosol applications will be considered, both based on the analysis of x-rays emitted by excited particulate matter.

The *x-ray fluorescence detector* consists of a monochromatic x-ray source, sample holder, and semiconductor detector (Goulding and Jaklevic, 1973). The data-processing system includes a computer for chemical analysis. A sample of particulate matter collected on a filter is excited by the x-ray source. The excited sample produces an x-ray spectrum characteristic of the atomic species present in the sample.

Enough material can be conveniently collected on a filter in a period of 1 or 2 hours for analysis of many of the species present in polluted city air. It is difficult to detect elements lighter than potassium because of their low fluorescence yields and the strong absorption of fluorescence x-rays by other matter including the particles. A semiautomatic version of the instrument has been developed in which the filters are transferred manually from the filter holder to the counter (Goulding and Jaklevic, 1973*a*). By modifying the sample holder to introduce a continuous tape filter, it may be possible to automate the system more completely.

6·10 Elemental Composition: Electron Microprobe

The electron microprobe can be used to measure the concentrations of elements with atomic numbers greater than about nine (fluorine) in individual particles larger than a few tenths of a micrometer. Hence for this size range, it permits the determination of the size–elemental composition probability density function, g. The electron microprobe was a natural outgrowth of the development of electron microscopy and x-ray spectrochemical analysis. In the electron microscope, a narrow beam of electrons is formed. When such a beam strikes a solid specimen, x-rays characteristic of the sample are generated. By measuring the wavelength and intensity of the emitted x-rays, the concentrations of various elemental species can be determined. The application of this technique to the analysis of small particles is discussed by Bayard (1973).

A schematic diagram of the electron microprobe is shown in Fig. 6·12. The particles are mounted on a metal substrate, evaporated carbon film, or electron microscope grid. They must be sufficiently far apart to avoid interference, but the concentration must be high enough to permit easy identification in the scanning process. As an example of the scale of operation of such systems, an electron beam with a diameter of 1 μm when passed through an acceleration field of 15 kV will

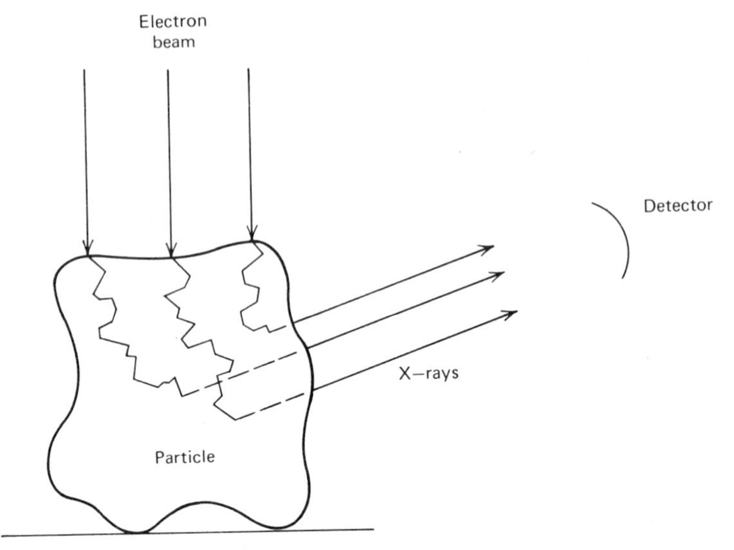

Fig. 6·12 Principle of the electron microprobe: Schematic diagram of the interaction of an electron beam with a deposited particle. Irregular lines represent the path of an electron before it produces ionization and the generation of x-rays characteristic of the emitting element. Straight lines represent the paths of x-rays headed in the direction of the detector. The irregular shape of the particle considerably complicates the interpretation of results (Armstrong and Buseck, 1975).

penetrate a volume of about 3 μm^3 in the sample. Scattering by particles smaller than this volume is difficult to interpret. This means that the limit of resolution for this instrument is only a little more than 1 μm.

Elemental composition is determined by x-ray spectrometry. The size of the smallest particle that can be analyzed is presently limited by spectrometric techniques. A computer connected to the system can be programmed to calculate the elemental concentrations. In this way, it is possible to obtain data on the composition of each area on which the beam is focused in a few minutes.

6·11 Summary Classification of Measurement Instruments

Aerosol measurement instruments can be conveniently classified according to the type and quantity of information they provide about aerosol properties. The physical principles on which some of the instruments are based are of secondary importance in this classification scheme, and indeed, the instruments can be considered "black boxes." This approach makes it possible to see how far measurement technology has developed and may indicate what direction it is likely to take in the future.

In Table 6·2, instruments discussed in previous sections are classified in terms of their performance characteristics. The second column shows whether the instrument classifies the particles according to size and also how fine the size resolution is. The third column shows the time response behavior of the system, that is, whether the instrument responds to single particles or to short- or long-term time averages. The fourth column indicates whether single particles are analyzed chemically or the average composition of collections of particles.

The first device listed, the nonexistent "single particle counter-analyzer" (SPCA), chemically sizes and analyzes each particle, thereby permitting the determination of g. The SPCA, operating perfectly, would classify the particles according to size, identifying each class separately with no "lumping" of classes. This perfect classification is

TABLE 6·2
Characteristics of Aerosol Measurement Instruments

Instrument	Resolution: Size	Resolution: Time	Resolution: Chemical Composition	Quantity Measured (Integrand × N_∞^{-1})
Perfect Single Particle Counter Analyzer	Resolution at single particle level	Resolution at single particle level	Resolution at single particle level	g
Optical Single Particle Counter	Discretizing process	Resolution at single particle level		$\int_{v_1}^{v_2}\int g\,dn_j\,dv$
Electrical Mobility Analyzer	Discretizing process	Discretizing process		$\int_{v_1}^{v_2}\int g\,dn_j\,dv$
Condensation Nuclei Counter	Averaging process	Discretizing process		$\int g\,dv\,dn_j = 1$
Impactor	Discretizing process	Averaging process		$\overline{\int_{v_1}^{v_2}\int g\,dn_j\,dv}$
Impactor Chemical Analyzer	Discretizing process	Averaging process	Discretizing process	$\overline{\int_{v_1}^{v_2}\int g\,n_j\,dn_j\,dv}$
Whole Sample Chemical Analyzer	Averaging process	Averaging process	Discretizing process	$\overline{\int\int g\,n_j\,dn_j\,dv}$

Key:

◁ Resolution at single particle level

Discretizing process

$\int$ Averaging process

represented by the open sector shown in the size column. As an example, a single class is shown passing to the time resolution column; since the counter responds to the individual particles, the time resolution is also perfect as indicated by the open sector in the time column. Complete chemical analysis of a single-time class is indicated by the open sector in the chemical composition column.

The electron microprobe provides some of the information necessary for the evlauation of g. Since the particles must be collected on a surface and then transferred to the instrument, it does not allow continuous, real-time reporting of data. Only particles larger than a few tenths of a micrometer can be analyzed. Certain chemical elements present are reported but not the form in which they are combined chemically, that is, the chemical speciation.

The other devices listed in the table are real instruments; they measure certain integral function of g as shown in the last column. The first few devices have a relatively fast time response. However, the single particle optical counter lumps the data over small but discrete size ranges. This lumping of size classes is represented by the striated pattern shown in the size sector. Optical counters currently available are limited to particles larger than a few tenths of a micrometer in diameter. Hence the size resolution covers a limited range as shown by the reduced sector in the table. Since the counter responds to single particles, the time resolution can be considered perfect as indicated by the open sector.

The striated pattern in the time resolution column for the mobility analyzer shows that particle counts are integrated over an interval of minutes.

The condensation nuclei counter responds to all particles larger than an ill-defined size probably of the order of 10 to 100 Å. Since it counts all such particles without distinguishing among them, there is no resolution with respect to size as indicated by the box containing the integral sign. The quantity measured is in principle $N_\infty(t)$, the total number of particles per unit volume, as a function of the time t. The response time is of the order of a few seconds so the instrument can be considered as continuous for applications to atmospheric monitoring.

Size distribution data obtained with the cascade impactor, the fifth device shown in the table, are lumped over size ranges corresponding to each stage as shown by the striated pattern in the size resolution column. The data are then averaged over the sampling volume or time of operation as indicated by the integrated symbol in the time resolution column.

The cascade impactor is frequently used to classify particles for chemical analysis. The quantity obtained is the concentration of various chemical species averaged over all the particles in a given size range and then averaged again with respect to time:

$$\overline{\Delta\rho}_i = \frac{M_i}{T}\int_0^T N_\infty \int_{v_1}^{v_2}\left[\int \cdots \int g n_i \, dn_2 \cdots dn_i \cdots dn_k\right] dv\, dt$$

where $\overline{\Delta\rho}_i$ is the mass of species i per unit volume of air in the size range v_1 to v_2 averaged over the time T, and M_i is the molecular weight of the ith species.

The last system shown in Table 6·2 corresponds to a filter in combination with a method of chemical analysis such as x-ray fluorescence. It provides data on the chemical composition of the entire aerosol. The particles are collected on a filter usually over a period of hours. Concentrations measured by whole sample chemical analysis can be represented by the following expression:

$$\bar{\rho}_i = \frac{M_i}{T}\int_0^T N_\infty \int\left[\int \cdots \int g n_i \, dn_2 \cdots dn_i \cdots dn_k\right] dv\, dt$$

where $\bar{\rho}_i$ is the mass of species i per unit volume of air averaged over the time T.

6·12 Monodisperse Aerosols: Condensation Generators

Monodisperse aerosols are almost always used to calibrate the instruments described previously. They are also important for performance studies of gas-cleaning devices and in investigations of fundamental aerosol behavior.

It has long been possible to generate aerosols of fairly uniform size using biological particles such as spores, pollen, bacteria, viruses, and bacteriophage. Interest in the characteristics of screening smokes and in the design of gas mask filters during World War I led to the development of condensation aerosol generators; those were easier to setup and operate in nonbiological laboratories. By seeding a condensable vapor with nuclei and then allowing condensation to take place under carefully controlled conditions, aerosols of nearly uniform size were produced. The method is of great practical importance in producing test aerosols from a variety of liquids and from solid materials, such as salts, as well.

A diagram of an apparatus of this type developed by Liu, Whitby, and Yu (1966) is shown in Fig. 6·13. The substance from which the aerosol is to be made, generally a high boiling material such as dioctyl

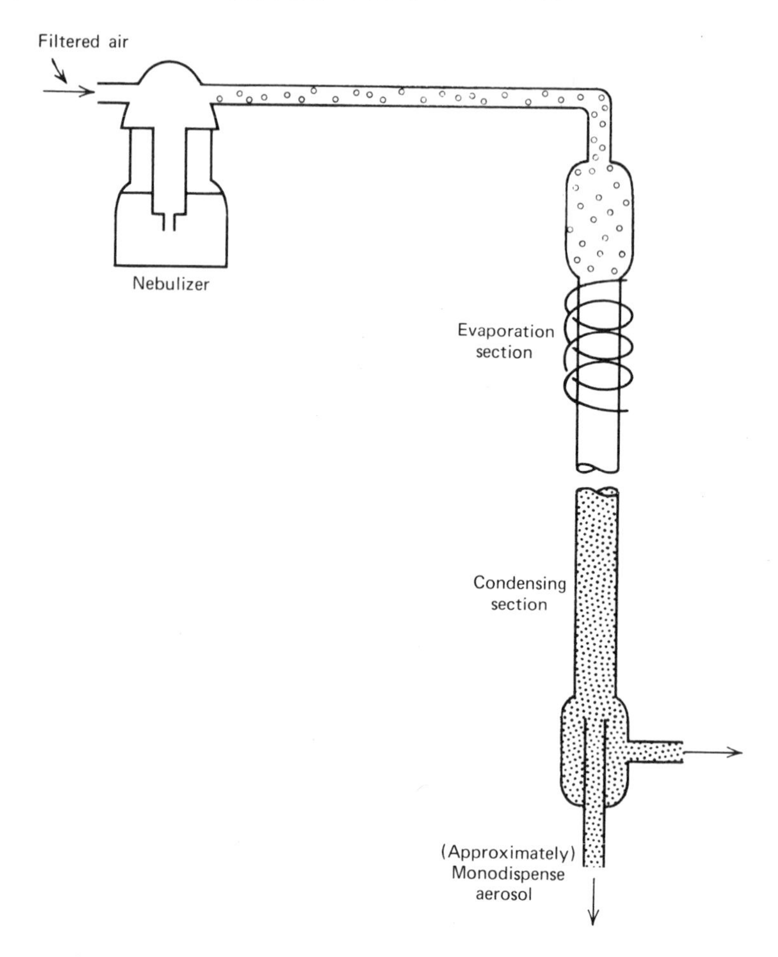

Fig. 6·13 Condensation aerosol generator (Liu, Whitby, and Yu, 1966).

phthalate or oleic or stearic acids, is atomized, and the resulting spray is then evaporated in an electrically heated glass tube. There is usually a sufficient amount of nonvolatile impurity present in the atomized liquid to leave tiny residual particles that can serve as condensation nuclei. Since one condensation nucleus is formed from each droplet and the atomizer can be operated in a stable manner, the generator provides a steady source of nuclei and condensable vapor.

After passing through the evaporation section, condensation takes place on the residual nuclei, forming an aerosol. The most uniform portion of the aerosol is the part flowing near the center of the tube where the temperature profile is flat. It is this portion of the stream that is sampled to provide an almost monodisperse aerosol. By diluting the aerosol substance material with alcohol, Liu, Whitby, and Yu were able to generate aerosols with diameters ranging from about 0.036 to 1.1 μm. The larger particles are more uniform than the smaller, the geometric standard deviation increasing from 1.22 at 0.6 μm to 1.50 at 0.036 μm. Several other condensation aerosol generators have been developed and these are reviewed by Mercer (1973). Aerosol flow rates in the range 0.1 to 1 liter/min and concentrations in the range 10^4 to 10^7 particles/cm^3 can be generated in this way.

6·13 Monodisperse Aerosols: Atomizing Generators

Monodisperse, spherical polystyrene latex particles in aqueous suspension are available commercially in sizes ranging from 0.088 to about 2 μm. Relative standard deviations are usually less than 10% and sometimes less than 1%. The suspensions are prepared industrially by carefully controlled emulsion polymerization. Monodisperse polyvinyltoluene particles of somewhat larger diameter, up to 3.5 μm, are also available. The properties of these systems are reviewed by Mercer (1973).

Monodisperse aerosols can be generated by atomizing a suspension of these particles using a commercial nebulizer. The solvent is then evaporated by mixing with dry air leaving an aerosol particle composed of the contents of the drop. Particle concentration and droplet size must be controlled to insure that no more than one particle is present in each droplet.

Small quantities of dissolved substances, including stabilizing agents to prevent coagulation of the particles, are present in the suspension. These also appear in the aerosol either coated on the latex sphere or as a very small particle when the original droplet contained no latex. Such aerosols are, therefore, composed of two types of particles, an (almost) monodisperse latex component and a secondary aerosol of much smaller particles, the residual solids originally present in the suspension. The secondary aerosol may cause problems in certain types of applications.

Several ingenious methods have been devised for the production of monodisperse aerosols, based on the break-up of suspensions, solutions, or pure liquids under carefully controlled conditions. In the case of the *spinning disk generator*, a liquid is fed continuously on to the center of a rotating disk, and a spray of droplets is formed. Uniform droplets result if the liquid wets the disk surface, and the flow is controlled between certain limits. Actually two sizes of droplets are produced. Most of the liquid goes into a set of larger droplets projected to a greater distance than the accompanying satellite droplets. The diameter of the satellite droplets is about one-fourth that of the primary droplets, and there are about four satellite droplets for each of the primary. Less than 10% of the liquid appears in the satellites for liquid flow rates below about 1 cm^3/min. This proportion increases with increasing liquid flow rates.

Separate large and small drop aerosols can be produced by taking advantage of the different stop distances of the primary and satellite droplets. Aerosols of the original pure liquids can be produced in this way with primary droplet diameters ranging from 6 to 3000 μm and liquid flow rates up to 168 cm^3/min. When solutions with volatile solvents are used, the solvents can be evaporated leaving behind particles whose size depends on the original droplet diameter and concentration. In this way, aerosols in the size range between 0.6 and 10 μm have been generated.

The *vibrating orifice generator* produces uniform droplets by the break-up of a jet of liquid forced through a vibrating orifice. The diameter of the orifice ranges from 5 to 20 μm, corresponding to droplet diameters ranging from 15 to 40 μm. Droplet diameters are continuously adjustable over an approximate 25% range by varying the vibration frequency, f. Each vibration cycle produces one droplet with a volume given by

$$v = \frac{Q}{f} \tag{6·2}$$

where Q is the volumetric rate at which liquid is fed to the orifice. The droplets are dispersed by turbulence in the air stream thereby minimizing collision and coalescence. If the liquid is a solution of a nonvolatile material in a volatile solvent, the solvent can be evaporated to produce particles as small as 0.6 μm conveniently. The size of the particles formed in this way can be calculated from (6·2) and the concentration of the nonvolatile substance. Particle diameters can be calculated in this way more accurately than they can be measured with conventional microscope techniques (Berglund and Liu, 1973). Concentrations of the order of 100 particles/cm^3 can be produced at aerosol flow rates of 100 liter/min.

Problems

1. It is proposed to filter a gas stream to obtain an aerosol sample for chemical analysis. A standard 30 mm commercial filter is to be used, rated at 0.02% penetration for 0.3 μm particles at a face velocity of 53 cm/sec. If the velocity of the gas stream is 6 m/sec, what should be the diameter of the sampling probe for isokinetic sampling?

2. It is planned to use a cascade impactor to sample the atmospheric aerosol for chemical analysis. The size distribution, measured by a single particle optical counter and an electrical mobility analyzer, is given in Problem 5, Chap. 1. The impactor is to have four stages followed by a 100% efficient afterfilter. The impactor stages can be characterized by the size corresponding to an efficiency of 50%. If the first stage collects all particles larger than 5 μm, determine the characteristic particle size for the other three stages that will provide equal mass on each stage and the after filter. Assume particle density is constant. What is the reason for designing for equal mass on each stage?

3. A round jet impactor is to be designed to sample automobile exhaust aerosol from the tailpipe of a car burning leaded gasoline. For a maximum jet velocity of 200 m/sec and a flow rate of 1 liter/min, determine the minimum particle size (aerodynamic diameter) that can be collected. Assume the pressure throughout the impactor is approximately atmospheric and the temperature is 20°cc. Estimate the smallest particle size (actual diameter), assuming the particles have the density of lead bromide ($\rho_p = 6.7$ g/cm^3). Note that it is necessary to account for the slip correction factor (Chap. 2).

4. For the optical particle counter of Fig. 6·6, the sensing volume is 1.5×1.5×1 mm. Determine the total energy scattered by the air molecules in the volume. Compare this with the scattering by *single* particles with diameters ranging from 0.05 to 5 μm and a refractive index of 1.5. Assume $\lambda = 0.5$ μm and express your answers in terms of the incident intensity. The temperature is 20° C.

5. It is desired to relate measured values of the light scattered by the atmospheric aerosol to the contributions of different ranges of the size distribution function. Discuss the components of a measurement system capable of providing the necessary information. Discuss any assumptions that must be made in carrying out calculations.

6. The integrating nephelometer provides information on total aerosol light scattering. Show how the characteristics of the instrument would appear within the framework of Table 6·2.

7. By rotating a flat disk about an axis perpendicular to its face, an air flow can be induced in the direction of the surface. Small particles diffuse to the surface of the disk at a rate given by the expression

$$|J| = 0.62 D^{2/3} \nu^{-1/6} \omega^{1/2} n_\infty$$

where J is the flux in particles per square centimeter per second, ω is the angular speed of the rotating disk in radians per second, and n_∞ is the particle concentration at large distances from the disk.

The particle deposition rate is independent of position on the surface as long as the boundary layer is laminar. By attaching an electron micrograph grid to the surface, a sample can be collected for examination under the electron microscope. If the atmospheric concentration is 10^5 particles/cm^3, determine the sampling time necessary to have 10 particles in an area 100 μm on a side. The speed of rotation is 20,000 rpm and the temperature is 20°C. To simplify the calculation, assume the particle size is 0.05 μm. (For an application of this method, see Friedlander and Pasceri (1964) *Surface Contamination*, Pergamon, Oxford and New York, p. 107.)

8. Select the components of an air monitoring and analysis system capable of detecting the properties of particulate pollution to which humans respond most sensitively. Include visibility degradation and health effects.

References

American Conference of Governmental Industrial Hygienists (1972) Air Sampling Instruments 4th ed. ACGIH, Cincinnati, Ohio. This is a good, practical guide to instruments for the measurement of the physical properties of aerosols. It includes many tables and figures showing operating ranges with limited discussion of fundamental principles. It is the closest thing available to a handbook on aerosol instrumentation, with different specialists writing about each type of instrument.

Armstrong, J. T. and Buseck, P. R. (1975) *Anal. Chem.* **47**, 2178.

Bayard, M. (1973) Application of the Electron Microprobe to the Analysis of Free Particulates in Andersen, C. A. (Ed.) *Microprobe Analysis*, Wiley, New York.

Berglund, R. N., and Liu, B. Y. H. (1973) *Environ. Sci. Technol.*, **7**, 147.

Butcher, S. S., and Charlson, R. J. (1972) *An Introduction to Air Chemistry*, Academic, New York.

Conner, W. D., and Hodkinson, J. R. (1967) Optical Properties and Visual Effects of Smoke-Stack Plumes, PHS Publication No. 999-AP-30, U.S. Dept. H.E.W., Cincinnati, Ohio.

Goulding, F. S., and Jaklevic, J. M. (1973) Photon-Excited Energy-Dispersive X-Ray Fluorescence Analysis for Trace Elements in Segre, G. (Ed.) *Annual Review of Nuclear Science*, vol. 23, Annual Reviews Inc., Palo Alto, Calif.

Goulding, F. S., and Jaklevic, J. M. (1973*a*) X-Ray Fluorescence Spectrometer for Airborne Particulate Monitoring, Office of Research and Monitoring Report EPA R2-73-182 (NTIS PB-225038), April 1973.

Husar, R. B. (1974) Recent Developments in *in Situ* Size Spectrum Measurement of Submicron Aerosol in *Instrumentation for Monitoring Air Quality*, A.S.T.M.S.T.P. 555, American Society for Testing Materials, Philadelphia.

Liu, B. Y. H., Whitby, K. T., and Yu, H. H. S. (1966) *J. Rech. Atmos.*, **2**, 397.

Liu, B. Y. H., Pui, D. Y. H., Hogan, A. W., and Rich, T. A. (1975) *J. Appl. Meteor.*, **14**, 46.

Lockhart, L. B., Patterson, R. L., Jr., and Anderson, W. L. (1964) Characteristics of Air Filter Media Used for Monitoring Airborne Radioactivity, Report No. 6054, U. S. Naval Research Laboratory, Washington, D. C. (March 20, 1964).

Data are given on penetration and resistance as a function of air velocity for 36 different filter materials. In all cases, an 0.3 μm monodisperse test aerosol was used.

May, K. R. (1967) Physical Aspects of Sampling Airborne Microbes in Gregory, P. H. and Monteith, J. L. (Eds.) *Airborne Microbes*, Cambridge Univ. Press, Cambridge.

McCrone, W. C., Draftz, R. G., and Delly, J. G. (1973) *The Particle Atlas*, Ann Arbor Sciences Publishers, Ann Arbor, Mich.

Mercer, T. T. (1973) *Aerosol Technology in Hazard Evaluation*, Academic, New York.

The title is somewhat misleading. This is primarily a review of instrumentation for the measurement of the physical properties of aerosols. It includes more theoretical background than the ACGIH reference and much useful information on the operating characteristics of commercial instrumentation.

Middleton, W. E. K. (1952) *Vision Through the Atmosphere*, Univ. of Toronto Press, Toronto.

Pollak, L. W., and Metnieks, A. L. (1957) *Geofis. Pura Appl.*, **37**, 174.

Ranz, W. E., and Wong, J. B. (1952) *Ind. Eng. Chem.*, **44**, 1371.

Silverman, L., Billings, C. E., and First, M. W. (1971) *Particle Size Analysis in Industrial Hygiene*, Academic, New York.

Spurny, K. R., Lodge, J. P., Jr., Frank, E. R., and Sheesley, D. C. (1969) *Environ. Sci. Technol.*, **3**, 453.

Stern, S. C., Zeller, H. W., and Schekman, A. I. (1962) *Ind. Eng. Chem.. Fund.*, **1**, 273.

Whitby, K. T., Liu, B. Y., Husar, R. B., and Barsic, N. J. (1972) *J. Colloid Interface Sci.*, **39**, 136.

Zworykin, V. K., Morton, G. A., Ramberg, E. G., Hillier, J., and Vance, A. W. (1945) *Electron Optics and the Electron Microscope*, Wiley, New York.

CHAPTER 7

Collision and Coagulation

In previous discussions of light scattering and deposition, the size distribution function, $n(v)$, was considered a given quantity. The deposition process itself results in loss of particles preferentially with respect to size thereby changing $n(v)$. In addition, processes occurring within the gas, including coagulation and gas-to-particle conversion, modify $n(v)$ and the size composition p.d.f. This occurs in the evolution of the atmospheric aerosol (Chap. 1) as well as at sources such as the automobile tailpipe. The rest of this text is concerned with the processes taking place in the aerosol that modify $n(v)$, leading eventually to the discussion of a general dynamic equation for the distribution function. In this chapter, we consider coagulation alone.

Aerosols are unstable with respect to coagulation. The reduction in surface area that accompanies particle attachment corresponds to a reduction in the Gibbs free energy under conditions of constant temperature and pressure. The theory of coagulation is composed of two parts. The first, primarily mathematical, is essentially a scheme for keeping count of particle collisions as a function of particle size; it incorporates a general expression for the collision frequency function. An expression for the collision frequency based on a physical model is then introduced into the equation keeping count of collisions. The collision mechanisms include Browian motion and fluid shear under the influence of interacting particle force fields. The process is basically nonlinear, and this leads to formidable difficulties in the mathematical theory.

In this chapter, we limit consideration to some relatively simple cases of practical importance. First, we consider the initial coagulation of monodisperse aerosols for which analytical solutions can be obtained rather easily. Then solutions approached asymptotically after long periods of time ("self-preserving" size distributions) are discussed. Numerical methods for solving the coagulation equations have been developed, but these are beyond the scope of the text.

Considerable effort has gone into the development of sonic agglomeration as an industrial process. The goal is to grow the particles to a sufficiently large size in an acoustic field and separate the agglomerates by relatively inexpensive equipment such as a cyclone separator. So far, a practical process has not been developed, and the method is not discussed in this chapter.

The availability of monodisperse aerosols and improvements in instrumentation for particle counting have led to important experimental advances. In certain cases, however, data for hydrosols are the best available for testing theory, and where appropriate are discussed in the chapter.

7·1 Collision Frequency Function

Particle collision and coagulation lead to a reduction in the total number of particles and an increase in the average size. An expression for the time rate of change of the particle size distribution function can be derived as follows:

Let N_{ij} be the number of collisions occurring per unit time per unit volume between the two classes of particles of volumes v_i and v_j. All particles are assumed to be spherical, which means that i and j are uniquely related to particle diameters. When two particles collide, according to this simplified model, they coalesce to form a third whose volume is equal to the sum of the original two. The collision frequency can be written as follows in terms of the concentrations of particles with volumes v_i and v_j:

$$N_{ij} = \beta(v_i, v_j) n_i n_j \tag{7·1}$$

where $\beta(v_i, v_j)$, the collision frequency function, depends on the sizes of the colliding particles and on such properties of the system as temperature and pressure.

In the case of a discrete spectrum (Chap. 1), the rate of formation of particles of size k by collision of particles of size i and j is given by $\frac{1}{2}\Sigma_{i+j=k} N_{ij}$ where the notation $i+j=k$ indicates that the summation is taken over those collisions for which

$$v_i + v_j = v_k$$

The factor of 1/2 is introduced because each collision is counted twice in the summation. The rate of loss of particles of size k by collision with all other particles is $\Sigma_{i=1}^{\infty} N_{ik}$. Hence the net rate of generation of particles of size k is given by the expression:

$$\frac{dn_k}{dt} = \frac{1}{2} \sum_{i+j=k} N_{ij} - \sum_{i=1}^{\infty} N_{ik} \tag{7·2}$$

Substituting (7·1) in (7·2), the result is

$$\frac{dn_k}{dt} = \frac{1}{2} \sum_{i+j=k} \beta(v_i, v_j) n_i n_j - n_k \sum_{i=1}^{\infty} \beta(v_i, v_k) n_i \qquad (7\cdot3)$$

which is the dynamic equation for the discrete size spectrum when coagulation alone is important. The solution to (7·3) depends on the form of $\beta(v_i, v_j)$, which is determined by the mechanism of particle collision as discussed in the sections that follow. The theory for the discrete spectrum, including expressions for the collision frequency function for Brownian coagulation and laminar shear is due to Smoluchowski (1917*a*, *b*).

7·2. Brownian Coagulation

Particles smaller than about 1 μm collide as a result of their Brownian motion; most of the theoretical and experimental studies of coagulation have been concerned with this mechanism. For particles much larger than the mean free path of the gas, there is experimental evidence that the collision process is diffusion limited. Consider a sphere of radius a_i, fixed at the origin of the coordinate system in an infinite medium containing suspended spheres of radius a_j. Particles of radius a_j are in Brownian motion and diffuse to the surface of a_i, which is a perfect sink. Hence the concentration of a_j particles vanishes at $r = a_i + a_j$. For the spherical symmetry, the equation of diffusion (Chap. 2) takes the form:

$$\frac{\partial n}{\partial t} = D \frac{\partial r^2 (\partial n / \partial r)}{r^2 \partial r} \qquad (7.4)$$

For this case, the initial and boundary conditions are

$$\text{at } r = a_i + a_j, \qquad n = 0 \text{ for all } t$$

$$r > a_i + a_j, \quad t = 0 \quad n = n_\infty$$

Let

$$w = \left(\frac{n_\infty - n}{n_\infty} \right) \left(\frac{r}{a_i + a_j} \right)$$

and

$$x = \frac{r - (a_i + a_j)}{a_i + a_j}$$

Then (7·4) can be transformed to

$$\frac{\partial w}{\partial t} = D\frac{\partial^2 w}{\partial x^2} \tag{7·5}$$

with the boundary conditions

$$\text{at } x=0, \quad w=1 \text{ for all } t$$

$$x>0, \quad t=0, \quad w=0$$

Equation (7·5) with these boundary conditions corresponds to one-diminsional diffusion in a semi-infinite medium for which the solution is

$$w = 1 - erf\frac{x}{2(Dt)^{1/2}} \tag{7·6}$$

where *erf* denotes the error function. As $t \to \infty, w \to 1 - erf(0) = 1$, and

$$\frac{n_\infty - n}{n_\infty} \to \frac{a_i + a_j}{r} \tag{7·7}$$

which is the steady-state solution for the concentration distribution, also obtained by setting $\partial n/\partial t = 0$ in (7·4) and solving for n.

The rate at which particles arrive at the surface ($r = a_i + a_j$) is found by differentiating (7·6)

$$F(t) = 4\pi D\left(r^2\frac{\partial n}{\partial r}\right)_{r=a_i+a_j}$$

$$= 4\pi D(a_i + a_j)n_\infty\left[1 + \frac{a_i + a_j}{(\pi D t)^{1/2}}\right] \tag{7·8}$$

This is the rate at which particles of size a_j collide with a fixed particle of size a_j (particles/sec). For sufficiently long times ($t \gg (a_i + a_j)^2/D$), this becomes

$$F = 4\pi D(a_i + a_j)n_\infty \tag{7·9}$$

which is equivalent to the steady-state solution (7·7).

If the central particle is also in Brownian motion, the diffusion constant, D, should describe the relative motion of two particles. The relative displacement is given by $x_i - x_j$ where x_i and x_j are the displacements of the two particles in the x direction measured from a given reference plane. The diffusion constant for the relative motion can be obtained from the Einstein equation for the diffusion coefficient (Chap. 2):

$$D_{ij} = \frac{\overline{(x_i - x_i)^2}}{2t}$$

$$= \frac{\overline{x_i^2}}{2t} - \frac{\overline{2x_i x_j}}{2t} + \frac{\overline{x_j^2}}{2t}$$

$$= D_i + D_j$$

The quantity $\overline{x_i x_j} = 0$ since the motion of the two particles is assumed to be independent. The collision frequency function is then obtained by substitution in (7·9).

$$\beta(v_i, v_j) = 4\pi(D_i + D_j)(a_i + a_j) \tag{7·10}$$

For particles 0.1 μm in radius, the characteristic time $(a_i + a_j)^2/(D_i + D_j)$ is about 10^{-3} sec, and the use of the steady-state solution is justified in most cases of practical interest. When the Stokes–Einstein relation holds for the diffusion coefficient and $d_p \gg l$, this expression becomes

$$\beta(v_i, v_j) = \frac{2kT}{3\mu}\left(\frac{1}{v_i^{1/3}} + \frac{1}{v_j^{1/3}}\right)\left(v_i^{1/3} + v_j^{1/3}\right) \tag{7·11}$$

For particles much smaller than the mean free path of the gas, less than about one-tenth, say, the collision frequency is obtained from the expression derived in the kinetic theory of gases for collision among molecules that behave as rigid elastic spheres:

$$\beta(v_i, v_j) = \left(\frac{3}{4\pi}\right)^{1/6}\left(\frac{6kT}{\rho_p}\right)^{1/2}\left(\frac{1}{v_i} + \frac{1}{v_j}\right)^{1/2}\left(v_i^{1/3} + v_j^{1/3}\right)^2 \tag{7·12}$$

where ρ_p is the particle density. Fuchs (1964) has proposed a general interpolation formula for β, which takes into account the transition from the free molecule regime (7·12) to the continuum range (7·11) and this problem is further discussed by Hidy and Brock (1970). Values of β as a function of particle size are given in Table 7·1.

TABLE 7·1
COLLISION FREQUENCY FUNCTION[a]
IN AIR AT 23°C AND 1 atm
(Based on Fuchs, 1964, p. 294)

	$10^{10}\beta$ cm^3/sec		
d_{p1} \ d_{p2}	0.01	0.1	1
0.01	18		
0.1	240	14.4	
1	3200	48	6.8

[a] Sometimes called the coagulation constant, although it is a function of particle size and the properties of the gas.

7·3 Brownian Coagulation: Dynamics of Discrete Distribution

A simple solution to the kinetic equation for Brownian coagulation can be obtain for nearly monodisperse systems. Setting $v_i = v_j$ in (7·11), the collision frequency function is given by

$$\beta(v_i = v_j) = \frac{8kT}{3\mu} = K$$

In this special case of $\beta = K$ independent of particle size, a simple, analytical solution can be obtained for the discrete size distribution of an initially monodisperse aerosol. Substituting in (7·3), the result is

$$\frac{dn_k}{dt} = \frac{K}{2} \sum_{i+j=k} n_i n_j - K n_k \sum_{i=1}^{\infty} n_i$$

Let $\sum_{i=1}^{\infty} n_i = N_\infty$ be the total number of particles per unit volume of fluid. Summing over all values of k, the result is

$$\frac{dN_\infty}{dt} = \frac{K}{2} \sum_{k=1}^{\infty} \sum_{i+j=k} n_i n_j - K N_\infty^2$$

It is not difficult to show by expanding the summation that the first term on the right-hand side is $(K/2)N_\infty^2$ so that the equation becomes

$$\frac{dN_\infty}{dt} = -\frac{K}{2} N_\infty^2 \tag{7·13}$$

Integrating once gives

$$N_\infty = \frac{N_\infty(0)}{1 + (K N_\infty(0) t/2)} \tag{7·13a}$$

where $N_\infty(0)$ is the total number of particles at $t=0$. For $k=1$, the kinetic equation is

$$\frac{dn_1}{dt} = -Kn_1N_\infty$$

Solving, we find

$$n_1 = \frac{N_\infty(0)}{(1+t/\tau)^2}$$

where $\tau = 2/KN_\infty(0) = 3\mu/4kTN_\infty(0)$ and for $k=2$

$$n_2 = \frac{N_\infty(0)t/\tau}{(1+t/\tau)^3}$$

In general,

$$n_k = \frac{N_\infty(0)(t/\tau)^{k-1}}{(1+t/\tau)^{k+1}}$$

which is the equation for the discrete size distribution, with an initially monodisperse aerosol and a collision frequency function independent of particle size. The variation in n_k with time is shown in Fig. 7·1.

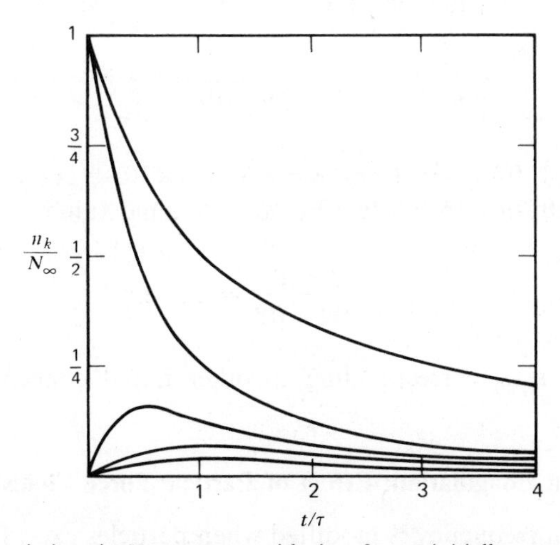

Fig. 7·1 The variations in $N_\infty, n_1, n_2, \ldots$ with time for an initially monodisperse aerosol. The total number concentration, N_∞, and the concentration of monomer, n_1, both decrease monotonically with increasing time. The concentrations of the polymers pass through a maximum.

Support for the theory of diffusion controlled coagulation comes primarily from experiments with polydisperse aerosols (Whytlaw-Gray and Patterson, 1932). The coagulation coefficient K was shown to be approximately independent of the chemical nature of the aerosol material with a value close to that predicted theoretically.

Tests of the theory have also been conducted by following the coagulation of monodisperse aerosols of dioctylphthalate generated by a condensation aerosol generator (Chap. 6). Experimentally measured values of the coagulation coefficient were compared with values calculated from theory (Devir, 1963). Taking into account wall losses, good agreement between theory and experiment was obtained. Experiments of this type support the assumption that small, uncharged particles in air adhere when they collide, since the theoretical rate constants are based on this assumption.

Example. Estimate the time for the concentration of a monodisperse aerosol to fall to 10% of its original value. The particle diameter is 0.1 μm, and the initial concentration is 10^8 cm^{-3}. The gas is air at 20°C.

SOLUTION. Rearranging (7·13*a*), the result is

$$t = \frac{N_\infty(0)/N_\infty - 1}{KN_\infty(0)/2}$$

For $N_\infty(0)/N_\infty = 10$, this becomes

$$t_{1/10} = \frac{18}{KN_\infty(0)}$$

From Table 7·1, for $d_p = 0.1$ μm, $K = \beta \approx 14.4 \times 10^{-10}$ cm^3/sec. The time for the concentration to fall to 10% of its original value is

$$t_{1/10} = \frac{18 \times 10^9}{1.44 \times 10^8} = 125 \text{ sec}$$

For values of $t_{1/10}$ corresponding to other initial concentrations, see Table 1·1.

7·4 Brownian Coagulation: Effect of Particle Force Fields

The collision frequency is modified when particles exert forces on one another. The fields of most interest result from van der Waals forces, which are always present, and Coulomb forces, which result when the particles are charged. Both are considered in the following sections.

Once again we consider a particle of radius a_i to which particles of radius a_j are diffusing. In this case, however, the a_j particle exerts a force $K(r)$ per unit mass (that is, it produces an acceleration $K(r)$) on the particles of radius a_j in Brownian motion in the surrounding fluid. By (2·19) and the assumption of spherical symmetry, the rate of diffusion of a_j particles to the surface of the a_i particle is given by

$$J(r) = -D\frac{\partial n}{\partial r} + \frac{K(r)}{f}n$$

In the steady state, the number of particles crossing each spherical surface concentric with the central particle is constant:

$$4\pi r^2 J = \text{const} = -F = -D4\pi r^2 \frac{\partial n}{\partial r} + \frac{4\pi r^2 K(r) n}{f}$$

Instead of dealing with the force $K(r)$, it is usually more convenient to deal with the potential energy of two particles as a function of their separation distance, $\Phi(r)$:

$$K(r) = -\frac{d\Phi(r)}{dr}$$

Substituting,

$$F = 4\pi r^2 D\left(\frac{dn}{dr} + \frac{n}{kT}\frac{d\Phi}{dr}\right)$$

The solution to this ordinary linear differential equation is

$$n = n_\infty \exp\left[\frac{-\Phi(r)}{kT}\right] + \frac{F\exp[-\Phi(r)/kt]}{4\pi D}\int_\infty^r \frac{\exp[\Phi(x)/kT]}{x^2}dx$$

where n_∞ is the concentration at $r=\infty$. If $n=0$ at $r=a_i+a_j$, the total flow of particles to the central sphere is given by

$$F = \frac{4\pi D n_\infty (a_i+a_j)}{(a_i+a_j)\int_{a_i+a_j}^\infty \left[\exp(\Phi(x)/kT)/x^2\right]dx} \tag{7·14a}$$

$$= \frac{4\pi D n_\infty (a_i+a_j)}{W} \tag{7·14b}$$

Comparing this equation with the flow in the absence of an external force field, we see that the result has been modified by a correction factor, W, which appears in the denominator. The potential energy term, $\Phi(r)$, may be positive or negative and depends in different ways on r for different types of force fields. When Φ is negative (attraction), the integral is less than $(a_i + a_j)^{-1}$, and the denominator is less than unity leading to an increase in the collision frequency over the rate for diffusion alone. When Φ is positive (repulsion), the denominator is greater than unity, and the collision frequency is reduced. The effect of specific forces on the rate of coagulation can be determined by evaluating the appropriate integrals as shown in the following sections.

7·5 Effect of van der Waals Forces

Attractive (van der Waals) forces between uncharged nonpolar molecules result from dipoles produced by fluctuations in the electron clouds (Section 2·8). The energy of attraction can be calculated from quantum theory, and depends on the molecular properties and the distance between the molecules. The attractive potential between like molecules can often be expressed in the Lennard–Jones form:

$$\Phi_m = -4\epsilon\left(\frac{\sigma}{r}\right)^6$$

where ϵ is a constant with the dimensions of energy and σ is a constant with the dimensions of length, of the order of the molecular diameter. The constants ϵ and σ have been obtained by semiempirical calculations for various substances, and some values are listed in Table 7·2.

TABLE 7·2
LENNARD–JONES FORCE CONSTANTS FOR SELECTED ORGANIC COMPOUNDS BASED ON VISCOSITY
(Hirschfelder, Curtiss, and Bird, 1954, p. 1212)

Substance	ϵ/k[a] (°K)	σ(Å)
Ethanol	415	4.37
Methyl acetate	417	5.05
Ethyl acetate	531	5.16
Cyclohexane	313	6.14
1,3,5 Trimethylbenzene	234	7.71
n-Octane	333	7.41
Benzene	335	5.63
Toluene	377	5.93

[a] k = Boltzmann's constant.

The energy of attraction between two particles is found by integrating over the interactions between pairs of molecules in the separate particles. For this purpose, it is convenient to introduce a new coupling constant between molecular pairs defined as follows:

$$Q = \frac{4\epsilon\sigma^6}{v_m^{\,2}}$$

where v_m is the molecular volume in the particles. If the density of the particle is ρ_p, then $v_m = M/\rho_p N_{av}$, where M is the molecular weight and N_{av} is Avogadro's number. Clearly Q, like ϵ, has the dimensions of energy.

The energy of attraction between two spherical particles of radii a_i and a_j was calculated by Hamaker (1937):

$$\Phi = -\frac{\pi^2 Q}{6}\left[\frac{2a_i a_j}{r^2-(a_i+a_j)^2} + \frac{2a_i a_j}{r^2-(a_i-a_j)^2} + \ln\frac{r^2-(a_i+a_j)^2}{r^2-(a_i-a_j)^2}\right]$$

where r is the distance between the centers of the spheres. For two spherical particles of the same radius, a, the energy of attraction is found by setting $a_i = a_j = a$:

$$\Phi = -\frac{\pi 2 Q}{6}\left\{2\left(\frac{a}{r}\right)^2 + \frac{2a^2}{r^2-4a^2} + \ln\left[1-\frac{4a^2}{r^2}\right]\right\}$$

where r is the distance separating the particle centers.

Substituting in the correction factor, W, which appears in the denominator of (7·14), the result is

$$W = \int_0^1 \exp\left[-\frac{\pi^2 Q f(x)}{6kT}\right] dx \tag{7·15}$$

where

$$f(x) = \left[\frac{x^2}{2} + \frac{x^2}{2(1-x^2)} + \ln(1-x^2)\right]$$

Hence for colliding particles of the same diameter the effect of the van der Waals forces on collision rate does not depend on the size but only on Q/kT. The integral has been evaluated and the result is shown in Fig. 7·2. The determination of the effect of the van der Waals forces on

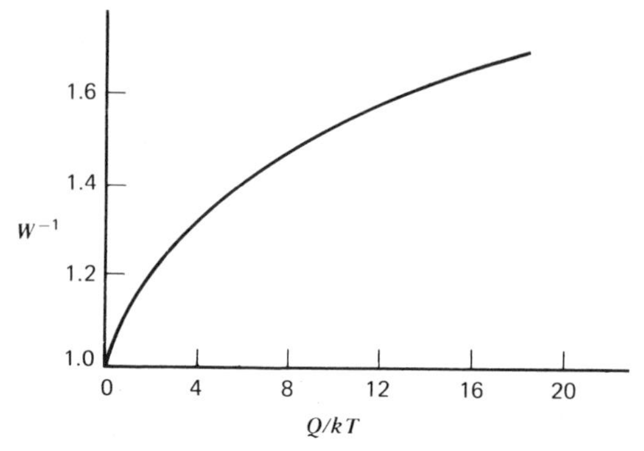

Fig. 7·2 Increase in rate of collision of particles of equal diameter resulting from the action of dispersion forces (Tikhomirov, Tunitskii, and Petrjanov, 1942).

the coagulation rate thus reduces to the evaluation of Q/kT, at least for particles of equal size.

Example. Estimate the effect of the dispersion forces on the rate of coagulation of ethyl acetate droplets.

SOLUTION. From Table 7·2, $\epsilon/k = 531°K$ and $\sigma = 5.16$ Å. The density at 20°C is 0.9 g/cm^3, and the molecular weight is 88. Hence $v_m = 183 \times 10^{-24}$cm^3 and $Q/kT = 3.6$, corresponding to an increase in the coagulation rate of about 25% (Fig. 7·2). In problems of practical importance, a factor of 25% is usually not of great significance because of uncertainties resulting from convection and deposition on the walls of containment vessels.

7·6 Effect of Coulomb Forces

In the case of electrically charged particles, the complete expression for the force of interaction between the particles includes terms for induction forces, in addition to the leading term for the Coulomb force. It is often possible to neglect the induction forces, and the potential energy of interaction then takes the form:

$$\Phi = \frac{z_i z_j e^2}{\epsilon r}$$

where z_i and z_j are the number of charges on the interacting particles, e is the electronic charge, and ϵ is the dielectric constant of the medium.

The integral correction factor to (7·14) can now be evaluated, and the result is

$$W = \frac{1}{y}(e^y - 1)$$

where the dimensionless parameter,

$$y = \frac{z_i z_j e^2}{\epsilon kT(a_i + a_j)}$$

represents the ratio of the electrostatic potential energy at contact to the thermal energy, kT. In the limiting case of uncharged particles, $y=0$, the correction factor becomes unity and (7·14) reduces to the field-free case. When the particles are of opposite sign, y is negative and the correction factor is positive and less than unity as can be seen by expanding the exponential. The result is that collisions occur more rapidly than in the case of uncharged particles. When the particles are of like sign, the correction factor is positive and greater than unity. The result is that the collision rate is smaller than for uncharged particles.

Depending on the charging mechanism, aerosols may be composed of particles of like charges (unipolar charging) or of unlike charges (bipolar charging), and the magnitudes of the charges may vary. For $|y| \ll 1$, the charging may be termed weak, and for $|y| \gg 1$, strong. The atmospheric aerosol has a weak bipolar charge, with roughly equal numbers of positively and negatively charged particles. For such an aerosol, the effect on the collision frequency can be estimated by calculating separately the collision rates for particles of the same and opposite signs and then averaging the rates. For $y = 1/2$, for example, the correction factor is about 1.3, and for $y = -1/2$, about 0.8. Hence the average rate is but little affected by charging over the range $y=0$ to $y = \pm 1/2$. This has been confirmed experimentally (Fuchs, 1964, p. 308).

For strong bipolar aerosols ($|y| \gg 1$), this compensation does not take place. The large increase in coagulation resulting from attractive forces strongly outweighs the decrease caused by repulsion.

7·7 Collision Frequency for Laminar Shear

Particles in a uniform, laminar shear flow collide because of their relative motion (Fig. 7·3). The streamlines are assumed to be straight, and the particle motion, rectilinear. This is an oversimplification of what really happens since the particles affect the shear flow, and their

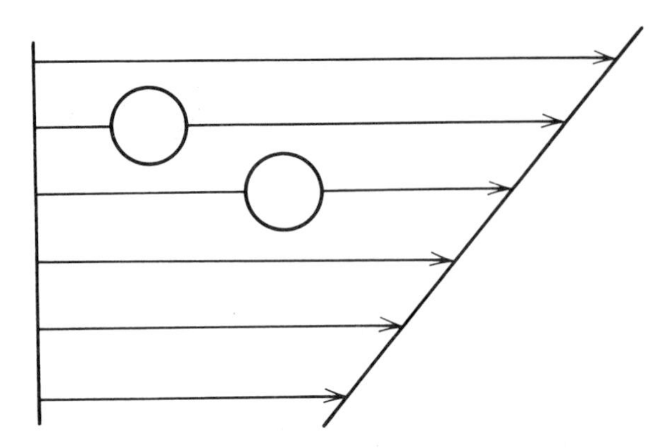

Fig. 7·3 Idealized model of particle collision in a shear field. The upper particle, moving at a higher velocity, overtakes and collides with the slower moving particle.

motion does not remain rectilinear. The model is useful, however, for an approximate calculation.

To derive an expression for the collision frequency function, we refer to Fig. 7·3, which shows a single particle in the shear field with radius a_i interacting with particles of radius a_j (Fig. 7·4). The velocity of the aerosol normal to the surface of the page, relative to the particle shown, is $x(du/dx)$. Hence the flow of particles into the shaded portion of the strip dx is given by

$$F = n_j x \frac{du}{dx} (a_i + a_j) \sin\theta \, dx$$

Since $x = (a_i + a_j) \cos\theta$, the total number of particles flowing into the

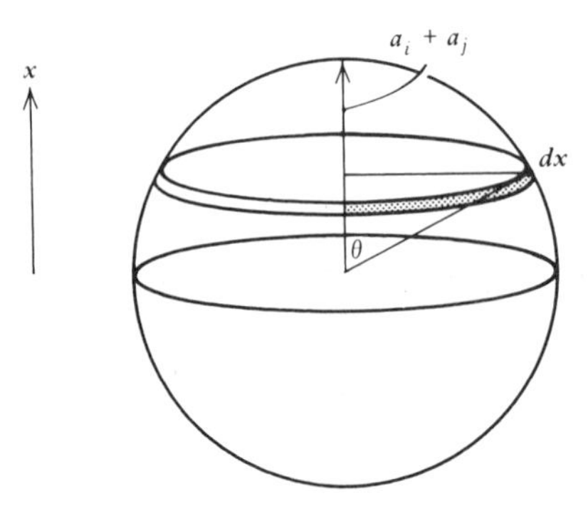

Fig. 7·4 Geometry for coagulation in a laminar shear field. The flow is normal to the paper. The particle of radius a_i has its origin at the center of the coordinate system. The velocity gradient du/dx is constant, and the velocity relative to the central sphere at x is $x(du/dx)$.

central sphere is given by the expression

$$F=2(2)n_j\int_0^{\pi/2}(a_i+a_j)^3\frac{du}{dx}\sin^2\theta\cos\theta\,d\theta$$

where the first factor 2 takes into account the flow into the upper hemisphere from this side of the paper plus the flow into the bottom hemisphere from the backside of the page. The second factor 2 is necessary because the integration from 0 to $\pi/2$ must be done twice. Carrying out the integration, the result is

$$F=\frac{4}{3}(a_i+a_j)^3\frac{du}{dx}n_j$$

Therefore,

$$N_{ij}=\frac{4}{3}(a_i+a_j)^3\frac{du}{dx}n_in_j$$

The collision frequency function for coagulation by laminar shear is, therefore,

$$\beta(v_i,v_j)=\frac{4}{3}(a_i+a_j)^3\frac{du}{dx} \tag{7·16}$$

Substituting in (7·3), the equation of coagulation by laminar shear for the discrete spectrum becomes

$$\frac{dn_k}{dt}=\frac{1}{2}\sum_{i+j=k}\left[\frac{4}{3}(a_i+a_j)^3\frac{du}{dx}n_in_j\right] - \sum_{i=1}^{\infty}\frac{4}{3}(a_i+a_k)^3\frac{du}{dx}n_in_k \tag{7·17}$$

If the system is composed of particles that are all of nearly the same size, $a_i \approx a$, and (7·17) becomes

$$\frac{dn_k}{dt}=\frac{1}{2}\sum_{i+j=k}\left[\tfrac{32}{3}a^3\frac{du}{dx}n_in_j\right]-\sum_{i=1}^{\infty}\tfrac{32}{3}a^3\frac{du}{dx}n_in_k$$

Summing over all k, the result is

$$\frac{dN_\infty}{dt}=-\frac{16}{3}\frac{du}{dx}a^3N_\infty^{\ 2}$$

But

$$\tfrac{4}{3}\pi a^3 N_\infty = V = \text{const}$$

Hence

$$\frac{dN_\infty}{dt} = -\frac{4V}{\pi}\frac{du}{dx}N_\infty \tag{7·18}$$

It should be noted that the decay rate is proportional N_∞ and not to $N_\infty^{\ 2}$, as in the case of coagulation by Brownian motion in the continuum range. Integrating from the initial state for which $N = N_\infty(0)$ at $t=0$ the result is

$$\ln\frac{N_\infty(0)}{N_\infty} = \frac{4V}{\pi}\frac{du}{dx}t \tag{7·19}$$

7·8 Simultaneous Laminar Shear and Brownian Motion

Swift and Friedlander (1964) carried out experiments on the coagulation of hydrosols in the presence of simultaneous laminar shear and Brownian movement. They used a Couette type apparatus consisting of an outside plastic cylindrical shell, and inside brass cylinder, and two brass end plates. The brass cylinder had an outside radius of 6.325 cm, and the annular space was 0.200 cm. The outer shell and end plates were fixed and the inner cylinder was free to rotate. Suspensions of polystyrene latex particles 0.871 μm in diameter were destabilized by the addition of sodium chloride solution and allowed to coagulate in the annular space; the shear field was varied by controlling the speed of rotation of the inner cylinder.

Assuming additivity of the rates of coagulation by shear and Brownian motion, the rate of change of the total particle concentration is given by

$$\frac{dN_\infty}{dt} = -\frac{4\alpha}{3}\frac{kT}{\mu}N_\infty^{\ 2} - \frac{4G\alpha_1 V}{\pi}N_\infty \tag{7·20}$$

where α is an empirical collision efficiency for Brownian motion, α_1 is the collision efficiency for shear flow, and $G = du/dx$. Integrating (7·20) assuming $\alpha_1 = \alpha$, the result is

$$\ln\left[\left(\frac{N_\infty + R}{N_\infty}\right)\left(\frac{N_\infty(0)}{N_\infty(0)+R}\right)\right] = \frac{4\alpha G V t}{\pi} \tag{7·21}$$

where $R=3GV\mu/\pi kT$. The value of α was found to be 0.375 in an experiment carried out in the absence of shear. Experiments were carried out with the latex dispersion subjected to shear rates of 1, 5, 20, 40, and 80 sec^{-1}, and the results were plotted in a form suggested by (7·21) as shown in Fig. 7·5. According to (7·21), the slopes of the lines of Fig. 7·5 should be $4\alpha GV/\pi$. When these slopes were, in turn, plotted as a function of G, the value of α_1 was found to be 0.364, consistent with the assumption that $\alpha_1=\alpha$.

At higher shear rates, the contribution of the Brownian motion to coagulation can be neglected and (7·21) reduces to (7·19). As shown in Fig. 7·6, this form of the equation satisfactorily correlated the experimental data.

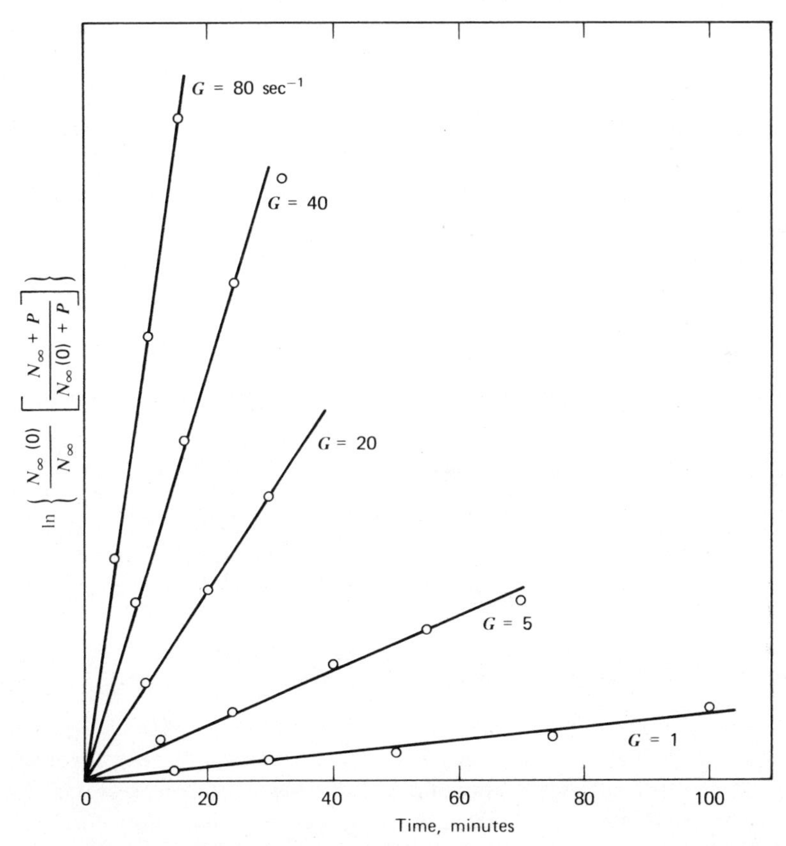

Fig. 7·5 Shear coagulation of a monodisperse latex dispersion. Straight lines were obtained for differing shear rates in accordance with a simple theory assuming additivity of the effects of Brownian motion and shear. The points are experimental results (Swift and Friedlander, 1964).

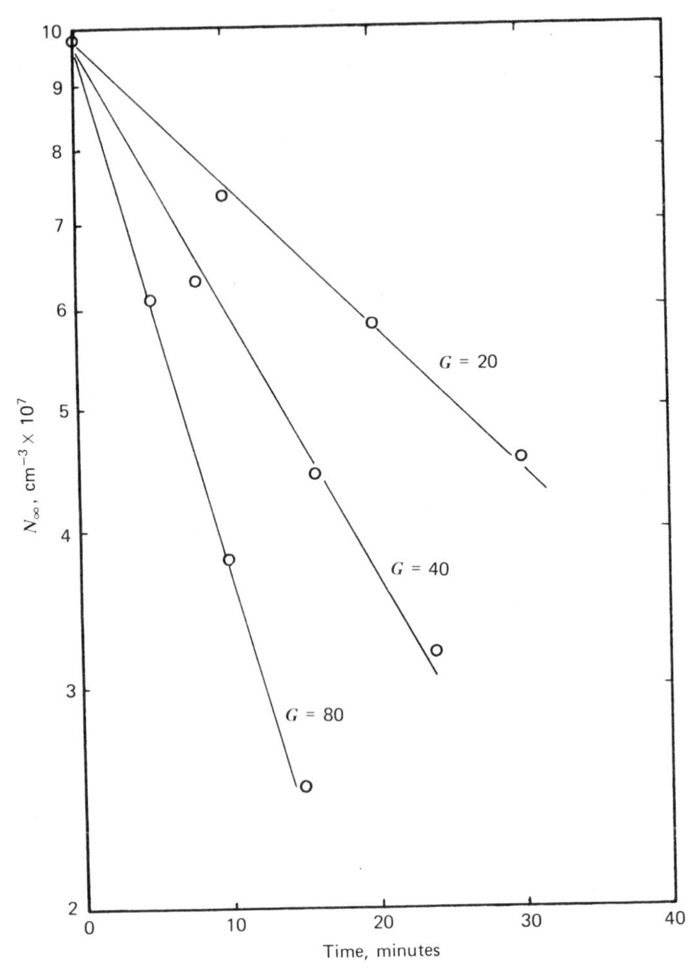

Fig. 7·6 When shear coagulation controls, a plot of $\log N_\infty$ versus t should give a straight line. This was confirmed experimentally at the higher shear rates with a monodisperse hydrosol (Swift and Friedlander, 1964).

7·9 Other Collision Mechanisms

Other collision mechanisms that operate in the atmosphere and in natural waters as well as in gas and water cleaning include turbulent shear and differential sedimentation. Approximate expressions for $\beta(v_i, v_j)$ have been derived for both cases. For coagulation by turbulent shear of particles much smaller than the size of the energy dissipating eddies, Saffman and Turner (1956) give

$$\beta = 1.30(a_1 + a_j)^3\left(\frac{\epsilon_d}{\nu}\right)^{\frac{1}{2}} \tag{7·22}$$

where ϵ_d is the energy dissipation by turbulence (cm^2/sec^3) and ν is the kinematic viscosity of the fluid. Note the similarity in form to (7·16) for coagulation by laminar shear with $(\epsilon_d/\nu)^{1/2}$ replacing the velocity gradient.

Differential sedimentation refers to the sweeping out of small particles by larger ones falling from above. The collision frequency factor

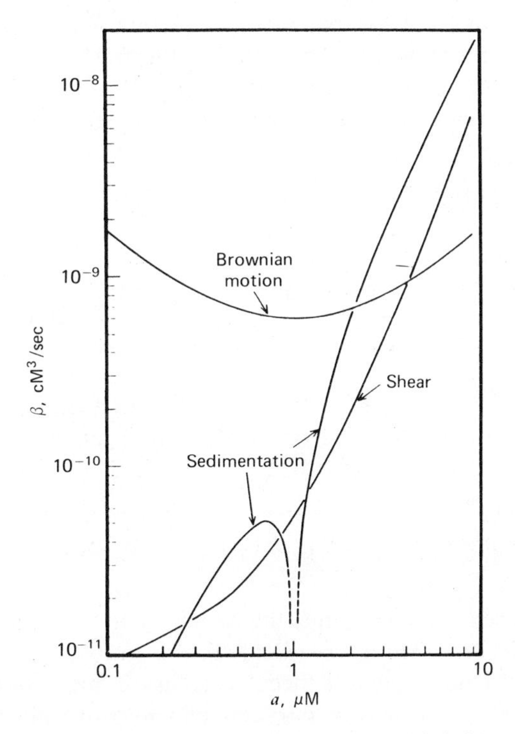

Fig. 7·7 Comparison of coagulation mechanisms for particles of 1 μm radius interacting with particles of radius between 0.1 and 10 μm. Coagulation by shear based on $\epsilon_d = 5\ cm^2/sec^3$. Differential sedimentation curves were obtained by an approximate calculation assuming Stokes flow around the larger of the falling spheres (Firedlander, 1964).

The value of ϵ_d corresponds to the open atmosphere at a height of about 100 m (Lumley and Panofsky, 1964). At a height of 1 m, $\epsilon_d \approx 1{,}000\ cm^2/sec^3$ and shear becomes the dominant mechanism of coagulation for larger particles.

For the core region of a turbulent pipe flow, the energy dissipation (based on Laufer, 1954) is given by:

$$\epsilon_d \approx \frac{4}{d}\left(\frac{f}{2}\right)^{3/2} U^3$$

where f is the Fanning friction factor. For a smooth pipe, 10 cm in diameter, with air at 20°C and a Reynolds number of 50.000, $\epsilon_d \approx 2 \times 10^4\ cm^2/sec^3$.

can be written in the form

$$\beta=\gamma(a_i,a_j)\pi(a_i+a_j)^2(v_i-v_j) \tag{7·23}$$

where γ is a collision efficiency that depends on the flow field, and the rest of the expression represents a particle flow through the effective collision cross section.

The relative importance of the various collision mechanisms can be compared as shown in Fig. 7·7. The figure shows the collision frequency function for 1 μm particles interacting with particles of any other size. At conditions corresponding to the free atmosphere ($\epsilon_d \approx 5$ cm^2/sec^3), either Brownian motion or differential sedimentation plays a dominant role. Brownian motion controls for particles smaller than 1 μm.

7·10 Equation of Coagulation: Continuous Distribution Function

For the continuous distribution function, the collision rate between particles of size v and $\tilde{v}$ is given by

$$\text{coll. rate}=\beta(v,\tilde{v})n(v)n(\tilde{v})\,dv\,d\tilde{v}$$

where the forms of the collision frequency function discussed in previous sections are applicable. The rate of formation of particles of size v by collision of smaller particles of size $v-\tilde{v}$ and $\tilde{v}$ is then given by

$$\text{formation in range } dv=\tfrac{1}{2}\left[\int_0^v \beta(\tilde{v},v-\tilde{v})n(\tilde{v})n(v-\tilde{v})\,d\tilde{v}\right]dv$$

Here we have used the result that the Jacobian for the transformation from the coordinate system $(\tilde{v},v-\tilde{v})$ to $(\tilde{v},v)$ is unity. The factor 1/2 is introduced as in the discrete case because collisions are counted twice in the integral. The rate of loss of particles of size v by collision with all other particles (except monomer) is given by

$$\text{loss in range } dv=\left[\int_0^\infty \beta(v,\tilde{v})n(v)n(\tilde{v})\,d\tilde{v}\right]dv$$

The net rate of formation of particles of size v is given by

$$\frac{\partial(n\,dv)}{\partial t}=\frac{1}{2}\left[\int_0^v \beta(\tilde{v},v-\tilde{v})n(\tilde{v})n(v-\tilde{v})\,d\tilde{v}\right]dv$$
$$-\left[\int_0^\infty \beta(v,\tilde{v})n(\tilde{v})n(v)\,d\tilde{v}\right]dv$$

Dividing through by dv, the result is

$$\frac{\partial n}{\partial t}=\frac{1}{2}\int_0^v \beta(\tilde{v},v-\tilde{v})n(\tilde{v})n(v-\tilde{v})\,d\tilde{v}$$

$$-\int_0^\infty \beta(v,\tilde{v})n(\tilde{v})n(v)\,d\tilde{v} \qquad (7\cdot 24)$$

which is the equation of coagulation for the continuous distribution function. Mathematical methods for solving this equation are reviewed by Drake (1972). Few solutions have been obtained for collision frequency functions of physical interest. Most of these were calculated by the similarity theory, discussed in the following sections, and by numerical methods.

Solutions to (7·24) are subject to two important physical interpretations. They represent the change with time of the aerosol in a box in the absence of convection or deposition on the walls. Alternatively, they can be interpreted as the steady-state solution for an aerosol in steady "plug" flow through a duct. In this case $\partial n/\partial t = U(\partial n/\partial x)$ where U is the uniform velocity in the duct and x is the distance in the direction of flow.

7·11 Similarity Solution for the Continuous Distribution Function

A method of solving certain coagulation and condensation problems has been developed based on the use of a similarity transformation for the size distribution function (Swift and Friedlander, 1964; Friedlander and Wang, 1966). Solutions found in this way are believed to be asymptotic forms approached after long times, and appear to be independent of the initial size distribution. Closed form solutions for the upper and lower ends of the distribution can sometimes be obtained in this way, and numerical methods can be used to match the solutions for intermediate size particles.

The similarity transformation for the particle size distribution is based on the assumption that the fraction of the particles in a given size range is a function only of particle volume normalized by the average particle volume:

$$\frac{n\,dv}{N_\infty}=\psi\left(\frac{v}{\bar{v}}\right)d\left(\frac{v}{\bar{v}}\right) \qquad (7\cdot 25)$$

where $\bar{v}=V/N_\infty$ is the average particle volume. Both sides of (7·25) are dimensionless. Rearranging, the result is

$$n(v,t)=\frac{N_{\infty}^{2}}{V}\psi(\eta) \tag{7·26}$$

where $\eta=v/\bar{v}=N_{\infty}v/V$. There are also the integral relations:

$$N_{\infty}=\int_{0}^{\infty} n\,dv \tag{7·27a}$$

and

$$V=\int_{0}^{\infty} nv\,dv \tag{7·27b}$$

In terms of the distribution function $n_a(a)$, the similarity transformation takes the form

$$n_a(a,t)=\frac{N_{\infty}^{4/3}}{V^{1/3}}\psi_a(\eta_a) \tag{7·28}$$

where $\eta_a=a(N_{\infty}/V)^{1/3}$. Both N_{∞} and V are in general functions of time. In the simplest case, no material is added or lost from the system, and V is constant. The number concentration N_{∞} decreases as coagulation takes place. If the size distribution corresponding to any value of N_{∞} and V is known, the distribution for any other value of N_{∞}, corresponding to a different time, can be determined from (7·26) if $\psi(\eta)$ is known. The shapes of the distribution at different times are similar when reduced by a scale factor. For this reason, the distribution is said to be "self-preserving."

The determination of the form of ψ is carried out in two steps. First, the special form of the distribution function (7·26) is tested by substitution in the equation of coagulation for the continuous distribution function (7·24) with the appropriate collision frequency function. If the transformation is consistent with the equation, an ordinary integro-differential equation for ψ as a function of η is obtained. The next step is to find a solution of this equation subject to the integral constraints (7·27a) and (7·27b).

7·12 Similarity Solution for Brownian Coagulation

For Brownian coagulation in the continuum range, the collision frequency function is given by (7·11). Substitution of the similarity form (7·26) reduces the coagulation equation for the continuous distribution

(7·24) with (7·11) to the following form:

$$\frac{1}{N_\infty^2}\frac{dN_\infty}{dt}\left[2\psi+\eta\frac{d\psi}{d\eta}\right]$$

$$=\frac{kT}{3\mu}\int_0^\eta \psi(\tilde{\eta})\psi(\eta-\tilde{\eta})\left[\tilde{\eta}^{1/3}+(\eta-\tilde{\eta})^{1/3}\right]\left[\frac{1}{\tilde{\eta}^{1/3}}+\frac{1}{(\eta-\tilde{\eta})^{1/3}}\right]d\tilde{\eta}$$

$$-\frac{2kT}{3\mu}\psi(\eta)\int_0^\infty \psi(\tilde{\eta})\left[\tilde{\eta}^{1/3}+\tilde{\eta}^{1/3}\right]\left[\frac{1}{\eta^{1/3}}+\frac{1}{\tilde{\eta}^{1/3}}\right]d\tilde{\eta}. \quad (7\cdot 29)$$

The change in the total number concentration with time is found by integrating over all collisions:

$$\frac{dN_\infty}{dt}=-\frac{1}{2}\int_0^\infty\int_0^\infty \beta(v,\tilde{v})n(v)n(\tilde{v})\,dv\,d\tilde{v} \quad (7\cdot 30)$$

The factor 1/2 is introduced because the double integral counts each collision twice. Substituting (7·26) and (7·11) in (7·30), the result is

$$\frac{dN_\infty}{dt}=-\frac{2kT}{3\mu}(1+ab)N_\infty^2 \quad (7\cdot 31)$$

where

$$a=\int_0^\infty \eta^{1/3}\psi\,d\eta \quad (7\cdot 32a)$$

and

$$b=\int_0^\infty \eta^{-1/3}\psi\,d\eta \quad (7\cdot 32b)$$

Equation (7·31) is of the same form as (7·13) for the decay of the total number concentration in a monodisperse system. However, the constant has a somewhat different value. Substituting (7·31) in (7·29) and consolidating terms, the result is

$$(1+ab)\eta\frac{d\psi}{d\eta}+(2ab-b\eta^{1/3}-a\eta^{1/3})\psi$$

$$+\int_0^\eta \psi(\eta-\tilde{\eta})\psi(\tilde{\eta})\left[1+\left(\frac{\eta-\tilde{\eta}}{\tilde{\eta}}\right)^{1/3}\right]d\tilde{\eta}=0 \quad (7\cdot 33)$$

which is an ordinary integro-differential equation for ψ with η the independent variable. Hence the similarity transformation (7·26) represents a possible particular solution to the coagulation equation with the Brownian motion coagulation mechanism.

It is still necessary to show that a solution can be found to the transformed equation (7·33) with the integral constraints, (7·27*a*) and (7·27*b*). Analytical solutions to (7·33) can be found for the upper and lower ends of the distribution by making suitable approximations (Friedlander and Wang, 1966). The complete distribution is obtained numerically by matching the distributions for the upper and lower ends, subject to the integral constraints which follow from (7·27*a*) and (7·27*b*):

$$\int_0^\infty \psi \, d\eta = 1 \tag{7·34a}$$

and

$$\int_0^\infty \eta\psi \, d\eta = 1 \tag{7·34b}$$

The results of the numerical calculation are shown in Fig. 7·8, where they are compared with numerical calculations carried out for the

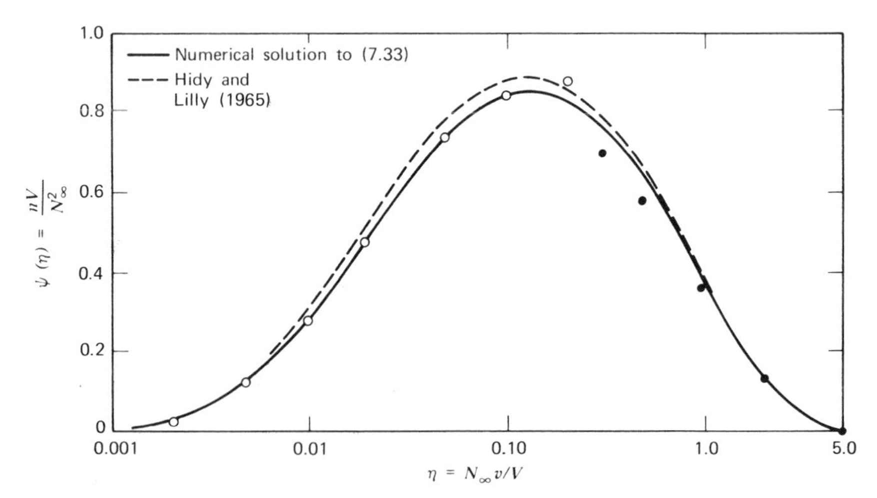

Fig. 7·8 Self-preserving particle size distribution for Brownian coagulation. The form is approximately lognormal. The result obtained by solution of the ordinary integro-differential equation for the continuous spectrum is compared with the limiting solution of Hidy and Lilly (1965) for the discrete spectrum, calculated from the discrete form of the coagulation equation. Shown also are points calculated from analytical solutions for the lower and upper ends of the distribution (Friedlander and Wang, 1966).

discrete spectrum starting with an initially monodisperse system. Good agreement is found between the two methods of calculation. Other calculations indicate that the similarity form is an asymptotic solution independent of the initial distributions so far studied. The values of a and b are found to be 0.9046 and 1.248, respectively. By (7·31) this corresponds to a 6.5% increase in the coagulation constant compared with the value for the monodisperse aerosol (7·13).

The time to reach the self-preserving form depends on the shape of the initial distribution—the closer the initial distribution to $(N_\infty{}^2/V)\psi(\eta)$, the faster the approach. For initially monodisperse aerosols, Hidy and Lilly (1965) found that the self-preserving form was reached after a time $\approx 9\mu/kTN_\infty(0)$ by numerical computations.

To predict the size distribution of a uniform aerosol coagulating in a chamber without deposition on the walls, the following procedure can be adopted: The volumetric concentration of aerosol is assumed constant and equal to its (known) initial value. The change in the number concentration with time is calculated from (7·31). The size distribution at any time can then be determined for each value of $v = V_\eta/N_\infty$ from the relation $n = (N_\infty^2/V)\Psi(\eta)$, using tabulated values (Table 7·3). The calculation is carried out for a range of values of η.

TABLE 7·3
VALUES OF $\psi(\eta)$ OBTAINED FROM (7·33) BY FINITE-DIFFERENCE METHOD
(Friedlander and Wang, 1966)

η	$\psi(\eta)$
0.0010	0.0030
0.0019	0.0165
0.0035	0.0593
0.0052	0.1154
0.0070	0.1750
0.0104	0.2764
0.0190	0.4593
0.0345	0.6409
0.0515	0.7394
0.0695	0.7943
0.1037	0.8373
0.1400	0.8451
0.2089	0.8215
0.3444	0.7371
0.5138	0.6279
0.6935	0.5256
1.0346	0.3729
1.8852	0.1572
3.4351	0.0317
5.1246	0.0050

7·13 Experimental Measurements in the Continuum Range: Comparison with Theory

The change in the particle size distribution function with time for coagulating cigarette smoke has been measured by Keith and Derrick (1960). Smoke issuing from a cigarette was rapidly mixed with clean air, and the mixture was then introduced into a 12 liter flask where coagulation took place. The dilution ratio was 294 volumes of air to 1 volume of raw smoke.

To follow the coagulation process, samples of the smoke were taken at intervals over a period of 4 min from the flask and passed into a conifuge, a centrifugal aerosol collector and classifier. The conifuge had been calibrated with several test aerosols including polystyrene latex particles, over the range from 0.08 to 5 μm. Size distribution curves were measured, together with values for the total number of particles per unit volume, obtained by the graphical integration of the size distribution curves. The volume fraction of aerosol material was $V = 1.11 \times 10^{-7}$, based on measured values for the mass concentrations of the tobacco smoke. This value was checked by numerical integration of the size distribution corresponding to a nominal aging time of 30 sec.

Size distribution data were reported in terms of a distribution function equivalent to n_x defined as follows:

$$dN = n_x\, dx \qquad (7\cdot35)$$

where dx represents an infinitesimal length along the outer cone. Values of n_x were calculated by Friedlander and Hidy (1969) from the self-preserving distribution using the relation

$$n_x = -n_v \frac{dv}{dx} = -\frac{N_\infty{}^2}{V}\psi(\eta)\frac{dv}{dx} \qquad (7\cdot36)$$

where dv/dx was obtained from the Keith and Derrick conifuge calibration curve and $\psi(\eta)$ from Friedlander and Wang (1966). Values of N_∞ and V were those reported in the reference. Theory and experiment are compared in Fig. 7·9 and 7·10. In Fig. 7·9, the experimental points for a nominal aging time of 30 sec are shown with the theoretical prediction. In Fig. 7·10, the results of measurements of the change in the distribution function with time are compared with theory. Agreement is fair; the experimental results appear to fall significantly higher than theory at the upper end of the spectrum (large particle sizes).

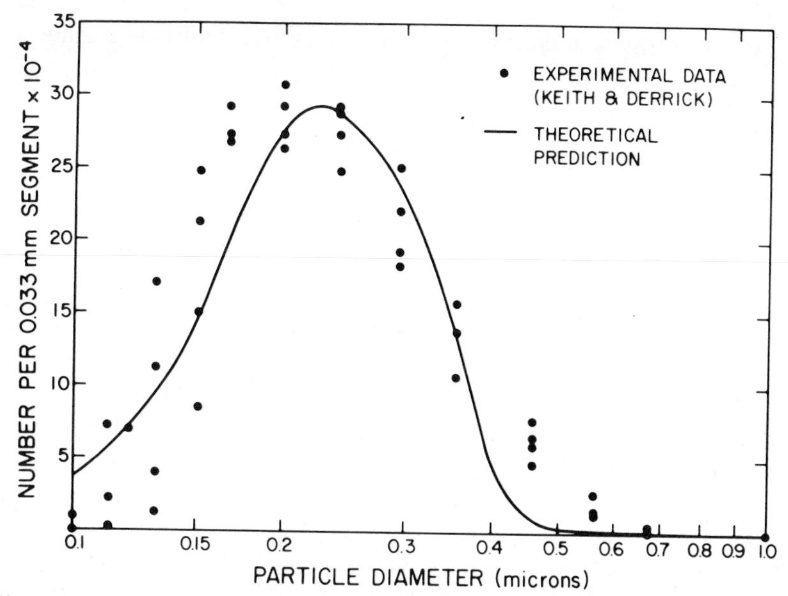

Fig. 7·9 Comparison of experimental size distribution data for tobacco smoke with prediction based on self-preserving size spectrum theory. $V=1.11\times10^{-7}$, $N_\infty=1.59\times10^7$ cm^{-3}. The peak in the number distribution measured in this way occurs at $d_p\approx0.2$ μm (Friedlander and Hidy, 1969).

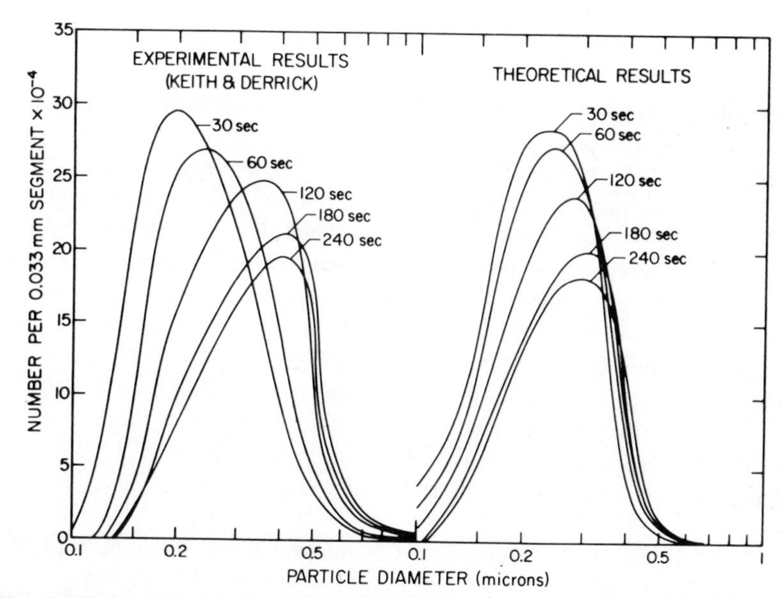

Fig. 7·10 Experiment and theory compared for an aging tobacco smoke aerosol. Calculation based on $V=1.11\times10^{-7}$ and experimental values of N_∞ (Friedlander and Hidy, 1969).

7·14 Similarity Solution: Coagulation in the Free Molecule Region

Is it possible to make the similarity transformation (7·26) for other collision mechanisms? In general, when the collision frequency $\beta(\tilde{v}, v)$ is a homogeneous function of particle volume, the transformation to an ordinary integro-differential equation can be made. However, even though the transformation is possible, a solution to the transformed equation may not exist that satisfies the boundary conditions and integral constraints.

When the particles are much smaller than the mean free path, the collision frequency function is given by (7·12):

$$\beta(v, \tilde{v}) = \left(\frac{3}{4}\pi\right)^{1/6}\left(\frac{6kT}{\rho_p}\right)^{1/2}\left[\frac{1}{v} + \frac{1}{\tilde{v}}\right]^{1/2}(v^{1/3} + \tilde{v}^{1/3})^2$$

which is a homogeneous function of order 1/6 in particle volume. The similarity transformation can be made and a solution can be found to the transformed equation in much the same way as in the previous sections (Lai et al., 1972). The change in the total number of particles with time is found to be

$$\frac{dN_\infty}{dt} = -\frac{\alpha}{2}\left(\frac{3}{4\pi}\right)^{1/6}\left(\frac{6kT}{\rho_p}\right)^{1/2} V^{1/6} N_\infty^{11/6} \tag{7·37}$$

The constant α is an integral function of ψ and is found to be about 6.67 by numerical analysis.

Husar (1971) studied the coagulation of the very small particles produced by a propane torch aerosol in a 90 m^3 polyethylene bag. The size distribution was measured as a function of time with an electrical mobility analyzer. The results of the experiments are shown in Fig. 7·11 in which the size distribution is plotted as a function of particle diameter and in Fig. 7·12 in which ψ_a is shown as a function of η_a (see (7·28)). Numerical calculations were carried out by a Monte Carlo method, and the results of the calculation are also shown in Fig. 7·12; the Monte Carlo calculations are in good agreement with direct calculations made from the transformed equation by Lai et al. (1972). The agreement between experiment and theory is quite satisfactory.

7·15 Light Scattering Dynamics: Coagulation in the Rayleigh Range

In the Rayleigh range, light scattering is proportional to the square of the particle volume; when two particles of the same size combine to

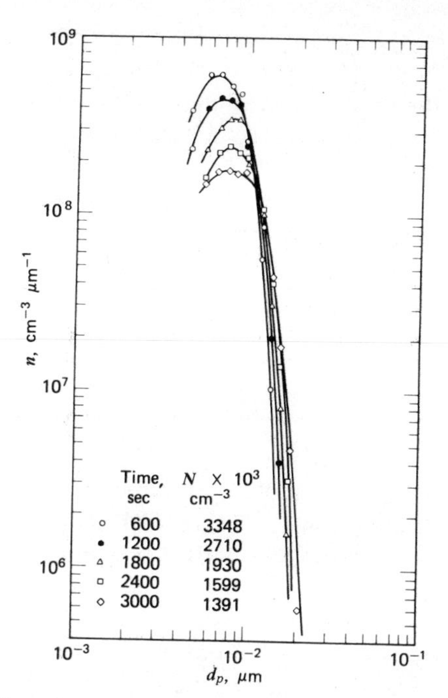

Fig. 7·11 Coagulation of aerosol particles much smaller than the mean free path. Size distributions measured with the electrical mobility analyzer (Husar, 1971).

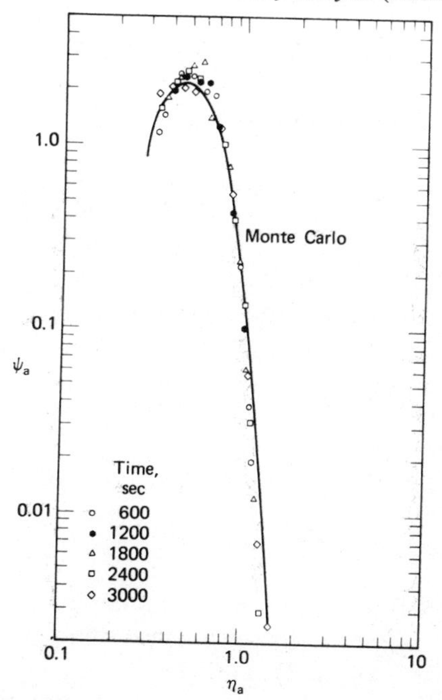

Fig. 7·12 Size distribution data of Fig. 7·11 for coagulation of small particles plotted in the coordinates of the similarity theory. Shown also is the result of a Monte Carlo calculation for the discrete spectrum (Husar, 1971).

form a larger one, the total light scattered doubles (Chap. 5). This is true so long as the two original particles are separated by a distance much greater than the wavelength of the incident light. In this case, the two particles scatter independently and out of phase, and the energy of the scattered light is the sum of the energies scattered separately by the two particles. When the two particles are combined and still much smaller than the wavelength of the light, the electric field scattered will be the sum of the two electric fields in phase. As a result, double the amplitude of the single particle or four times the energy of a single particle will be scattered. Hence the light scattered by a coagulating small particle aerosol increases with time.

Light scattering by a coagulating aerosol in the Rayleigh size range has been measured experimentally by Graham and Homer (1973). The aerosol was generated by passing a shock wave through argon containing tetramethyl lead (TML). The TML decomposes behind the shock forming a supersaturated lead vapor. As a result, small lead droplets form and subsequently coagulate. Light scattering during coagulation was followed experimentally with an argon–ion laser.

In the Rayleigh range, the total scattering is given by a relationship of the form:

$$b_{scat} \sim \int_{\infty} n(v) v^2 \, dv \tag{7·38}$$

If the aerosol size distribution is in the self-preserving form, (7·38) becomes

$$b_{scat} \sim V\bar{v} \int_0^{\infty} \psi(\eta)\eta^2 \, d\eta \tag{7·39}$$

That is, for an aerosol with a constant particulate volume concentration, V, the scattered light intensity is proportional to the instantaneous mean particle volume, $\bar{v} = V/N_{\infty}$, which for the self-preserving theory takes the form:

$$\bar{v} = \left[\frac{5}{12}\left(\frac{3}{4\pi}\right)^{1/3}\left(\frac{6kT}{\rho_p}\right)^{1/2} H\alpha V \right]^{6/5} t^{6/5} \tag{7·40}$$

by integrating (7·37). The constant H is a correction factor introduced to account for the attractive forces between the particles (Graham and Homer, 1973). Conbining (7·39) and (7·40), $b_{scat} \sim t^{6/5}$, that is, the total light scattered by a coagulating small particle aerosol increases as $t^{6/5}$.

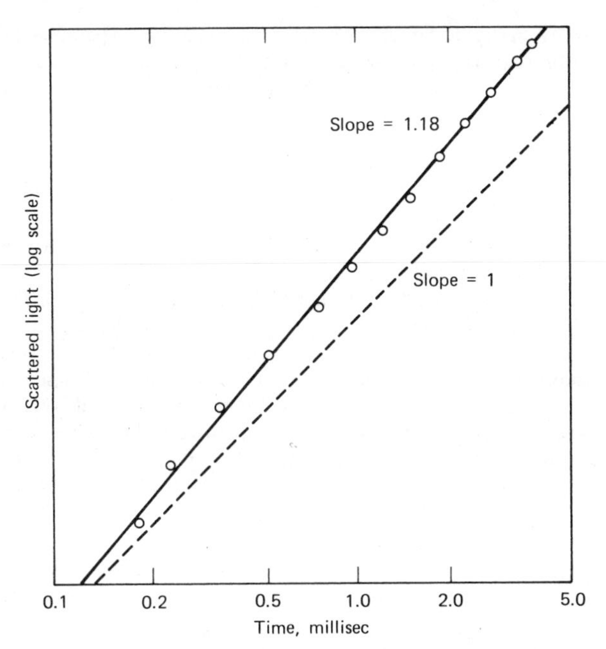

Fig. 7·13 Increase, with time, of light scattered by coagulating lead particles generated by the decomposition of tetramethyl lead. The light source was an argon laser (Graham and Homer, 1973).

The output voltage of the photomultiplier tube measuring the light scattered perpendicular to the incident beam is proportional to b_{scat}; the experimental dependence on time for a case in which coagulation is controlling is shown in Fig. 7·13. The slope in logarithmic coordinates is very close to the theoretical value of $6/5$. Hence the experimental result is consistent with the model of an aerosol coagulating according to the self-preserving theory and scattering light according to the Rayleigh theory. Absolute values of mean particle size and coagulation rates can be determined by proper calibration of the system.

For larger particles out of the Rayleigh scattering range, it is, in general, necessary to calculate the light scattering numerically based on experimental or theoretical expressions for the variation of the size distribution with time.

Problems

1. It is desired to quench the coagulation of an aerosol composed of very small particles ($d_p \ll l$). If the rate of coagulation is to be reduced to 1% of its original value by isothermal, constant pressure dilution with particle-free gas, determine the dilution ratio. The rate is to be reduced by the same factor by a reversible adiabatic expansion. Determine the volume expansion ratio assuming the gas is ideal.

2. (*a*) How many electronic charges (opposite sign) must two 0.5 μm particles have to produce an (i) 1% and (ii) 10% increase in collision rate? The temperature is 20°C and pressure 1 atm.

(*b*) Estimate the number of charges of like sign that two 0.5 μm droplets of cyclohexane must have to just balance the effect of van der Waals type attractive forces. The temperature is 20°C and pressure 1 atm.

3. Whytlaw-Gray and his co-workers experimentally demonstrated the applicability of the theory of diffusion controlled coagulation to aerosols. Many of their results were summarized in the book *Smoke* by Whytlaw-Gray and Patterson (1932). In a later set of experiments (Whytlaw-Gray, Cawood, and Patterson, 1936), the particle concentration was determined as a function of time by trapping a known volume of smoke from the center of a smoke chamber in a shallow box with a glass bottom. These sampling boxes were fixed to a support in the center of the smoke chamber and were withdrawn at definite intervals by means of strings passing through corks in the side of the chamber. The particles that settled on the glass plate were counted optically. In one set of experiments with a cadmium oxide smoke, the following results were obtained:

Time from start (min)	Number per $cm^3 x 10^{-6}$
8	0.92
24	0.47
43	0.33
62	0.24
84	0.21

Determine the coagulation constant from these data and compare with theory for monodisperse and self-preserving aerosols.

4. Consider the flow of an aerosol through a 4 in. duct at a velocity of 50 ft/sec. Compare the coagulation rate by Brownian motion and laminar shear in the viscous sublayer, near the wall. Present your results by plotting the collision frequency function for particles with $d_p = 1$ μm colliding with particles of other sizes. Assume a temperature of 20°C.

Hint: In the viscous sublayer, the velocity distribution is given by the relation:

$$u = \frac{yfU^2}{2\nu}$$

where y is the distance from the wall, f is the Fanning friction factor, U is the mainstream velocity, and ν is the kinematic viscosity.

5. As an idealized model for the automobile exhaust aerosol, assume the distribution is self-preserving and calculate the total particle concentration and average diameter at the tailpipe exit. The mass loading of particles is 1 mg/cu ft (STP) and the average temperature 400°F. The density of the aerosol material is 6 g/cm^3. Assume steady flow at a velocity of 90 ft/sec with all particles forming by condensation at a point 15 ft upstream from the exit. Neglect deposition on the walls of the pipe and assume the initial number concentration is effectively infinite.

6. Let A be the surface area per unit volume of gas of a coagulating aerosol. Assume that at $t=0$, both N_∞ and A are infinite. Show that for a self-preserving aerosol composed of particles much larger than the mean free path of the gas

$$A = \text{const } V^{2/3} t^{-1/3}$$

where t is the time. Find an expression for the constant.

7. Using the self-preserving transformation, derive an expression for the dynamics of the size distribution function when laminar shear is the controlling mechanism of coagulation. Find the corresponding expression for the change with time of the total number concentration. (Note that a solution satisfying the condition $\psi(\eta)\to 0$ as $\eta \to 0$ does not seem to exist in this case.)

References

Devir, S. F. (1963) *J. Colloid Sci.*, **18**, 744.

Drake, R. L. (1972) A General Mathematical Survey of the Coagulation Equation in Hidy, G. M., and Brock, J. R., (Eds.) *Topics in Current Aerosol Research*, vol. 3 (Part 2).

Friedlander, S. K. (1964) The Similarity Theory of the Particle Size Distribution of the Atmospheric Aerosol in *Proceedings of the First National Conference on Aerosols,* Publishing House of the Czechoslovak Academy of Sciences.

Friedlander, S. K., and Wang, C. S. (1966), *J. Colloid Interface Sci.*, **22**, 126.

Friedlander, S. K., and Hidy, G. M. (1969) New Concepts in Aerosol Size Spectrum Theory in Podzimek, J. (Ed.) *Proceedings of the 7th International Conference on Condensation and Ice Nuclei,* Academia, Prague.

Fuchs, N. A. (1964) *Mechanics of Aerosols*, Pergamon, New York.

Graham, S. C, and Homer, J. B. (1973), *Faraday Symp.*, **7**, 85.

Hamaker, H. C. (1937) *Physica,* **4**, 1058.

Hidy, G. M., and Lilly, D. K. (1965) *J. Colloid Sci.*, **20**, 867.

Hidy, G. M., and Brock, J. R. (1970) *Dynamics of Aerocolloidal Systems*, Pergamon, New York, p. 308–309.

Hirschfelder, J. O., Curtiss, C. F., and Bird, R. B. (1954) *Molecular Theory of Gases and Liquids*, Wiley, New York.

Husar, R. B. (1971) Coagulation of Knudsen Aerosols, Ph.D. thesis, Department of Mechanical Engineering, Univ. of Minnesota.

Keith, C. H., and Derrick, J. E. (1960) *J. Colloid Sci.*, **15**, 340.

Lai, F. S., Friedlander, S. K., Pich, J., and Hidy, G. M. (1972) *J. Colloid Interface Sci.*, **39**, 395.

Landau, L. D., and Lifshitz, E. M. (1959) *Fluid Mechanics,* Addison-Wesley, Reading, Mass.

Laufer, J. (1954) NACA Report 1174.

Lumley, J. L. and Panofsky, H. A. (1964) *The Structure of Atmospheric Turbulence*, p. 123, Interscience, New York.

Saffman, P., and Turner, J. (1956) *J. Fluid Mech.*, **1**, 16.

Smoluchowski, M. (1917*a*) *Z. Physik Chem.*, **92**, 129.

Smoluchowski, M. (1917*b*) *Z. Physik Chem.*, **92,** 155.

Swift, D. L., and Friedlander, S. K. (1964) *J. Colloid Sci.*, **19**, 621.

Tikhomirov, M. V., Tunitskii, N.N., and Petrjanov, I.V. (1942) *Acta Phys.-Chim. U.R.S.S.*, **17**, 185.

Whytlaw-Gray, R., and Patterson, H. S. (1932) *Smoke: A Study of Aerial Disperse Systems*, Edward Arnold, London.

In this classic monograph, the authors describe experiments on the coagulation of smokes composed of substances of many different types. The studies were carried out before the development of monodisperse aerosol generators and the electron microscope. The coagulation rate was shown to be approximately independent of the chemical nature of the aerosol material and consistent with the predictions of the diffusion limited theory of Smoluchowski. Included are many interesting photomicrographs of agglomerated particles.

Whytlaw-Gray, R., Cawood, W., and Patterson, H. S. (1936) *Disperse Systems in Gases; Dust, Smoke and Fog,* A General Discussion of the Faraday Society, April 1936.

CHAPTER 8

Thermodynamic Properties

Aerosols are by their nature multiphase, and equilibrium thermodynamics provide constraints and limiting conditions on their properties. This is particularly important for systems in which material exchange can occur between the dispersed and continuous phases. Sulfate containing aerosols formed by the oxidation of SO_2 in power plant plumes are an important example in air pollution. In nature, the marine aerosol, primarily a sodium chloride solution, is the most important case. Changing conditions of temperature and humidity affect the behavior of these systems.

Thermodynamic considerations play an important role in the behavior of the lead compounds added to gasoline to prevent "knock." These additives are introduced in an organic form, lead tetraethyl, sufficiently volatile to evaporate with the gasoline hydrocarbons in the cylinders. The mechanisms by which the lead compounds act in suppressing knock are not well understood. After completion of combustion, however, lead oxide tends to form and precipitate as a solid in the cylinders. To prevent fouling of the cylinders, ethylene bromide and chloride are also added to gasoline. Lead bromochloride, a compound of relatively high vapor pressure forms and remains in the gas phase until it condenses later in the automobile tailpipe.

As shown in this chapter, equilibrium thermodynamics provides information on the vapor pressures of particles and droplets and on the conditions under which condensation can occur. The distributions of various species between gas and aerosol phases with simultaneous chemical reaction is thermodynamically determined, under limiting conditions. Usually, aerosols are not in equilibrium or only partially so with respect to certain constituents. Particle size and chemical species distributions depend on the interactions of transport processes and equilibrium thermodynamics (Chap. 10).

8·1 The Vapor Pressure Curve: Saturation Ratio

For a single component, two phase system such as a liquid and vapor of the same substance, the relationship between the vapor pressure and

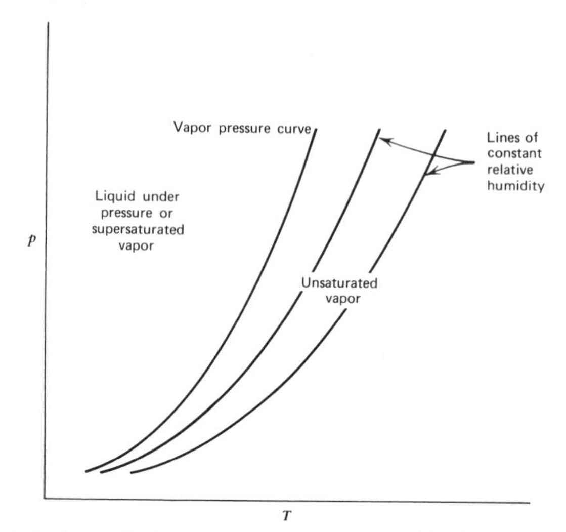

Fig. 8·1 Typical form of the vapor pressure curve with lines of constant relative humidity.

temperature is usually of the form shown in Fig. 8·1. The region to the right of the curve represents unsaturated vapor, whereas the region to the left represents liquid under pressure. Along the curve, vapor and liquid co-exist in equilibrium.

An expression for the slope of the vapor pressure curve can be derived from fundamental thermodynamic considerations (Denbigh, 1966):

$$\frac{dp_s}{dT}=\frac{\Delta H}{T\Delta v}$$

where ΔH is the molar heat of vaporization and Δv the volume change per mole accompanying vaporization. Away from the critical point, which is represented by the end point of the vapor pressure curve in Fig. 8·1, the molar volume of the gas is much larger than that of the liquid. If the gas is ideal, the slope of the vapor pressure curve is given by

$$\frac{dp_s}{dT}=\frac{\Delta H p_s}{RT^2} \tag{8·1}$$

which is known as the Clapeyron equation. Since the heat of vaporization is approximately constant over a wide range of temperatures, the vapor pressure can be represented approximately by the expression

$$\ln p_s \approx -\frac{\Delta H}{RT}+\text{const}$$

These results hold to a close approximation even in the presence of an inert gas, such as air, with the total pressure near an atmosphere.

How does an initially unsaturated vapor, represented by a point to the right of the vapor pressure curve, reach conditions under which condensation can occur? Any number of paths on the (p, T) diagram are imaginable but two are of particular interest: reversible adiabatic expansion and mixing with cooler air at a lower concentration. Both processes may lead to the formation of an aerosol composed of small liquid droplets. The paths followed through the unsaturated state up to the equilibrium curve can be followed approximately from theoretical considerations in both cases as shown in the following sections.

If insufficient condensation nuclei and/or surface are available, condensation is delayed and the system passes into a metastable state, even though it is on the liquid side of the equilibrium curve. The ratio of the actual pressure to the equilibrium vapor pressure at the temperature in question is the *saturation ratio* (or *relative humidity*):

$$S = \frac{p}{p_s(T)} \tag{8·2}$$

The saturation ratio is greater than unity when p and T correspond to a state on the liquid side of the vapor–liquid equilibrium curve. For $S > 1$, the system is said to be supersaturated. The supersaturation is given by $(p - p_s)/p_s$.

The behavior of a condensing system is determined by the interaction of thermodynamic and rate processes. This interaction, which is quite complex, is discussed in more detail in the next chapter.

8·2 Adiabatic Expansion

Adiabatic expansion may be carried out as a batch process in a Wilson-type cloud chamber or as a steady-flow process in the diverging section of the nozzle of a steam turbine or supersonic wind tunnel. If the process is carried out reversibly, and this is usually a good approximation, the conditions along the path are related by the expression:

$$\frac{p_2}{p_1} = \left(\frac{T_2}{T_1} \right)^{\gamma/(\gamma - 1)} \tag{8·3}$$

where p_1, p_2 and T_1, T_2 are the total pressures and temperatures before and after the expansion, and γ is the ratio of the specific heat at constant pressure to the specific heat at constant volume. Since $\gamma > 1$, reversible adiabatic expansion leads to a decrease in both temperature and pressure of an initially unsaturated gas up to the point of condensation. In the absence of condensation, the partial pressure is proportional to the pressure so (8·3) also represents the partial pressure ratio. The path of the expansion process is shown in Fig. 8·2 with the vapor pressure curve.

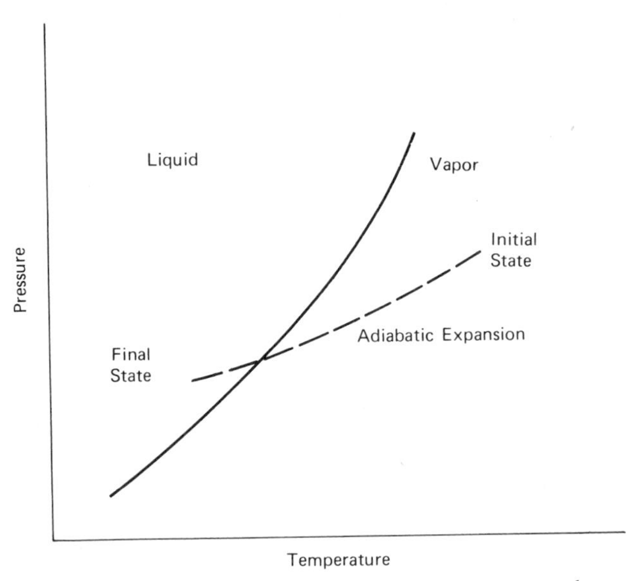

Fig. 8·2 Reversible, adiabatic expansion from an initially unsaturated state carries the vapor across the saturation curve into a region where the stable state is a liquid.

The original chamber experiments of Wilson are discussed in the next chapter. These led to the development of the theory of homogeneous nucleation. Experiments on condensation taking place in a reversible adiabatic expansion under steady-flow conditions were stimulated first by concern with droplet formation in turbine nozzles and the erosion of the turbine blades. When wind tunnels were designed for the study of supersonic flows, it was found that moisture condensed in the diverging section, altering the predicted Mach number and pressure and producing nonuniformities in the flow at the nozzle exit. This led to studies of the phenomenon in an attempt to reduce interference with wind tunnel operation.

More recently, studies of condensation have been carried out in converging–diverging nozzles as a means of investigating nucleation kinetics (Wegener and Pouring, 1964). In a typical experiment, air and water vapor are expanded through the nozzle as shown in Fig. 8·3. Both pressure and temperature drop isentropically, and saturation conditions are reached in the subsonic section upstream of the throat. Cooling rates of 10^5 to 10^6°C/sec are attainable in such flows. No condensate appears when saturation conditions are reached, or even considerably further downstream, because the rate of condensation is not sufficiently rapid for visible fog to form. In supersonic flows, the pressure and temperature downstream of the throat continue to decrease as the Mach number increases. At some point in the diverging section, nuclei generated in the

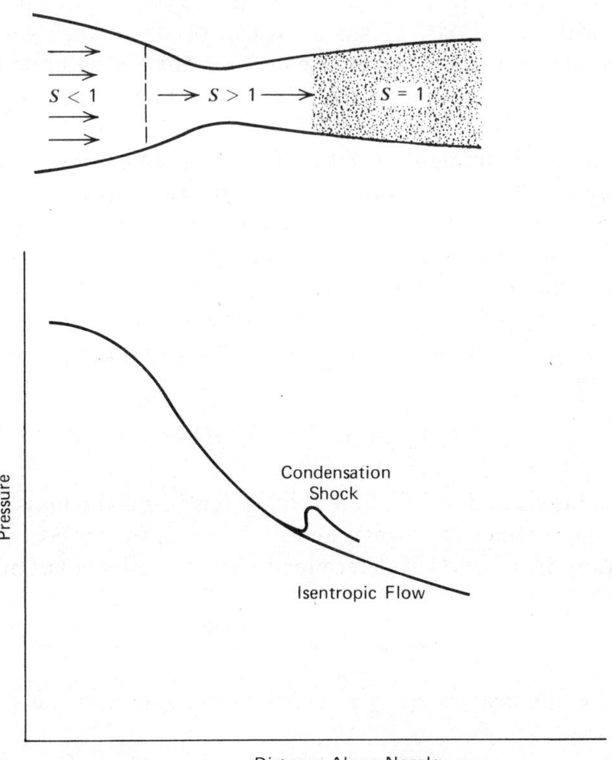

Fig. 8·3 Variation of the pressure in a converging–diverging nozzle. Condensation in the diverging section leads to an increase in pressure as a result of the release of latent heat. This pressure increase is known as a condensation shock. The onset of condensation also leads to a local increase in temperature, compared with the isentropic flow without condensation. Increasing humidity causes the location of the condensation shock to move upstream.

gas grow to a sufficiently large size and visible fog appears.

Fog formation is accompanied by the release of the latent heat of vaporization of the condensate; as a result both pressure and temperature (Fig. 8·3) rise above the curve for isentropic flow. The pressure rise has been termed a condensation shock but the name is misleading since the pressure and temperature of the gas do not show the almost discontinuous behavior of a mechanical shock. By measuring the pressure rise and calculating the heat release, it is possible to estimate the rate of condensation and to test condensation theory (Chap. 9).

8·3 Condensation by Mixing

Condensation can result when a hot gas carrying a condensable vapor is mixed with a cool gas. This process can occur in stack gases as they mix with ambient air or with exhaled air that is saturated at body temperature when it comes from the lungs. What determines whether condensation occurs in such systems?

As mixing with ambient air takes place, the temperature drops favoring condensation, but dilution tends to discourage condensation. Whether saturation conditions are reached depends on the relative rates of cooling and dilution during the mixing process. The situation can be analyzed in the following way:

In the absence of condensation, the concentration distribution in the fluid is determined by the equation of convective diffusion for a binary gas mixture:

$$\rho \frac{\partial c}{\partial t} + \rho \mathbf{v} \cdot \nabla c = \nabla \cdot \rho D \nabla c \tag{8·4}$$

where ρ is the mass density of the fluid (g/cm^3), c is the mass fraction of the diffusing species (g/g gas), and D is the diffusion coefficient. The temperature distribution is determined by the energy equation

$$\rho C_p \frac{\partial T}{\partial t} + \rho C_p \mathbf{v} \cdot \nabla T = \nabla \cdot \kappa \nabla T \tag{8·5}$$

where C_p is the heat capacity at constant pressure and κ is the thermal conductivity.

The mixing system that has received the most careful experimental study is the hot jet of a condensable vapor–air mixture, which mixes with air at a lower temperature. The boundary conditions for the jet geometry can be written:

$$c = c_0, \quad T = T_0 \text{ at the orifice of the jet}$$

$$c = c_\infty, \quad T = T_\infty \text{ in the ambient air}$$

When C_p is constant, the equations for the concentration and temperature fields and the boundary conditions are satisfied by the relation

$$\frac{c-c_{\infty}}{c_0-c_{\infty}}=\frac{T-T_{\infty}}{T_0-T_{\infty}} \tag{8·6}$$

provided $\kappa/C_p=\rho D$ or $\kappa/\rho C_p D=1$. The dimensionless group $\kappa/\rho C_p D$, known as the Lewis number, is usually of order unity for gas mixtures. Table 8·1 shows values of $\kappa/\rho C_p D$ for air and water vapor as a function of temperature.

TABLE 8·1
Lewis Number for Trace Amounts of Water Vapor in Air

T(°K)	Schmidt Number, ν/D	Prandtl Number, $C_p\mu/\kappa$	Lewis Number, $\kappa/\rho C_p D$
300	0.604	0.708	0.854
400	0.650	0.689	0.945
500	0.594	0.680	0.873
600	0.559	0.680	0.822
700	0.533	0.684	0.780

For $\kappa/\rho C_p D=1$, the relation between concentration and temperature, (8·6), is independent of the nature of the flow, laminar or turbulent. It applies both to the instantaneous and time average concentration and temperature fields but only in regions in which condensation has not yet occurred.

According to (8·6), the path of the condensing system on a mass fraction versus temperature diagram is a straight line determined by the conditions at the orifice and in the ambient atmosphere. For $c \ll 1$, the partial pressure is approximately proportional to the mass fraction. The path is shown in Fig. 8·4 with the vapor pressure curve. From this relationship, it is possible to place limits on the concentrations and temperatures that must exist at the jet orifice for condensation to occur. The mixing line must be at least tangent to the vapor pressure curve. For $\kappa/\rho C_p D \neq 1$, the relationship between c and T will in general depend on the flow field.

An experimental study of the behavior of a condensing jet has been carried out by Hidy and Friedlander (1964). These investigators passed a collimated light beam (2 mm slit width) through the center line of a condensing jet to reveal the internal structure of the fog. The cross section of the fog-filled region was roughly that of a wedge with its tip located vertically downstream from the edge of the nozzle. The region

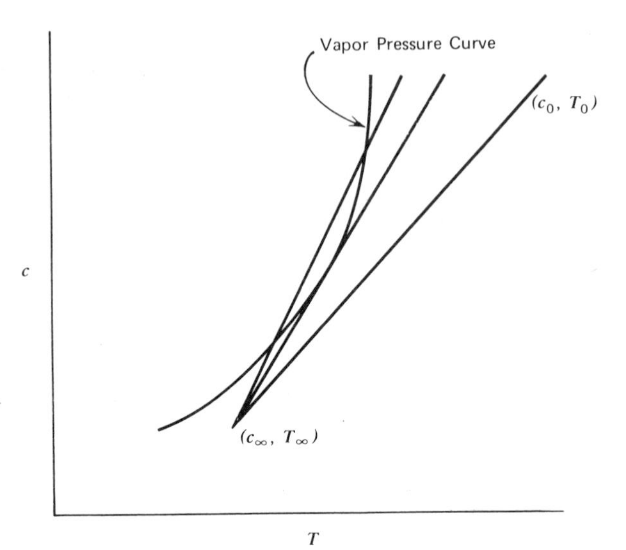

Fig. 8·4 Air–vapor mixtures at three different source conditions (c_0, T_0) mixing with air of the same ambient conditions (c_∞, T_∞). No condensation occurs for the source on the right while condensation can occur, depending on availability of nuclei and mixing rates, for the one on the left. The middle line shows a limiting situation. Case of the Lewis number $= \kappa/\rho C_p D = 1$.

where condensation first appeared fell within the turbulent mixing zone between the emerging jet and the ambient air. No fog was observed in the mixing zone near the edge of the nozzle, and it was concluded that this region was supersaturated. The extent of the fog-free region could be increased by decreasing concentration or velocity or by increasing temperature. Observable condensation could be produced in this region by inserting fine wires or by introducing sulfuric acid nuclei. Temperature measurements were made in a hot jet with and without condensable vapor. The release of latent heat by the fog produced a temperature increase across the condensation zone when compared with the fog-free jet (Fig. 8·5).

It was concluded that homogeneous nucleation in the supersaturated region near the nozzle initiates fog formation. The nuclei formed in this way then mix with fresh saturated and supersaturated gas and may serve as foreign nuclei in other areas of the mixing zone. Individual nuclei will, in general, possess different temperature and concentration histories, and this makes the system a difficult one to use in basic studies of nucleation kinetics.

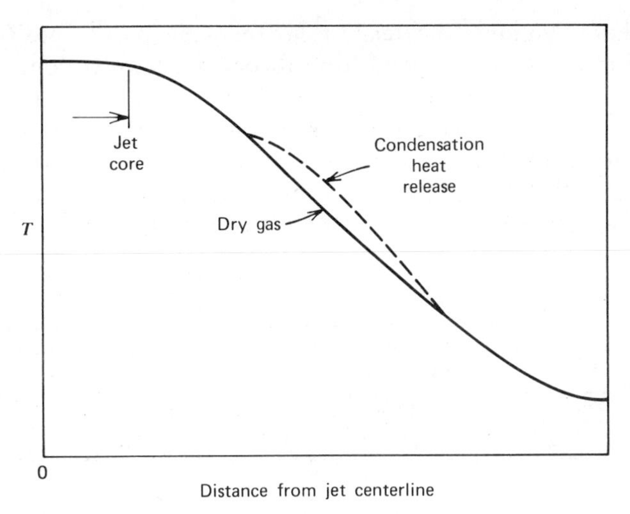

Fig. 8·5 Temperature increase across the condensation zone for a hot jet of glycerine in air injected into cool air (Hidy and Friedlander, 1964).

8·4 Effect of Solutes on Vapor Pressure

The presence of a nonvolatile solute in an aqueous solution tends to reduce its vapor pressure to an extent that depends on the nature and concentration of the solute. On a purely geometric basis, there are fewer solvent molecules in the surface layer than in the case of a pure solvent drop. This would lead to a vapor pressure reduction proportional to concentration, and this is observed for *ideal solutions*. Specific chemical effects of an attractive nature between solute and solvent may lead to a further reduction in vapor pressure. The reduction of vapor pressure makes it possible for aqueous solution droplets to exist in equilibrium with air whose relative humidity is significantly less than 100%. This effect contributes to visibility degradation by air pollution.

An important example is the droplets containing dissolved sulfates that form in power plant stack plumes as a result of the oxidation of SO_2. The concentration of SO_2 depends on the sulfur content of the fuel. The sulfates may be present as sulfuric acid or in a partially neutralized form as ammonium salts or salts of the metal oxides in flyash. The droplet size distribution and chemical composition are determined by a combination of thermodynamic and kinetic factors. In this section, we consider only equilibrium thermodynamics as it affects the vapor pressure of the drop.

For dilute solutions, the relationship between partial pressure and composition can be determined from theory; over wider concentration ranges, it is in general necessary to determine the relationship by experiment. This has been done for solutions of certain salts and acids; data for the equilibrium vapor pressure of water over solutions of sulfuric acid 25°C are shown in Table 8·2. As the concentration of sulfuric acid increases, the vapor pressure of water over the solution drops sharply.

TABLE 8·2
WATER VAPOR PRESSURE OVER SULFURIC ACID SOLUTIONS AT 25°C (PERRY, 1950)

wt.%	Density (g/cm^3)	Vapor Pressure (mm Hg)
0	0.997	23.8
10	1.064	22.4
20	1.137	20.8
30	1.215	17.8
40	1.299	13.5
50	1.391	8.45
60	1.494	3.97
70	1.606	1.03
80	1.722	0.124
90	1.809	0.00765

For a binary solution at constant composition, an expression of the form (8·1) is found for the slope of the vapor pressure curve as a function of temperature in which the latent heat of vaporization is the value for the solution. Solution vapor pressure curves can be represented as a set of parametric curves at constant composition on the vapor pressure diagram.

For binary solutions such as sulfuric acid and water, droplets may be distributed with respect to size, but at equilibrium all have the same composition unless the Kelvin effect is important as discussed in a later section. For ternary mixtures, the situation is more complicated; the same droplet size may result from different chemical compositions in equilibrium at a given relative humidity.

When equilibrium between the bulk of the gas and the droplet phase does not exist for a chemical species, it is usually assumed that there is local equilibrium between the phases at the interface. From the transport rates in the gas and droplet phases and the equilibrium boundary condition, the transfer rate can be calculated as shown in the next chapter.

8·5 Multicomponent Equilibrium: Reacting Droplets

Gases such as SO_2, NH_3, CO_2, and Cl_2 dissolve in water and hydrolyze to produce a variety of ionic species. The sulfur dioxide–water system can be represented by the following set of reactions at equilibrium (Fig. 8·6):

$$SO_2(\text{gas}) \rightleftharpoons H_2SO_3 \tag{8·7}$$

$$H_2SO_3 \rightleftharpoons H^+ + HSO_3^- \tag{8·8}$$

$$HSO_3^- \rightleftharpoons H^+ + SO_3^{2-} \tag{8·9}$$

$$H_2O \rightleftharpoons H^+ + OH^- \tag{8·10}$$

The corresponding equilibrium relationships can be written as follows:

$$p_{SO_2} = \mathrm{K}[H_2SO_3] \tag{8·11}$$

$$K_1 = \frac{[H^+][HSO_3^-]}{[H_2SO_3]} \tag{8·12}$$

$$K_2 = \frac{[H^+][SO_3^{2-}]}{[HSO_3^-]} \tag{8·13}$$

$$K_3 = [H^+][OH^-] \tag{8·14}$$

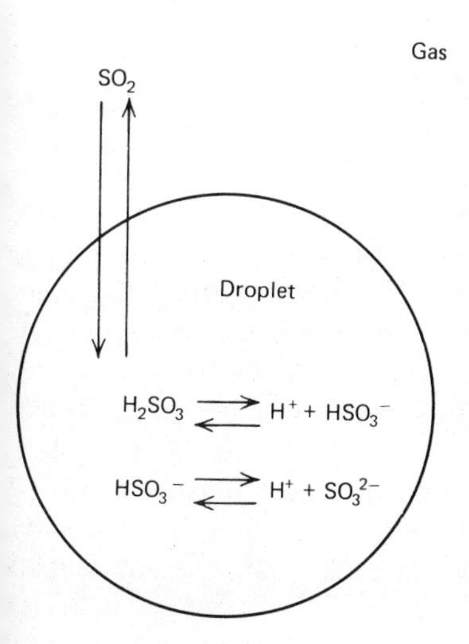

Fig. 8·6 Equilibrium relationships in the distribution of SO_2 between the gas and aqueous solution phases.

where K is the Henry's law constant and K_1, K_2, and K_3 are equilibrium constants. With p_{SO_2} given, the five concentrations $[H_2SO_3]$, $[H^+]$, $[HSO_3^-]$, $[SO_3^{2-}]$, and $[OH^-]$ are determined by these four equations and the electroneutrality requirement that the sums of positive and negative ions are equal.

Combining (8·11) through (8·13), an expression for the sulfite ion concentration is obtained:

$$[SO_3^{2-}] = \frac{K_1 K_2}{K} \frac{p_{SO_2}}{[H^+]^2} \tag{8·15}$$

This result can be applied to a collection of small droplets suspended in a gas containing SO_2. If the droplet concentration is not too high, the partial pressure of SO_2 in the gas is approximately constant despite SO_2 exchange between the gas and solution phases. The hydrogen ion concentration or pH can be varied independently by addition of NH_3 or other basic (or acidic) species to the solution phase. Reducing $[H^+]$ increases the equilibrium concentration of SO_3^{2-}.

This effect is of importance in atmospheric processes since SO_3^{2-} is oxidized to sulfate at a rate that is usually sufficiently slow for reactions (8·7) through (8·10) to remain in equilibrium. The reaction rate depends on $[SO_3^{2-}]$ so reducing $[H^+]$ increases the rate of conversion by increasing $[SO_3^{2-}]$.

Both NH_3 and CO_2 are likely to be present in the atmosphere along with SO_2. Ammonia is produced in combustion processes and by the breakdown of natural products including animal urine. Both of these gases affect pH, and the following additional equilibrium reactions must be considered along with (8·7) through (8·10):

$$NH_4OH \rightleftharpoons NH_4^+ + OH^- \tag{8·16}$$

$$H_2CO_3 \rightleftharpoons H^+ + HCO_3^- \tag{8·17}$$

$$HCO_3^- \rightleftharpoons H^+ + CO_3^{2-} \tag{8·18}$$

There are, in addition, Henry's law relationships between NH_3 and NH_4OH, and CO_2 and H_2CO_3. Finally, there is the electroneutrality requirement. The result is a set of ten equations in ten unknown solution concentrations and the partial pressures of SO_2, NH_3, and CO_2, which must be given. These equations can be solved to give the concentrations of the various ionic species (Scott and Hobbs, 1967).

If, in addition, there is a slow oxidation of dissolved SO_2 to SO_4^{2-}, the rest of the reactions remain approximately in equilibrium. If only SO_4^{2-}

(and not HSO_4^-) is present, this new species must be included in the electroneutrality balance as another unknown. Scott and Hobbs (1967) assume that the corresponding additional equation is a rate expression for the conversion of SO_3^{2-} to SO_4^{2-}. This new set of equations can then be solved simultaneously for the increase in $[SO_4^{2-}]$ with time (Chap. 9).

8·6 Stability of Small Droplets: The Kelvin Effect

The vapor pressure of a liquid is determined by the energy necessary to separate a molecule from the attractive force exerted by its near neighbors and bring it into the gas phase. In the case of a small droplet, there are fewer molecules in the layers adjacent to the surface than for a plane surface. As a result, it is easier for the molecules on the surface of a small drop to escape into the vapor phase, and the vapor pressure over a drop is greater than that over a plane surface.

Thermodynamic reasoning can be used to derive an expression for the equilibrium vapor pressure, p_d, of a drop of diameter d_p:

$$\ln \frac{p_d}{p_s} = \frac{4\bar{v}\sigma}{d_p RT} = \frac{4\sigma v_m}{d_p kT} \tag{8·19}$$

where p_s is the vapor pressure above a flat surface and σ is the surface tension. The molar volume of the liquid, $\bar{v}$, is related to the molecular volume (volume per molecule of liquid) by $v_m = \bar{v}/N_{av}$ where N_{av} is Avogadro's number. Equation (8·19) is usually called the Kelvin relationship.

The percentage increase in vapor pressure as a function of drop diameter is shown in Fig. 8·7 for several different liquids. In making these calculations, it is assumed that the surface tension is constant, independent of drop size. This assumption is satisfactory for relatively large droplets of pure substances, but breaks down as the number of molecules composing the droplet becomes very small.

The Kelvin effect appears often in the field of aerosol growth and condensation. Despite its importance, direct experimental verification is difficult, and few such tests have been made (LaMer and Gruen, 1952).

The Kelvin effect sets a lower limit on the particle size of a polydisperse aerosol that can serve as condensation nuclei. In the atmosphere, the water vapor supersaturation rarely exceeds a few percent. From Fig. 8·7, it is clear that nuclei smaller than 0.2 μm will not be activated if the supersaturation is less than 1%. For condensing organic vapors, still larger particles must be activated at the same supersaturation.

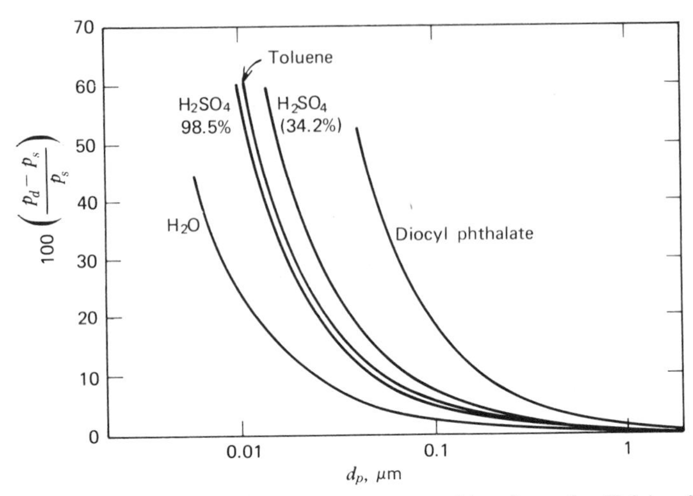

Fig. 8·7 Percentage increase in vapor pressure resulting from the Kelvin effect for various liquids. Sulfuric acid at a concentration of 34.2% has a water vapor pressure p_s corresponding to a relative humidity of 75%. Water shows the smallest effect over the size range shown because of the large value of σv_m.

8·7 Drops Containing Nonvolatile Solute

As a good approximation, the vapor pressure of the solvent over a drop containing a nonvolatile solute is given by an expression of the same form as the Kelvin relationship (8·19) (Defay and Prigogine, 1966). The partial molar volume, $\bar{v}$, is that of the solvent, and p_s is the vapor pressure of the solvent over a solution with a flat surface.

The curve of vapor pressure versus drop size is significantly different in shape from that for a droplet of pure solvent (Fig. 8·7). A droplet of pure solvent is always unstable at vapor pressures below saturation; a solution droplet may be stable because of the vapor pressure lowering of the solute. The vapor pressure of the solvent can be expressed by the relation:

$$p_s = \gamma x p_{s0} \tag{8·20}$$

where γ is the activity coefficient, x is the mole fraction of the solvent, and p_{s0} the vapor pressure of the pure solvent at the temperature of the system. There are thus two competing effects. The Kelvin effect tends to increase the vapor pressure, whereas the solute tends to reduce it. Consider a droplet containing a fixed amount of nonvolatile solute. The volume of the droplet can be expressed in terms of the partial molar volumes of the solvent and solute:

$$\frac{\pi d_p^3}{6} = n_1 \bar{v}_1 + n_2 \bar{v}_2 \tag{8·21}$$

where n_1 and n_2 represent the number of moles of solvent and solute, respectively. Rearranging (8·21) in terms of the mole fraction of solvent

$$\frac{1}{x_1} = 1 + \frac{n_2}{n_1} = 1 + \frac{n_2 \bar{v}_1}{\pi d_p^3/6 - n_2 \bar{v}_2} \tag{8·22}$$

Substituting (8·20) and (8·22) in (8·19), the result is

$$\ln \frac{p_d}{p_{s0}} = \frac{4\sigma \bar{v}_1}{d_p RT} + \ln \gamma_1 - \ln \left[1 + \frac{n_2 \bar{v}_1}{\pi d_p^3/6 - n_2 \bar{v}_2} \right]$$

Since the number of moles of solute is fixed, this expression relates the saturation ratio to particle size at equilibrium.

In the case of a dilute, ideal solution of a non-surface active solute, $\gamma_1 = 1$ and $n_2 \bar{v}_2 \ll \pi d_p^3/6$. Expanding the last term on the right-hand side of (8·4) and keeping only the first term, the result is

$$\ln \frac{p_d}{p_{s0}} = \frac{4\sigma \bar{v}_1}{d_p RT} - \frac{6 n_2 \bar{v}_1}{\pi d_p^3} \tag{8·23}$$

Since the solution is ideal, $\bar{v}_1 = v_{10}$. As a good approximation, it can be assumed that the surface tension is independent of concentration. The first term on the right corresponds to the Kelvin effect for the pure solvent. The second is the contribution resulting from the vapor pressure lowering of the solute. For small droplets, the second term dominates giving the lower branch of the curve shown in Fig. 8·8. For large values of d_p, the first term of (8·23), corresponding to the Kevin effect, is of controlling importance.

This analysis can be applied to a small dry salt particle exposed to increasing relative humidity. The particle remains solid until, if it is hygroscopic, a characteristic relative humidity less than 100% at which it absorbs water and dissolves forming a saturated solution. The relative humidities at which this occurs for bulk samples of various salts are shown in Table 8·3. These values will vary with crystal size because of the Kelvin effect. For sodium chloride, solution takes place at a relative humidity of 75% at which the diameter about doubles. With increasing

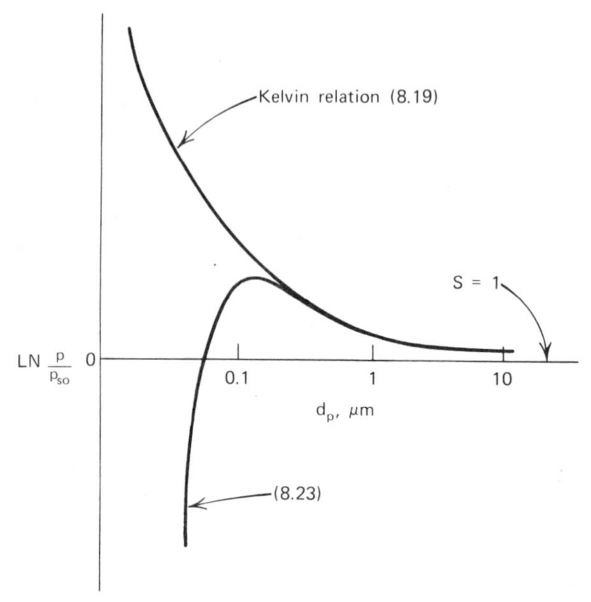

Fig. 8·8 Equilibrium vapor pressure curves for droplets composed of solvent alone (Kelvin relation) and of a solvent with a fixed mass of non-volatile solute.

TABLE 8·3
RELATIVE HUMIDITY AND CONCENTRATION FOR SATURATED SOLUTIONS AT 20°C[a]

Salt	Relative Humidity (%)	Solubility (g/100 g H_2O)
$(NH_4)_2SO_4$	81	75.4
NaCl	75.7	36
NH_4NO_3	62 (25°C)	192
$CaCl_2 \cdot 6H_2O$	32	74.5

[a] Based on Lange's Handbook of Chemistry (1973) and Stokes and Robinson (1949).

relative humidity, the equilibrium relationship between drop size and vapor pressure is determined by the interaction of the Kelvin effect and vapor pressure lowering.

As humidity is *decreased* in the range below 100%, the sodium chloride droplet shrinks following the path, in reverse, of increasing humidity. However, instead of crystallizing at 75%, the droplet evaporates while remaining as a supersaturated solution until a humidity of about 40% at which crystallization does take place. Droplet diameter changes relatively little. The failure to crystallize probably

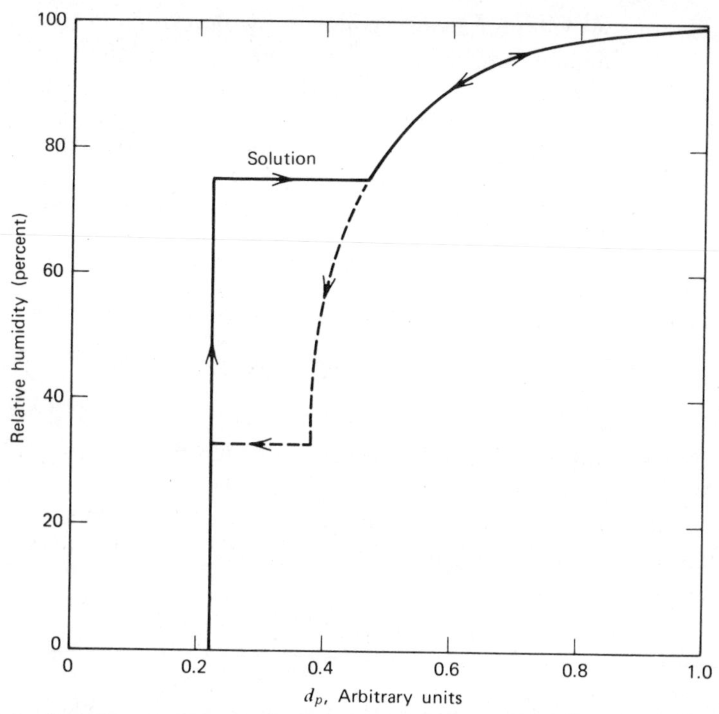

Fig. 8·9 Variation in particle (droplet) size with relative humidity. The solid curve shows the effect of *increasing* humidity on a salt crystal. At a relative humidity of about 75%, the crystal absorbs water and goes into solution; the droplet then continues to grow with increasing humidity. As humidity *decreases*, below 100%, the evaporation curve initially follows the condensation curve. As shown by the dashed curve, however, the droplet does not crystallize at a relative humidity of 75% but remains supersaturated until a much lower humidity (hysteresis effect) (Junge, 1963).

results from a lack of crystallization nuclei in the solution. Hence there is a hysteresis effect for small salt crystals exposed to varying relative humidities as shown in Fig. 8·9.

The size of the droplet formed when a salt crystal dissolves depends on the concentration of the saturated solution, that is, the solubility. Salts that absorb much water form dilute solutions and relatively large droplets. This is the case for sodium chloride (and other sodium salts) as shown in Table 8·3. Ammonium salts tend to form more concentrated solutions corresponding to smaller droplets for the same mass concentration.

Example. Taking the Kelvin effect into account, determine the percentage increase in the sulfuric acid concentration of an 0.05-μm diameter aqueous solution droplet compared with a solution with a flat surface. The temperature is 25°C and the relative humidity 40%. Assume the sulfuric acid is nonvolatile. This particle size falls within the size range of sulfuric acid droplets emitted by the catalytic converters with which automobiles have been equipped in recent years. Other data:

$$\sigma = 72 \text{ ergs/cm}^2$$

$$R = 8.3 \times 10^7 \text{ ergs/mole}^\circ\text{K}$$

$$\bar{v} = 18 \text{ cm}^3/\text{mole}$$

$$\text{M.W. of } H_2SO_4 = 98$$

SOLUTION. From Table 8·2, the vapor pressure of water, p_{s0}, is 23.8 mm Hg. At a relative humidity of 40%, $p=0.4$ (23.8)=9.52 mm. By linear interpolation in the table, this corresponds to a 47.9% solution of sulfuric acid with infinite radius of curvature (flat surface). By (8·19)

$$p_s = p_d \exp\left(-\frac{4\sigma\bar{v}}{d_p RT}\right)$$

where p_s is the vapor pressure over a flat surface of the same composition as a droplet with vapor pressure p_d. Now $p_d = 9.52$ mm and substituting the data given previously, $p_s = 9.52(0.96) = 9.14$ mm. By interpolation in Table 8·2, this corresponds to a droplet with a composition of 48.6% sulfuric acid. Thus the percentage increase in composition is $[(48.6-47.9)/47.9]100 \approx 1.5\%$.

8·8 Molecular Clusters

Even in a thermodynamically stable system, such as an unsaturated vapor, attractive forces between the molecules lead to the formation of molecular clusters. These are the London dispersion or van der Waals forces responsible for the deviations from the ideal gas laws.

There is now a great deal of experimental evidence for the existence of clusters in vapors. For example, Miller and Kusch (1956) reported experiments on the velocity distribution of the molecules in a beam produced as the vapor effused through a small slit in a source of heated alkali halide vapor. The measured velocity spectrum could be interpreted as the sum of the distributions of the individual molecular components, including monomers, dimers and in some gases higher polymers. The relative abundance of the different polymeric species was determined from the experimental velocity distributions. (Fig. 8·10).

Concentrations of the various cluster species have been measured directly by crossing an electron beam with the molecular beam extracted from an argon or carbon dioxide jet, and using a mass spectrometer to measure the cluster distribution. The results of such measurements for an argon jet are shown in Fig. 8·11. Other such studies are reviewed by Andres (1969).

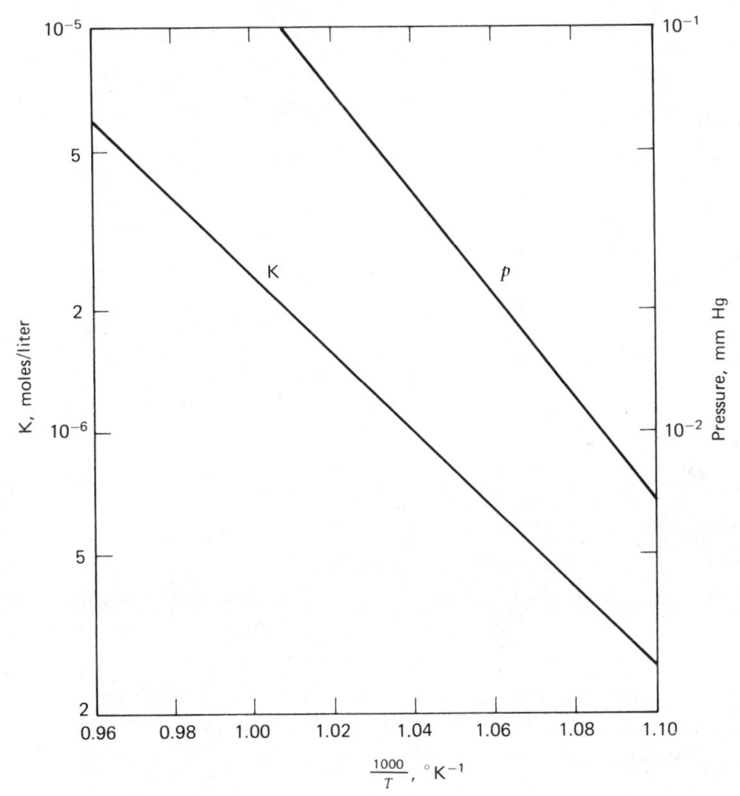

Fig. 8·10 The vapor pressure of NaCl and the equilibrium constant K for the reaction

$$[NaCl]_2 \rightleftharpoons 2NaCl$$

$$\text{with } K = [NaCl]^2/[NaCl]_2$$

Over this temperature range, 931 to 1037° K, between 25 and 35% of the total NaCl was present as dimer, $[NaCl]_2$. The data are those of Miller and Kusch (1956).

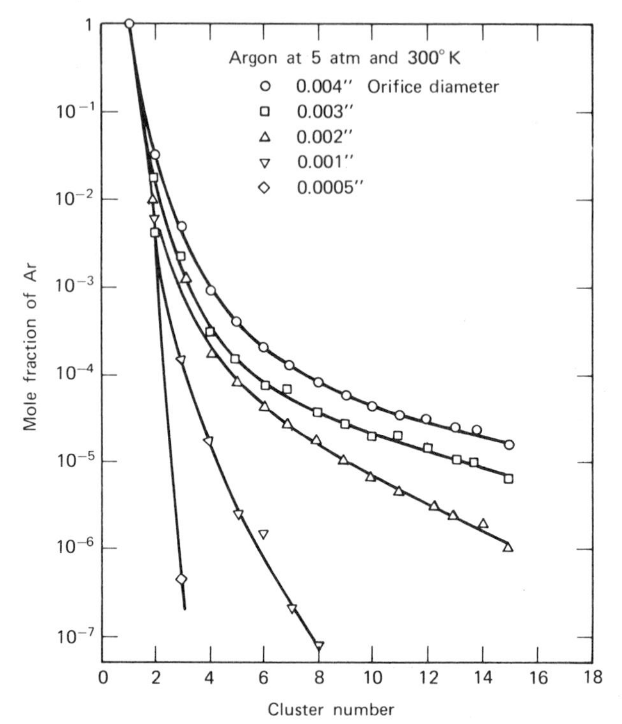

Fig. 8·11 Cluster size distributions in a free jet of argon for orifices of varying diameter (Milne and Green, 1967).

In the absence of foreign nuclei, such as smoke and dust particles, these molecular clusters serve as the nuclei of condensation in supersaturated gases. An approximate theory for the equilibrium size distribution ($S < 1$) is discussed in the next section. The theory is based on the Kelvin relationship and is not applicable to very small clusters since the surface tension concept is no longer applicable. For this reason and because of experimental difficulties, satisfactory agreement between theory and experiment has not been obtained.

8·9 Equilibrium Size Distribution of Clusters

Consider a system composed of single molecules of a condensable vapor (monomers), clusters of molecules of a condensable vapor distributed in size, and an inert carrier gas such as air. When the saturation ratio is less than unity, no net growth occurs, and the rates of formation and decay of clusters of any size are equal. This statement can be written using the formalism for a chemical reaction:

$$A_{g-1} + A_1 \rightleftarrows A_g \tag{8·24}$$

where A_g is a cluster containing g molecules and A_1 is a monomer molecule. The rate of formation of A_g by condensation of monomer on A_{g-1} is equal to the rate of loss of A_g by evaporation. The equilibrium relationship can also be written as follows:

$$\beta s_{g-1} n_{g-1} = \alpha_g s_g n_g \tag{8·25}$$

where n_g is the concentration of clusters containing g molecules. The flux of monomers (molecules per unit time per unit area) condensing on clusters of class $g-1$ is β and s_{g-1} is the effective area for condensation of the clusters of this class. The evaporative flux from class g is α_g and the effective area for evaporation is s_g. As an approximation, it can be assumed that $s_{g-1} \approx s_g$. The flux of condensing monomers (molecules/cm^2 sec) is assumed to be given by the following relation obtained from the kinetic theory of gases:

$$\beta = \frac{p_1}{(2\pi mkT)^{1/2}}$$

where p_1 is the monomer partial pressure and m is the molecular mass. It is assumed that all molecules that strike the surface of the nucleus stick. There is an evaporative flux, however, given by the Kelvin relation for the vapor pressure above a curved surface:

$$\alpha_g = \frac{p_s}{(2\pi mkT)^{1/2}} \exp\left[\frac{4\sigma v_m}{d_p kT}\right] \tag{8·26}$$

where p_s is the vapor pressure above a plane surface of the liquid, σ, the surface tension, and v_m the molecular volume of the liquid. Substituting in (8·25), the result is

$$\frac{n_{g-1}}{n_g} = \frac{1}{S} \exp\left[\frac{2\sigma v_m\left(\frac{4}{3}\pi/v_m\right)^{1/3}}{g^{1/3}kT}\right] \tag{8·27}$$

where the saturation ratio S has been set equal to p_1/p_s. Multiplying equations of this form for successively smaller values of g down to $g=2$, the result is

$$\frac{n_1}{n_2}\frac{n_2}{n_3}\cdots\frac{n_{g-2}}{n_{g-1}}\frac{n_{g-1}}{n_g}=\frac{n_1}{n_g}=\frac{1}{S^{g-1}}\exp\left[\frac{2\sigma v_m\left(\frac{4}{3}\pi/v_m\right)^{1/3}}{kT}\sum_{g=2}^{g}g^{-1/3}\right]$$

For sufficiently large values of g,

$$\sum_{g=2}^{g}g^{-1/3}\approx\int_0^g\frac{dg}{g^{1/3}}=\tfrac{3}{2}g^{2/3}\tag{8·28}$$

Hence the equilibrium distribution of nuclei (discrete spectrum) is given by the following relation:

$$n_g=n_sS^g\exp\left[\frac{-3\sigma v_m\left(\frac{4}{3}\pi/v_m\right)^{1/3}g^{2/3}}{kT}\right]\tag{8·29}$$

where $S=p_1/p_s$ and $n_s=p_s/kT$. This is one of the few cases in which an analytical solution can be obtained for the size distribution of a particulate system. For small g, the approximation (8·28) leads to considerable error in the value of n_g. If the gas is unsaturated, $S<1$ and n_g is a monotonically decreasing function of g since the exponential always decreases with g. For $S>1$, n_g passes through a minimum at a cluster size given by

$$d_p^*=\frac{4\sigma v_m}{kT\ln S}$$

determined by differentiating (8·29) with respect to g and setting the derivative equal to zero. This value of d_p is designated the critical nucleus size. Smaller nuclei tend to evaporate while larger ones grow

(Section 8·7). The number of nuclei of critical size is given by the equation:

$$n_g^* = n_1 \exp\left[\frac{-16\pi\sigma^3 v_m{}^2}{3(kT)^3(\ln S)^2}\right] \tag{8·30}$$

The shapes of the equilibrium distributions are sketched in Fig. 8·12 for unsaturated and supersaturated cases. Equilibrium over the entire distribution is unattainable in supersaturated cases. Such a state would require an infinite amount of condensable vapor.

These clusters play an important role in the growth of a dispersed phase from a continuous phase when foreign nuclei such as smoke or dust particles are not present in sufficient quantity. Their concentration and size distribution can be estimated by modifying the equilibrium theory for $S > 1$ as described in the next chapter.

When smoke and dust are present, the size distribution is composed of these foreign particles and the clusters of equilibrium theory. The mass concentration of foreign particles is normally many times greater than that of the clusters.

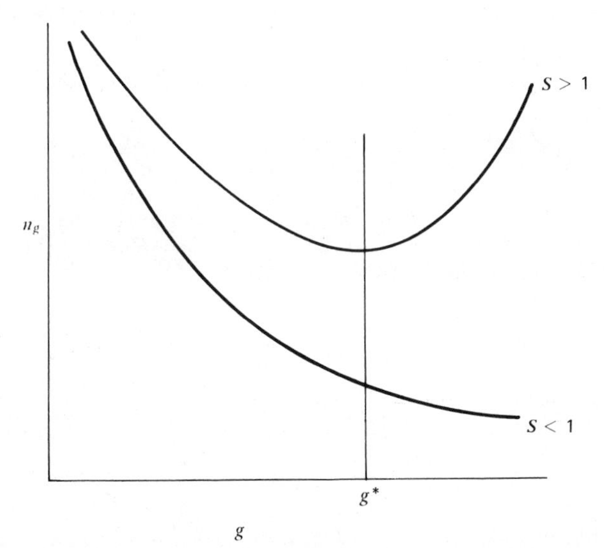

Fig. 8·12 Discrete size distribution at equilibrium for clusters formed by homogeneous nucleation. For $S > 1$, an infinite mass of material must be present in the cluster phase at equilibrium.

Problems

1. Estimate the humidity (% saturation) of the ambient air at which you would expect your breath to condense for ambient air temperatures of 10 and 20°C.

2. A gas is discharged from a stack at a temperature of 200°F and a relative humidity of 90%. In the ambient atmosphere, the temperature is 60°F and the relative humidity is 80%.

(*a*) Estimate the maximum possible mass concentration of condensed water in the plume. Express your answer in micrograms per cubic meter.

(*b*) To what temperature would the stack gases have to be heated to prevent formation of a visible plume?

3. In the special case of a Lewis number, $Le = \kappa/\rho C_p D = 1$, the path describing mixing on a concentration versus temperature diagram is a straight line. For real gases, the Lewis number is less than unity. Show on the c versus T diagram how the path for a real binary gas mixture deviates from the ideal path ($Le = 1$), starting at a given initial state.

4. Consider a gas containing 0.1 ppm SO_2 and an aerosol composed of water droplets with a total concentration of 200 $\mu g/m^3$. Calculate the amount of SO_2 in the droplet phase at equilibrium. Express your answer in micrograms per cubic meter. Do not consider oxidation to sulfate.

5. At a certain point downwind from a stack, the SO_2 concentration calculated from a stack plume dispersion model is 0.1 ppm on a volume basis. For SO_2, 1 ppm corresponds to about 2600 $\mu g/m^3$ under atmospheric conditions. To estimate worst case plume turbidity, assume all the SO_2 is converted to 0.4 μm sulfuric acid droplets in equilibrium with air at a relative humidity of 50%. Calculate the number concentration of such particles per cubic centimeter of air.

6. Referring to Section 8·5 show that a cubic algebraic expression is obtained for $[H^+]$ with coefficients that depend on the various equilibrium and Henry's law coefficients and the gas phase composition (Scott and Hobbs, 1967).

7. Estimate the increase in diameter of a crystal fragment of $(NH_4)_2SO_4$, which dissolves to form a saturated solution droplet at 20°C. The density of crystalline $(NH_4)_2SO_4$ is 1.77 g/cm^3.

8. Calculate the fraction of the total mass of water in the vapor phase present as clusters ($g \geqslant 2$) at a relative humidity of 50% and a temperature of 20°C.

9. Plot the number distribution of Problem 5, Chap. 1, together with the cluster size distribution for water vapor at a relative humidity of 50% and temperature of 20°C. Show your result on a log–log coordinate system.

References

Andres, R. P. (1969) Homogeneous Nucleation in a Vapor, in Zettlemoyer, A. C. (Ed.) *Nucleation*, Dekker, New York.

Defay, R., and Prigogine, I. (1966) *Surface Tension and Adsorption*, Wiley, New York.

Denbigh, K. G. (1966) *The Principles of Chemical Equilibrium*, Cambridge Univ. Press, Cambridge, 2nd Ed.

Hidy, G. M., and Friedlander, S. K. (1964) *AIChE J.* **10**, 115.

Junge, C. E. (1963) *Air Chemistry and Radioactivity*, Academic, New York, p. 133.

LaMer, V. K., and Gruen, R. (1952) *Trans. Faraday Soc.*, **48**, 410.

Mason, B. J. (1971) *The Physics of Clouds*, Clarendon Press, Oxford, 2nd ed.

Miller, R. C., and Kusch, P. (1956) *J. Chem. Phys.*, **11**, 860.

Milne, T. A., and Greene, F. T. (1967) *J. Chem. Phys.*, **47**, 4095.
Perry, J. H. (Ed.) (1950) *Chemical Engineers' Handbook*, McGraw-Hill, New York.
Robinson, R. A., and Stokes, R. H. (1959) *Electrolyte Solutions*, Butterworths, London, 2nd ed.
This reference includes much information in tabular form on the thermodynamic properties of electrolyte solutions.
Scott, W. D., and Hobbs, P. V. (1967) *J. Atmos. Sci.*, **24**, 54.
Stokes, R. H., and Robinson, R. A. (1949) *Ind. Eng, Chem.*, **41**, 2013.
Wegener, P. P., and Pouring, A. A. (1964) *Physics of Fluids*, **7**, 352.

C H A P T E R 9

Gas-to-Particle Conversion

Although coagulation (Chap. 7) modifies the size distribution of an aerosol, it causes no change in the mass concentration. The other important *internal* process that shapes the size distribution, gas-to-particle conversion, results in an increase in the mass concentration. In studying this process we are interested in the mechanisms by which gases are converted to particulate matter (or vice versa) and the rates at which conversion takes place.

Gas-to-particle conversion may result from homogeneous gas phase processes, or it may be controlled by processes in the particulate phase. Gas-phase processes, either physical or chemical, can produce a supersaturated state which then collapses by aerosol formation. Physical processes include adiabatic expansion or mixing with cool air—discussed in the last chapter—or radiative or conductive cooling.

Gas-phase chemical reactions such as the oxidation of SO_2 to sulfuric acid or the reaction of ozone with certain olefins may also generate condensable products. Even if the gas is not saturated with respect to one of the products, condensation may take place by the formation of droplets composed of binary solutions. The details of this process, *heteromolecular nucleation*, are beyond the scope of this text.

Once a condensable species has been formed in the gas phase, the system is in a nonequilibrium state. It may pass toward the equilibrium condition either by the generation of new nuclei or by condensation on existing particles. When condensation takes place on existing particles, self-nucleation is suppressed. This is important so far as air pollution effects are concerned, and a criterion is derived in this chapter for determining when self-nucleation occurs in the presence of foreign particles.

Alternatively, molecules from the gas may react on the particle surface or in a droplet. The process can be considered to consist of two steps in series. If the gas-to-particle transport step is rapid compared with the conversion step, the rate of particle growth will be controlled by the rate of chemical conversion in the particulate phase. This is probably the case for SO_2 conversion in stack plumes when solar radiation is not strong. Table 9·1 summarizes the various gas-to-particle conversion processes discussed in this chapter.

TABLE 9·1

EXAMPLES OF GAS-TO-PARTICLE CONVERSION MECHANISMS

- I. Homogeneous nucleation
  - A. Physical processes producing supersaturation
    1. Adiabatic expansion
    2. Mixing
    3. Conductive cooling
    4. Radiative cooling
  - B. Gas phase chemical reaction
    1. Single condensable species (classical theory)
    2. Multicomponent condensation (heteromolecular theory)
- II. Heterogeneous condensation
  - A. Transport limited
    1. Diffusion, $d_p \gg l$
    2. Molecular bombardment, $d_p \ll l$
  - B. Surface-controlled chemical reaction
  - C. Particulate phase-controlled chemical reaction

Measurements of the change in the size distribution function with time can be used to determine the form of particle growth laws. Inferences can then be drawn concerning the mechanism of growth.

9·1 Condensation by Adiabatic Expansion: The Experiments of C. T. R. Wilson

Cloud chamber experiments of the type carried out by Wilson at the end of the nineteenth century (summarized in his Nobel Lecture, 1927) demonstrate the nature of the condensation process at various saturation ratios with and without foreign nuclei. The air in a chamber is first saturated with water vapor. By rapid expansion of the chamber contents, both pressure and temperature fall carrying the system into a supersaturated state (Fig. 8·1). At first, condensation takes place on small particles initially present in the air. Concentrations of such particles in urban atmospheres range from 10^4 to 10^5 cm^{-3}. By repeatedly expanding the chamber contents and allowing the drops to settle, the vapor–air mixture can be cleared of these particles.

With the clean system, no aerosol forms unless the expansion exceeds a limit corresponding to a saturation ratio of about four. At this critical value, a shower of drops forms and falls. The number of drops in the shower remains about the same no matter how often the expansion process is repeated, indicating that these condensation nuclei are regenerated.

Further experiments show a second critical expansion ratio corresponding to a saturation ratio of about eight. At higher saturation ratios, dense clouds of fine drops form, the number increasing with the supersaturation. The number of drops produced between the two critical values of the saturation ratio is small compared with the number produced above the second limit.

Wilson interpreted these results in the following way: The nuclei that act between the critical saturation limits are air ions normally present in a concentration of about $1000/cm^3$. We know now that these result largely from cosmic rays and the decay of radioactive gases emitted by the soil. Wilson supported this interpretation by inducing condensation at saturation ratios between the saturation limits by exposing the chamber to x-rays that produced large numbers of air ions. Wilson proposed that the vapor molecules themselves serve as condensation nuclei when the second limit is exceeded, leading to the formation of very high concentrations of very small particles.

The original experiments were carried out with water vapor. Similar results were found with other condensable vapors but the value of the critical saturation ratio changed with the nature of the vapor.

Wilson used the droplet tracks generated in the cloud chamber at the lower condensation limit to determine the energy of atomic and subatomic species. Other workers subsequently became interested in the phenomena occurring at the upper condensation limit when the molecules themselves were believed to be serving as condensation nuclei.

9·2 Kinetics of Homogeneous Nucleation

According to the Kelvin relation (8·19), the higher the saturation ratio, the smaller the radius of the droplet that can serve as a stable nucleus for condensation. However, calculations based on the observations of Wilson and subsequent measurements by many other experimenters indicate values of d_p^* many times greater than the diameter of a single water molecule, about 2.8 Å. How, then, does condensation take place in systems that have been freed from condensation nuclei?

As shown in the previous chapter, molecular clusters are always present even in an unsaturated gas. When a system becomes supersaturated, these clusters increase in concentration and pass through the critical size d_p^* by attachment of single molecules. The formation of stable nuclei relieves the supersaturation in the gas. Since condensation nuclei are generated by the vapor itself, the process is known as *homogeneous nucleation* or *self-nucleation*.

When condensation occurs, the equilibrium relation $A_{g-1}+A_1 \rightleftarrows A_g$ no longer holds. With the nonequilibrium cluster distribution function now given by n_g, the difference

$$I_g = n_{g-1}s_{g-1}\beta - n_g s_g \alpha_g \tag{9·1}$$

is equal to the excess rate at which nuclei pass from the size $g-1$ to g by condensation over the rate of passage from g to $g-1$ by evaporation. The quantity, I_g, known as the *droplet current*, has c.g.s. dimensions of $\text{cm}^{-3}\ \text{sec}^{-1}$.

Eliminating α_g by substituting (8·25) for the equilibrium distribution n_g^e, the result is

$$I_g = n_{g-1}^e s_{g-1}\beta\left[\frac{n_{g-1}}{n_{g-1}^e} - \frac{n_g}{n_g^e}\right] \tag{9·2}$$

The rate of change of the number of clusters in a given class is given by

$$\frac{\partial n_g}{\partial t} = \underset{\text{condensation}}{\text{in from } g-1} + \underset{\text{evaporation}}{\text{in from } g+1} - \underset{\text{evaporation}}{\text{out from } g} - \underset{\text{condensation}}{\text{out from } g}$$

$$= I_g - I_{g+1} \tag{9·3}$$

For sufficiently large values of $g(g>10$ say) we can treat the variables appearing in these equations as continuous functions of g, and replace the difference equations by differential equations that are easier to handle analytically. In this way, (9·2) becomes

$$I(g) = -\beta n^e s\frac{\partial(n/n^e)}{\partial g} \tag{9·4a}$$

$$= -\beta s\frac{\partial n}{\partial g} + \beta sn\frac{\partial n^e}{n^e\,\partial g} \tag{9·4b}$$

Substitution of (8·29) for the equilibrium distribution n^e in (9·4*b*) gives

$$I(g) = -\beta s\frac{\partial n}{\partial g} - \frac{\beta sn}{kT}\frac{\partial \Delta\Phi}{\partial g} \tag{9·5}$$

where $\Delta\Phi/kT = (36\pi)^{1/3}\sigma v_m^{2/3}g^{2/3}/kT - g\ln S$. The first term of the right-hand side of $(9\cdot5)$ is proportional to the concentration gradient in g space. It can be interpreted as a diffusion of clusters through g (or v) space with βs playing the part of a spatially dependent diffusion coefficient. The second term represents the transport of droplets through g space under the influence of an external force field corresponding to a potential energy $\Delta\Phi$. The migration velocity is given by $-(\beta s/kT)(\partial\Delta\Phi/\partial g)$.

The kinetic equation $(9\cdot3)$ for the continuous distribution function becomes

$$\frac{\partial n}{\partial t} = -\frac{\partial I}{\partial g} \tag{9·6}$$

where $I_g - I_{g+1}$ has been replaced by $-\partial I/\partial g$. This expression represents a continuity (or Liouville) relation for particle transport through the g space.

An approximate solution to $(9\cdot6)$ with $(9\cdot5)$ can be obtained by making the following assumptions:

1. For $g \to 0$, the nonequilibrium distribution function approaches the equilibrium distribution, that is,

$$\frac{n}{n^e} \to 1 \text{ as } g \to 0$$

This is equivalent to the assumption that the time for the lower end of the spectrum to reach the equilibrium distribution is much shorter than the upper end.

2. For very large values of g, $n/n^e \to 0$ since the nuclei concentration for the nonequilibrium distribution is much smaller than for the equilibrium distribution.

3. A quasi-steady state exists such that as many nuclei enter a size range as leave. This means that the droplet current is independent of the size, that is,

$$\frac{\partial I}{\partial g} = -\frac{\partial n}{\partial t} \approx 0$$

or

$$I(g) = \text{const} = I$$

Integrating $(9\cdot4a)$ between limits with these assumptions, the result is

$$\int_1^0 d\left(\frac{n}{n^e}\right) = \frac{-I}{\beta}\int_0^\infty \frac{dg}{n^e s}$$

or

$$I=\beta/\int_0^{\infty}\frac{dg}{n^e s} \tag{9·7}$$

To calculate I, the droplet current, it is necessary to evaluate the integral in the denominator. The equilibrium distribution is given by (8·29), which can be written in the form:

$$\frac{1}{n^e}=\frac{1}{n_1}\exp\left[-\frac{\Delta\Phi}{kT}\right] \tag{9·8}$$

Now $1/n^e$ has a sharp maximum $1/n^{e*}$ when $g=g^*$ so in this region, we can replace $\Delta\Phi$ by its expansion about $g=g^*$:

$$\Delta\Phi\approx\Delta\Phi_{max}+\frac{1}{2}\left(\frac{\partial^2\Delta\Phi}{\partial g^2}\right)_{g=g^*}(g-g^*)^2$$

$$=\Delta\Phi_{max}-\frac{\pi}{9g^{*2}}\sigma d_p^{*2}(g-g^*)^2$$

Substituting in (9·8), the result is

$$\frac{1}{n^e}=\frac{1}{n_1}\exp\left[\frac{\Delta\Phi_{max}}{kT}\right]\exp\left[-\frac{\pi\sigma d_p^{*2}}{9kTg^{*2}}(g-g^*)^2\right] \tag{9·9}$$

Substituting (9·9) in the integral in (9·7), the result is

$$\int_{g\to 0}^{\infty}\frac{dg}{n^e s}=\frac{1}{s^* n^*}\int_{g\to 0}^{\infty}\exp\left[-\frac{\gamma}{2}(g-g^*)^2\right]dg$$

where

$$\gamma=\frac{2\pi\sigma d_p^{*2}}{9kTg^{*2}}$$

and

$$n^*=n_1\exp\left[-\frac{\Delta\Phi_{max}}{kT}\right]$$

As a good approximation, the lower limit can be changed to $-\infty$ with the following result:

$$\int_{-\infty}^{\infty} \exp\left[-\frac{\gamma}{2}(g-g^*)^2\right]dg = \int_{-\infty}^{\infty} \exp\left(-\frac{\gamma}{2}z^2\right)dz = \left[\frac{2\pi}{\gamma}\right]^{1/2}$$

Then the droplet current is given by

$$I = \beta \Big/ \int_0^{\infty} \frac{dg}{ns} = \frac{p_1 \pi d_p^{*2} n_1 \exp(-\Delta\Phi_{max}/kT)}{(2\pi mkT)^{1/2}[2\pi/\gamma]^{1/2}}$$

$$= 2\left[\frac{p_1}{(2\pi mkT)^{1/2}}\right]\left(n_1 v_m{}^{2/3}\right)\left[\frac{\sigma v_m{}^{2/3}}{kT}\right]^{1/2} \exp\left[-\frac{16\pi\sigma^3 v_m^2}{3(kT)^3(\ln S)^2}\right] \tag{9·10}$$

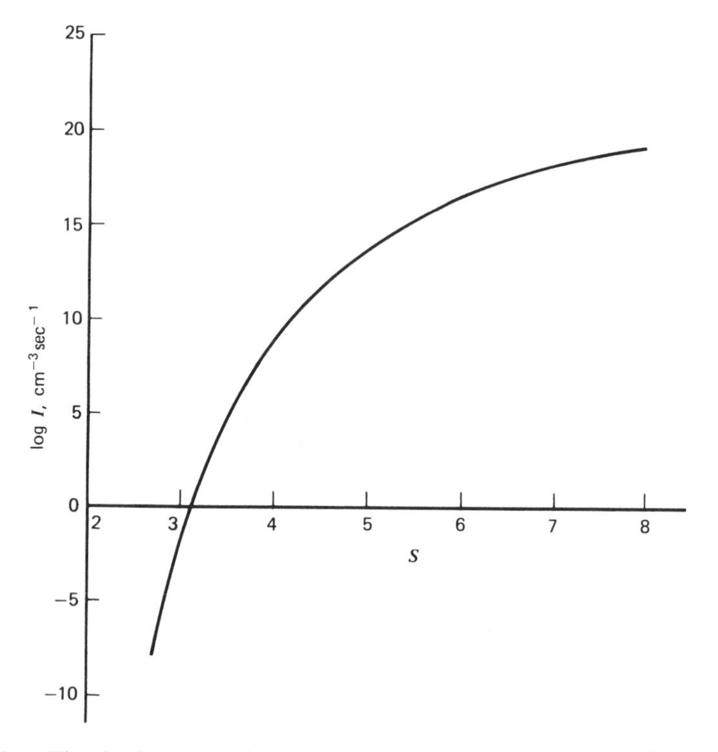

Fig. 9·1 The droplet current for supersaturated water vapor at $T = 300°K$ calculated from (9·10). The critical saturation ratio, corresponding to $I = 1$ cm^{-3} sec, is about 3.1 (see, also, Fig. 9·4).

The first term in brackets is the monomer flux (molecules per unit area per unit time) and the second is proportional to the monomer surface area per unit volume of gas. Their product has the same dimensions as I, number per unit volume per unit time. The group $\sigma v_m^{2/3}/kT$ is dimensionless. The droplet current calculated from (9·10) is shown in Fig. 9·1 for water vapor at a temperature of 300°K. Order of magnitude changes in I result from small changes in S, primarily because of the dependence on $\ln S$ in the argument of the exponential function. *Although a supersaturated vapor is always unstable*, the rate of generation of stable nuclei is negligible for small values of S. When $I = 1$ drop/cm^3 sec, particle formation can be conveniently observed experimentally. The corresponding value of S is designated the critical saturation ratio, S_{crit}.

9·3 Experimental Test of the Theory

It is difficult to carry out experimental studies that will verify the theory. Measurements of the nuclei size spectrum would constitute a sensitive check, but instruments capable of measurement in the 10 to 100 Å size range have not been available. Most experimental tests have involved measurements of the saturation ratio at which condensation occurs using an expansion (Wilson) cloud chamber. Data collected with the chamber are difficult to interpret because of the unsteady nature of the expansion process. These investigations are reviewed by Mason (1971). Reversible adiabatic expansion can be carried out as a steady process (Chap. 8), and such systems have been used to study nucleation (Wegener and Pouring, 1964).

The diffusion cloud chamber is a particularly attractive experimental system for the study of nucleation kinetics; it is compact and produces a well-defined, steady supersaturation field. The chamber is cylindrical in shape, perhaps 30 cm in diameter and 4 cm high. A heated pool of liquid at the bottom of the chamber evaporates into a stationary carrier gas, usually hydrogen or helium. The vapor diffuses to the top of the chamber, which cools, condenses, and drains back into the pool at the bottom. Since the vapor is denser than the carrier gas, the density is greatest at the bottom of the chamber, and the system is stable with respect to convection. Both diffusion and heat transfer are one dimensional, with transport occurring from the bottom to the top of the chamber. At some position in the chamber, the temperature and vapor concentrations reach levels corresponding to supersaturation. The variation in the properties of the system are calculated by a computer solution of the one dimensional equations for heat conduction and mass diffusion (Fig. 9·2). The saturation ratio is calculated from the computed local partial pressure and vapor pressure.

The goal of an experiment is to set up a "critical" chamber state, that is, a state that just produces nucleation at some height in the chamber where the vapor is critically supersaturated and droplets are visible. This occurs when the temperature difference across the chamber has been increased to the point where a rain of drops forms at an approximately constant height. Drop formation in this way must be distinguished from condensation on ions generated by cosmic rays passing through the chamber. An electrical field is applied to sweep out such ions which appear as a trail of drops.

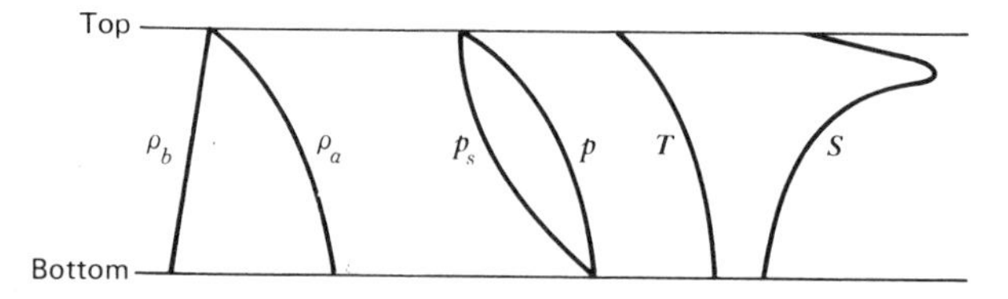

Fig. 9·2 Variation with height of the properties of a mixture in the diffusion cloud chamber. Shown are the mass densities of the carrier gas, ρ_b, and the vapor, ρ_a; the equilibrium vapor pressure, p_s; the partial pressure of the vapor, p; the temperature, T; and the saturation ratio, S. The highest temperature, vapor pressure, and gas density are at the chamber bottom, above the heated pool. The distributions with respect to chamber height are calculated by integrating expressions for the steady state fluxes of heat and mass through the chamber.

For each critical chamber state, the distribution of the saturation ratio and temperature can be calculated as shown in Fig. 9·2. The set of curves for the critical chamber states based on measurements with toluene is shown in Fig. 9·3. The experimental saturation ratio passes through a maximum with respect to temperature in the chamber. Condensation occurs not at the peak supersaturation but at a value on the high temperature side because the critical supersaturation decreases with increasing temperature. Hence the family of experimental curves should be tangent to the theoretical curve.

Good agreement between theory and experiment has been obtained in this way for toluene (Fig. 9·3) and other organic compounds. For water, agreement is poorer (Heist and Reiss, 1973). Such experiments offer the most convincing support yet developed for the theory of homogeneous nucleation.

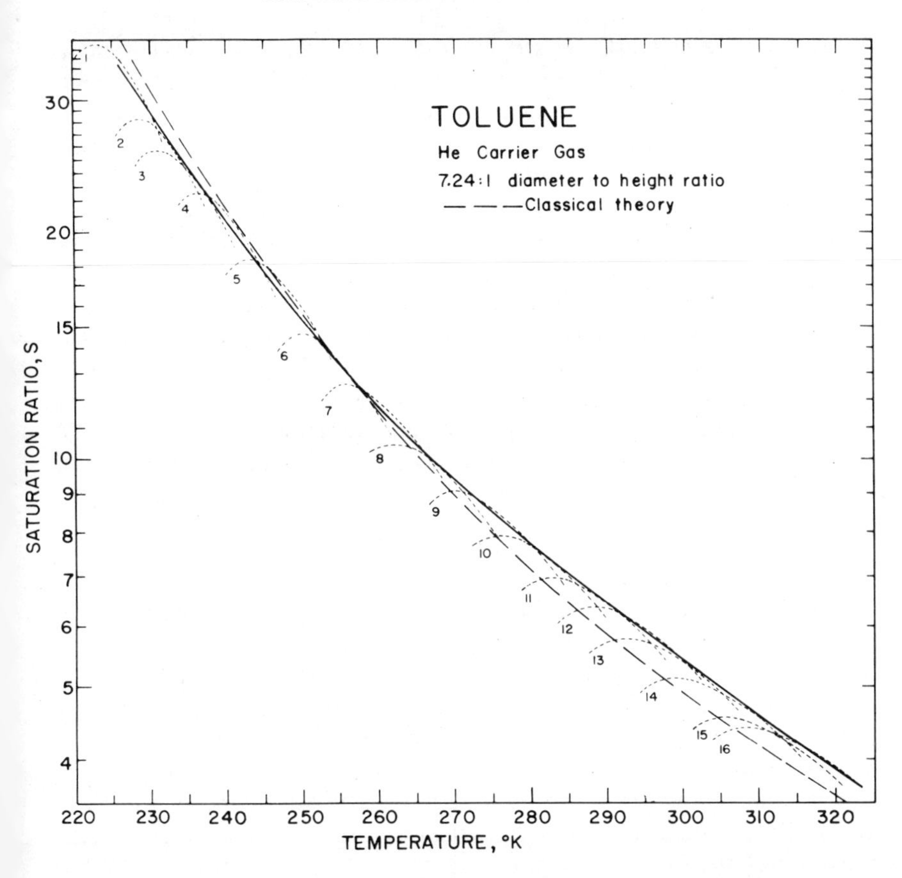

Fig. 9·3 Comparison of theoretical and experimental critical saturation ratios for toluene. The dashed line is the theoretical prediction ((9·10) with $I = 1\ \mathrm{cm}^{-3}\mathrm{sec}^{-1}$, and the solid line is the experimental result, the envelope to the numbered individual chamber state curves. (Katz, Scoppa, Kumar and Mirabel, 1975). Data for many other *n*-alkyl benzenes are given in this reference.

9·4 Chemically Generated Aerosols: Gas Phase Reaction

In the experiments discussed in Section 9·3, the supersaturated state was produced physically for a single condensable substance. Gas phase chemical reactions may also lead to the formation of condensable species, and several are usually present simultaneously. Of special air pollution interest are reactions involving SO_2, NO–NO_2, and certain organic compounds.

The rate of SO_2 conversion to particulate sulfur in the Los Angeles atmosphere, heavily polluted by the organic vapors and nitrogen oxides in automobile exhaust, ranged from 1 to 13% hr^{-1} in measurements reported by Roberts and Friedlander (1975). The SO_2 probably reacts with oxidizing agents generated by ozone–olefin reactions. The details of the mechanism including the nature of the oxidizing species remain uncertain (Roberts and Friedlander, 1976).

Automobiles emit nitrogen oxides and organic compounds that are converted in the atmosphere to particulate phase nitrate and organic compounds. These contribute a major share of the total particulate burden in regions where photochemical air pollution is important.

Particulate *nitrates* may be formed in the atmosphere by the gas phase oxidation of NO to NO_2 and then to nitric acid. Nitrogen oxides are produced in combustion processes and emitted into the atmosphere in large quantities by power plants and automobiles. Calvert (1973) considers oxidation by the hydroxyl radical the principal mode of formation:

$$\mathrm{HO} + \mathrm{NO_2} + M \rightarrow \mathrm{HONO_2} + M$$

where M represents a third body. The hydroxyl radical is probably a product of reactions involving oxygenated organic compounds, such as aldehydes, or carbon monoxide. The aerosol nitrate is usually present as the ammonium salt in polluted atmospheres. This form may result from gas phase reaction of nitric acid with ammonia or from reaction in the droplet phase.

Particulate *organics* are produced by gas phase ozone–olefin reactions. Rate data for these reactions can often be correlated by an equation of the form:

$$\frac{d[\mathrm{Ol}]}{dt} = -k[\mathrm{Ol}][\mathrm{O_3}]$$

Values of the bimolecular rate coefficient k for selected olefins are listed in Table 9·2. Those showing negligible conversion to aerosol have been listed for purposes of comparison. The cyclic olefins and α-pinene react rapidly forming large amounts of condensable species.

The cyclic olefins are present both in gasoline and in automobile exhaust and probably contribute significantly to the organics present in the photochemical smog aerosol. Forests produce α-pinene in significant quantities, and it is believed that aerosols generated from this material constitute a major portion of the total global aerosol burden.

If only a single condensable species is formed by chemical reaction, it may condense by homogeneous nucleation in the absence of existing particles. In other cases, two or more condensable species may be

TABLE 9·2
OZONE–OLEFIN REACTIONS LEADING TO CONDENSABLE SPECIES
(Grosjean, 1975)

Hydrocarbon	k_{O_3}[a]	Major Products	Product Volatility	Gas-to-Particle[b] Conversion, % as C
Ethylene	1.9	Formaldehyde		0
Propylene	13.0	Acetaldehyde, acetic acid	Very high	0
Tetramethyl ethylene	1510	Acetone		0
1-Hexene	11.1	Pentenoic acid	Medium	0–5
1-Heptene	~11	Hexanoic acid		
Cyclopentene	813	Difunctional compounds	Very low	15–30
Cyclohexene	169	(diacids [$COOH-(CH_2)_n-COOH$], acid nitrates, etc.)	Vapor pressure $\sim 10^{-6}$ mm Hg	
α-Pinene	330			>30

[a] k in 10^{-18} cm^3 $molecule^{-1}$ sec^{-1} from S. M. Japar, C. H. Wu, and H. Niki, (1974) *J. Phys. Chem.*, **78**, 2318 and *Environmental Letters* (1974), **74**, 245.
[b] Unpublished data, Grosjean (1975).

present simultaneously in the gas. If these are strongly interacting, nucleation can take place at partial pressures much lower than those required for the nucleation of the pure vapors. The most important example is the water vapor–sulfuric acid system. The details of this process, known as *heteromolecular* nucleation are beyond the scope of this text (Reiss, 1950). In general, the process of aerosol formation from chemically reacting gases, while important, is only poorly understood. (See, also, Section 9·10.)

9·5 Heterogeneous Condensation

When large concentrations of particles are present and the supersaturation is low, condensation takes place on the existing particles without formation of new nuclei. We call this process *heterogeneous condensation*. Cloud formation in the atmosphere takes place in this way since supersaturations are usually less than a few percent and nuclei concentrations are high.

The rate of heterogeneous condensation depends on the exchange of matter and heat between a particle and the continuous phase. The extreme cases of a particle much larger or much smaller than the mean free path of the suspending gas are easy to analyze. In the continuum range, ($d_p \gg l$) diffusion theory can be used to calculate the transport rate. For a single sphere in an infinite medium, the steady-state equation of diffusion in spherical coordinates takes the form:

$$\frac{\partial n}{\partial t} = \frac{D\partial r^2(\partial n/\partial r)}{r^2 \partial r} = 0$$

where n is the molecular concentration of the condensing species and D is its coefficient of diffusion. The solution that satisfies the boundary conditions—$n = n_d$, the concentration in equilibrium with the surface at $r = d_p/2$, and $n = n_1$ at $r = \infty$—is

$$\frac{n - n_d}{n_1 - n_d} = 1 - \frac{d_p}{2r}$$

The rate of diffusional condensation is given by

$$F = D\left(\frac{\partial n}{\partial r}\right)_{r = d_p/2} \pi d_p^{\,2} = 2\pi d_p D\,(n_1 - n_d)$$

$$= \frac{2\pi d_p D\,(p_1 - p_d)}{kT} \qquad (9\cdot 11)$$

where F is the flow of molecules (number per unit time) to the surface of the particle. The surface concentration, n_d, is determined by the surface temperature and curvature. It is assumed that the condensation rate is sufficiently slow for the latent heat of condensation to be dissipated without changing droplet temperature.

For particles much smaller than the mean free path of the gas, l, the rate of condensation can be calculated from kinetic theory: The net flow of molecules ($\sec^{-1}$) at a surface of area πd_p^2 is given by

$$F = \frac{\alpha(p_1 - p_d)\pi d_p^2}{(2\pi mkT)^{1/2}} \tag{9·12}$$

Separate accomodation coefficients, α, are often introduced for the condensation, $p_1/(2\pi mkT)^{1/2}$, and evaporation, $p_d/(2\pi mkT)^{1/2}$, fluxes, but the values are assumed equal in (9·12). In general, these coefficients must be determined experimentally. An approximate interpolation formula for the entire range of mean free path has been proposed by Fuchs and Sutugin (1971):

$$F = 2\pi D d_p (n_1 - n_d)\left\{\frac{1 + Kn}{1 + 1.71\ Kn + 1.333\ Kn^2}\right\} \tag{9·13}$$

where the Knudsen number is equal to l/a_p and l is the mean free path for collision of the condensing species. When $Kn \ll 1$, (9·13) reduces to (9·11) for the continuum range. When $Kn \gg 1$, (9·13) is about 1.2 times (9·12), with $\alpha = 1$ for rigid elastic spheres.

If the Kelvin effect is important, the partial pressure driving force for condensation takes the form:

$$\begin{aligned}(p_1 - p_d) &= p_1 - p_s \exp\left[\frac{4\sigma v_m}{d_p kT}\right] \\ &= p_1 - p_s \exp\left[\frac{d_p^* \ln S}{d_p}\right]\end{aligned} \tag{9·14}$$

where p_s is the vapor pressure over a flat surface or pool of liquid and d_p^* is the critical droplet diameter. It has been assumed that the nucleus behaves like a pure drop of the condensing species. If the surface of the particle is composed of a material different from that of the condensing vapor, this result must be modified to account for surface wetting effects.

Expanding the exponential of (9·14)

$$\Delta p = p_s\left[S - 1 - \frac{d_p^*}{d_p}\right]\ln S - \frac{1}{2}\left(\frac{d_p^*}{d_p}\ln S\right)^2 - \cdots$$

where $S = p_1/p_s$. For small values of $S-1$, this takes the approximate form:

$$\Delta p = \frac{p_s}{d_p}(S-1)(d_p - d_p^*) \tag{9·15}$$

As an example, this result can be substituted in (9·11) for growth by diffusion in the continuum range:

$$F = \left(\frac{2\pi D}{kT}\right)p_s(S-1)(d_p - d_p^*) \qquad (S - 1 \ll 1)$$

The rate of condensation is proportional to the difference between the particle diameter and the critical particle diameter.

9·6 Homogeneous Nucleation: Effect of Foreign Particles

Molecules of a condensable species may self-nucleate or condense out on existing nuclei. The existing nuclei either may have been introduced into the system initially or may themselves have resulted from a previous homogeneous nucleation process. The two condensation mechanisms result in quite different size spectra. Homogeneous nucleation generates high concentrations of very small particles while the number of particles is conserved in heterogeneous condensation. It is useful to derive a criterion for determining which mechanism controls a condensation process.

We consider the case of a single species of condensable molecules generated by a chemical reaction. When the available nuclei are smaller than the mean free path of the gas, the rate of loss of condensable species from the gas phase can be calculated from the expression

$$\frac{dn_1}{dt} = -\frac{(p_1 - p_s)A}{(2\pi mkT)^{1/2}} + g_1 \tag{9·16}$$

where g_1 is the rate at which monomer molecules are generated by chemical reaction and A is the surface area of aerosol per unit volume of gas. (The coefficient α has been set equal to unity, and the Kelvin effect neglected.) Assuming constant surface area, an assumption that holds so long as the total amount of material that condenses is small by comparison with the existing aerosol material—the solution to (9·16) is given by

$$S-1=\frac{g_1(2\pi mkT)^{1/2}}{p_s A}+\left(S_0-1-\frac{g_1(2\pi mkT)^{1/2}}{p_s A}\right)\exp\left[-A\left(\frac{kT}{2\pi m}\right)^{1/2}t\right]$$

where S_0 is the saturation ratio at $t=0$. Hence the saturation ratio asymptotically approaches a constant value given by the steady-state relationship between formation and condensation:

$$S_\infty=1+\frac{g_1(2\pi mkT)^{1/2}}{p_s A} \tag{9·17}$$

This value is the maximum attainable in the system since it has been calculated by neglecting homogeneous nucleation which would provide an alternate route for condensation to take place. Provided that $S_\infty < S_{crit}$, few particles will be generated by homogeneous nucleation.

If $S_\infty > S_{crit}$, homogeneous nucleation will occur, leading to the formation of high concentrations of small particles. However, the total *mass* of material condensing out by homogeneous nucleation may be small compared with that condensing on the foreign particles. The relative amounts will depend on the concentrations of foreign particles and the saturation ratio.

9·7 Droplet Phase Reactions

The solution phase oxidation of SO_2 is probably of importance in power plant stack plumes and in cloud processes related to "acid rain." A partial equilibrium model based on Section 8·6 offers a convenient theoretical framework. The SO_2, NH_3, and CO_2 present in the gas phase are assumed to be in equilibrium with H_2SO_3, NH_4OH, and H_2CO_3, respectively, in the solution phase. Based on the equilibrium reactions in solution, the concentration of SO_3^{2-} is given by (8·15):

$$[SO_3^{2-}]=\text{const}\frac{p_{SO_2}}{[H^+]^2}$$

If the rate of oxidation is assumed proportional to the concentration of SO_3^{2-}, the rate of formation of SO_4^{2-} is given by

$$\frac{1}{v}\frac{dn_{SO_4^{2-}}}{dt} = k[SO_3^{2-}] \tag{9·18}$$

where n_{SO_4} is the number of moles of sulfate in a droplet of volume v. The pseudo-first-order rate constant, k, includes the concentration of oxygen and is affected by trace metals, including iron and manganese, in solution (Junge and Ryan, 1958). These serve as catalysts and are present in urban and industrial areas in emissions from power plants and steel mills.

The partial equilibrium SO_2 model was set up for application to dilute solutions such as cloud droplets. Application to pollution aerosols, which are much more concentrated, is at best qualitative. Many other ionic species are present and concentrations are too high for ideal solution theory to be applicable.

A general growth law for conversion controlled by droplet phase reaction can be derived as follows: Consider an aerosol composed of small droplets all of the same composition but distributed with respect to size. The same chemical reactions take place in all droplets, leading to the conversion of molecules from gas to particle phases. The process is limited by the droplet phase reactions. As fast as material is consumed by reaction in the droplet phase, it is replenished by transport from the gas. Then the fractional rate of growth of all droplets *must be the same* provided that the Kelvin effect does not intervene. The rate of chemical conversion *per unit volume of droplet* is independent of size. That this must be true becomes clear if it is considered that the rate of conversion per unit volume is the same whether the solution is present as a large volume in a beaker or dispersed as an aerosol.

The result can be expressed mathematically. The change in mass of a droplet is

$$\frac{dm}{dt} = \sum_i \frac{dm_i}{dt} \tag{9·19}$$

where m_i is the mass of species i absorbed by the droplet. If the rate of uptake is equal to the rate of conversion by chemical reaction (quasi-stationary state),

$$\frac{dm_i}{dt} = M_i\frac{dn_i}{dt} = \nu_i M_i v\left(\frac{1}{\nu_i}\frac{dn_i}{v\,dt}\right) = \nu_i M_i v r \tag{9·20}$$

where n_i is the number of moles of species i, M_i is the molecular weight, and ν_i is the stoichiometric coefficient for species i in the reaction. The reaction rate

$$r = \frac{1}{\nu_i v}\frac{dn_i}{dt} \qquad (9\cdot 21)$$

is the same for all chemical species and can often be expressed by the power law forms of chemical kinetics (Denbigh, 1966). The droplet growth law is obtained by combining (9·19) through (9·21):

$$\frac{dm}{dt} = \rho_p \frac{dv}{dt} = v\left(\sum_i M_i \nu_i\right) r \qquad (9\cdot 22)$$

where the density of the droplet ρ_p has been assumed constant. Since the composition is the same in all droplets, the rate of reaction per unit volume, r, is independent of droplet size. Thus the growth rate is proportional to droplet volume. While the derivation was carried out for a single chemical reaction, the result can easily be generalized to the case of a multireaction system.

It may seem strange that the droplet growth law is so different in form from the transport limited law. After all, the gas phase species must be transported to the droplets. Actually, both laws are obeyed. The explanation is that the reactive species are *nearly* in equilibrium in the gas and droplet phases. Their small displacement from equilibrium differs, however, *depending on droplet size*, but not sufficiently to affect the rate of reaction in solution.

9·8 Growth Laws

Growth laws are expressions for dv/dt or $d(d_p)/dt$ as functions of particle size and the appropriate chemical and physical properties of the system. Such expressions are necessary for the calculation of changes in the size distribution function with time as shown in this and the next chapter. When growth is transport limited, the rate can be determined from the expressions derived in the previous section. For the continuum range, the growth law based on (9·11) is given by

$$\frac{dv}{dt} = \frac{2\pi D d_p v_m}{kT}(p_1 - p_d) \qquad (9\cdot 23)$$

where v_m is the molecular volume of the condensing species. The effect of the moving boundary of the growing particle is neglected.

For nuclei smaller than the mean free path of the gas, the growth law, based on (9·12), is given by

$$\frac{dv}{dt} = \frac{\pi d_p^2 v_m (p_1 - p_d)}{(2\pi mkT)^{1/2}} \tag{9·24}$$

where the accomodation coefficient α has been set equal to unity.

Chemical reactions at particle surfaces may also lead to particle growth. Such reactions are likely to be important near aerosol sources where the particle surfaces are fresh and their catalytic activity high. (In the atmosphere, however, contamination probably destroys the specific catalytic activity of aerosol surfaces.) Particles will grow if the products of reaction accumulate at the surface. When reaction rates are fast compared with transport, growth laws are of the same form as the transport limited laws. When reaction rates are slow compared with transport, the concentration of the reactive species in the gas near the surface is practically the same as in the bulk of the gas, and the rate of conversion is given by:

$$\frac{dv}{dt} = \frac{\alpha p_1 \pi d_p^2 v_m}{(2\pi mkT)^{1/2}} \tag{9·25}$$

where α, the fraction of effective collisions with the surface, is usually much less than unity.

Examples of growth laws including those limited by chemical reaction in the particulate phase are summarized in Table 9·3. The dependence

TABLE 9·3
GROWTH LAWS FOR GAS-TO-PARTICLE CONVERSION

Mechanism	Growth Law, dv/dt	Equation
Diffusion ($d_p \gg l$)	$\frac{2\pi D d_p v_m (p_1 - p_d)}{kT}$	(9·23)
Molecular bombardment ($d_p \ll l$)	$\frac{\alpha \pi d_p^2 v_m (p_1 - p_d)}{(2\pi mkT)^{1/2}}$	(9·24)
Surface reaction (all sizes)	$\frac{\alpha \pi d_p^2 v_m p_1}{(2\pi mkT)^{1/2}}$ $(\alpha \ll 1)$	(9·25)
Droplet phase reaction	$\frac{\pi d_p^3}{6\rho_p}\left(\sum M_i \nu_i\right) r$	(9·22)

of dv/dt varies from d_p for diffusion in the continuum range to d_p^3 for droplet phase chemical reaction. Different forms for the growth can lead to markedly different changes in the size distribution function with time and to the distribution of chemical species with respect to size.

Example 1. Derive an expression for the variation of particle diameter with time for a particle growing by transport from the gas phase for the case $d_p \ll l$. Neglect the Kelvin effect.

SOLUTION. The growth law is given by (9·24). Substituting $v = \pi d_p^3/6$ and rearranging, the result for $p_d = p_s$ is

$$d(d_p) = \frac{2v_m(p_1 - p_s)dt}{(2\pi mkT)^{1/2}}$$

Integrating from the initial condition $d_p = d_{p0}$ at $t=0$,

$$d_p = d_{p0} + \int_0^t \frac{2v_m(p_1 - p_s)}{(2\pi mkT)^{1/2}} dt'$$

The partial pressure driving force for growth, $p_1 - p_s$, is a function of time determined by the conditions of the system. For example, in the condensation aerosol generator (Chap. 6), it is determined by the cooling rate in the chimney. None of the quantities in the second term on the right is a function of drop radius. If $d_p \gg d_{p0}$ for all values of d_{p0}, the drops will all be roughly of the same size provided that the variation of $p_1 - p_s$ with time is the same for all particles. Hence the condensation generator produces almost monodisperse aerosols even through the original nuclei are nonuniform in size. In the atmosphere or in process gases, the temperature–time histories of the various gas parcels vary. As a result, $(p_1 - p_s)$ varies and the resulting size distributions are polydisperse.

Example 2. Ammonium sulfate molecules, formed by homogeneous gas phase reactions, deposit on small suspended aqueous solution droplets. Derive an expression for the droplet growth law, assuming the solution is dilute and ideal.

SOLUTION. The water vapor pressure above a droplet containing a nonvolatile salt such as ammonium sulfate is given by (8·23):

$$\ln \frac{p_d}{p_{s0}} = \frac{4\sigma \bar{v}_{solvent}}{d_p RT} - \frac{6 n_{solute} \bar{v}_{solvent}}{\pi d_p^3}$$

where p_d is the vapor pressure of the water in equilibrium with the drop, p_{s0} is the vapor pressure of pure water at the temperature T, $\bar{v}_{solvent}$ is the partial molal volume of the solvent, and n_{solute} is the number of moles of solute in the droplet. Rearranging with $v = \pi d_p^3/6$,

$$v \ln \frac{p_d}{p_{s0}} - 4\left(\frac{\pi}{6}\right)^{1/3} \frac{\sigma \bar{v}}{RT} v^{2/3} = -n\bar{v} \tag{1}$$

where the subscripts on n and $\bar{v}$ have been dropped for convenience.

As ammonium sulfate deposits and dissolves in the droplets, water is also transferred from the droplet to the gas phase to maintain the vapor pressure above the droplet. Since there is relatively little water in the aerosol phase compared with the gas phase, this transfer has little affect on the value of p_d. Hence it can be assumed the droplets are always in equilibrium with the gas so far as water vapor is concerned and that p_d is constant. Differentiating the equilibrium relationship (1) with respect to time, the result is

$$\frac{dv}{dt} = \frac{\bar{v}(dn/dt)}{\frac{8}{3}(\pi/6)^{1/3}(\sigma\bar{v}/RT)v^{-1/3} + \ln(p_{s0}/p_d)} \tag{2}$$

Thus the growth law depends on dn/dt, the rate at which ammonium sulfate molecules deposit on the droplets. For droplets much larger than the mean free path of the air, the deposition is controlled by diffusion:

$$\frac{dn}{dt} = 2\pi D d_p n_{gas}$$

where D and n_{gas} refer to the diffusion coefficient and concentration of ammonium sulfate (molecules/cm^3) in the gas phase. Substituting in (2), the result for the growth law is

$$\frac{dv}{dt} = \frac{2\pi(6/\pi)^{1/3} D v^{1/3} n_{gas}}{\frac{8}{3}(\pi/6)^{1/3}(\sigma\bar{v}/RT)v^{-1/3} + \ln(p_{s0}/p_d)}$$

9·9 Dynamics of Growth: Continuity Relation in v Space

We consider a polydisperse aerosol growing by gas-to-particle conversion. The system is spatially uniform in composition—a growing aerosol in a box. As growth occurs, the size distribution function changes with time; we wish to derive an expression for $\partial n/\partial t$. Let $I(v,t)$ be the

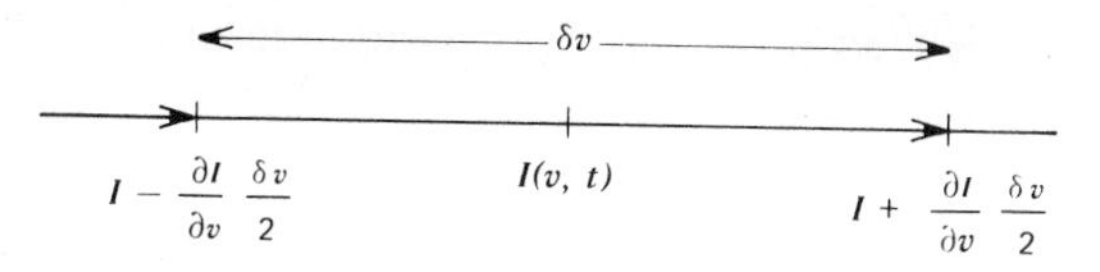

Fig. 9·4 The flow of particles through v space as a result of growth processes.

particle current or number of particles per unit time per unit volume of gas passing the point v. The rate at which particles enter the small element of length δv in v space (Fig. 9·4) is given by

$$I-\frac{\partial I}{\partial v}\frac{\delta v}{2}$$

The rate at which particles leave δv is given by

$$I+\frac{\partial I}{\partial v}\frac{\delta v}{2}$$

The net rate of change in particle number in δv is

$$\frac{\partial[n\,\delta v]}{\partial t}=I-\frac{\partial I}{\partial v}\frac{\delta v}{2}-\left[I+\frac{\partial I}{\partial v}\frac{\delta v}{2}\right]$$

$$=-\frac{\partial I}{\partial v}\delta v$$

Dividing both sides by δv, the result is

$$\frac{\partial n}{\partial t}=-\frac{\partial I}{\partial v} \tag{9·26}$$

which is the continuity relation for the v space equivalent to (9·6) for the discrete distribution.

Multiplying both sides of (9·26) by v and integrating over the range between v_1 and v_2, the result is

$$\int_{v_1}^{v_2} v\frac{\partial n}{\partial t}\,dv=\frac{\partial\int_{v_1}^{v_2} nv\,dv}{\partial t}=-\int_{v_1}^{v_2} v\frac{\partial I}{\partial v}\,dv \tag{9·27}$$

The last term can also be written as follows:

$$-\int_{v_1}^{v_2} v\frac{\partial I}{\partial v}\,dv=-\int_{v_1}^{v_2}\frac{\partial Iv}{\partial v}\,dv+\int_{v_1}^{v_2} I\,dv \tag{9·28}$$

Integrating the first term on the right-hand side and combining (9·27) with (9·28),

$$\frac{\partial \int_{v_1}^{v_2} nv\,dv}{\partial t} = [Iv]_1 - [Iv]_2 + \int_{v_1}^{v_2} I\,dv \tag{9·29}$$

The term on the left-hand side represents the rate of change in the volume in the size range bounded by v_1 and v_2. The first and second terms on the right are the flow of volume into and out of the range between v_1 and v_2. Hence by difference, the third term on the right represents the flow from the gas phase into the range. This leads to an alternate interpretation of I. The particle current also represents the volume of material converted from the gas phase per unit v space in unit volume of gas and unit time.

In general, the particle current can be expressed as the sum of two terms, one representing diffusion and the other migration in v space (9·5). Diffusion leads to a spread in v space of a group of particles initially of the same size. The diffusion term is proportional to $\partial n/\partial v$ (or $\partial n/\partial g$), which is very large for homogeneous nucleation. For the growth of larger particles, diffusion can be neglected in comparison with migration because $\partial n/\partial v$ is relatively small. The particle current is then given by the relation:

$$I(v,t) \approx n\frac{dv}{dt} \tag{9·30}$$

where dv/dt is the growth law (Table 9·3). According to (9·30), all particles of the same initial size grow to the same final size. By substitution of (9·30) with the appropriate growth law, (9·26) can be solved for $n(v,t)$.

Example. We consider the case of a growth law

$$\frac{dv}{dt} = F(t)v$$

which would hold for reaction in a droplet phase. Derive an expression for $n(v,t)$.

SOLUTION. Substitution in (9·26) gives

$$\frac{\partial n}{\partial t} = -F(t)\frac{\partial nv}{\partial v}$$

which can be written

$$\frac{\partial nv}{\partial \tau} + \frac{\partial nv}{\partial \ln v} = 0 \tag{1}$$

where $\tau = \int_0 F\,dt$. The solution to this equation, obtained by the method of characteristics is given by

$$nv = f(\ln v - \tau) \tag{2}$$

where f represents a functional relationship that is determined by the known distribution at any time. The result (2) can be checked by substitution in (1). If at $t = 0$, which corresponds to $\tau = 0$, the distribution function is given by a power law

$$n(v,0) = \text{const}\, v^p$$

then the functional form f, which satisfies (2), is

$$nv = \text{const}\, e^{(p+1)(\ln v - \tau)}$$

or

$$n = \text{const}\, v^p e^{-(p+1)\tau}$$

Thus if the distribution begins as a power law form, it will retain the same dependence on particle size as growth continues if $dv/dt \sim v$ (Brock, 1971).

The distribution with respect to size of a chemical species converted from the gas phase depends on the mechanism of conversion. In general, species that form by gas phase reaction and then diffuse to the particle surface are found in the smaller size range; species that form in a droplet phase tend to accumulate in the larger size range.

9·10 Measurement of Growth Rates: Homogeneous Gas Phase Reactions

The growth law for a polydisperse aerosol can be determined by measuring the change in the size distribution function with time. In experiments carried out by Heisler (1975), small quantities of organic vapors that served as aerosol precursors were added to a sample of the normal atmospheric aerosol contained in a 80 m^3 bag exposed to solar radiation. The bag was made of a polymer film almost transparent to solar radiation in the uv range and relatively unreactive with ozone and other species. Chemical reaction led to the formation of condensable species and to aerosol growth. The change with time of the size distribution function was measured with a single particle optical counter. (See frontispiece.).

The number of particles per unit volume larger than a given particle size d_p, $\int_{d_p}^{\infty} n_d(\tilde{d}_p)d(\tilde{d}_p)$ is shown in Fig. 9·5 for an experiment with cyclohexene. Consider a horizontal line on the figure corresponding to constant values of this integral. In the absence of homogeneous nucleation, each such line corresponds to the growth with time of a particle of size given initially by the curve for $t=0$. No particle can move across such a line because the total number larger is conserved.

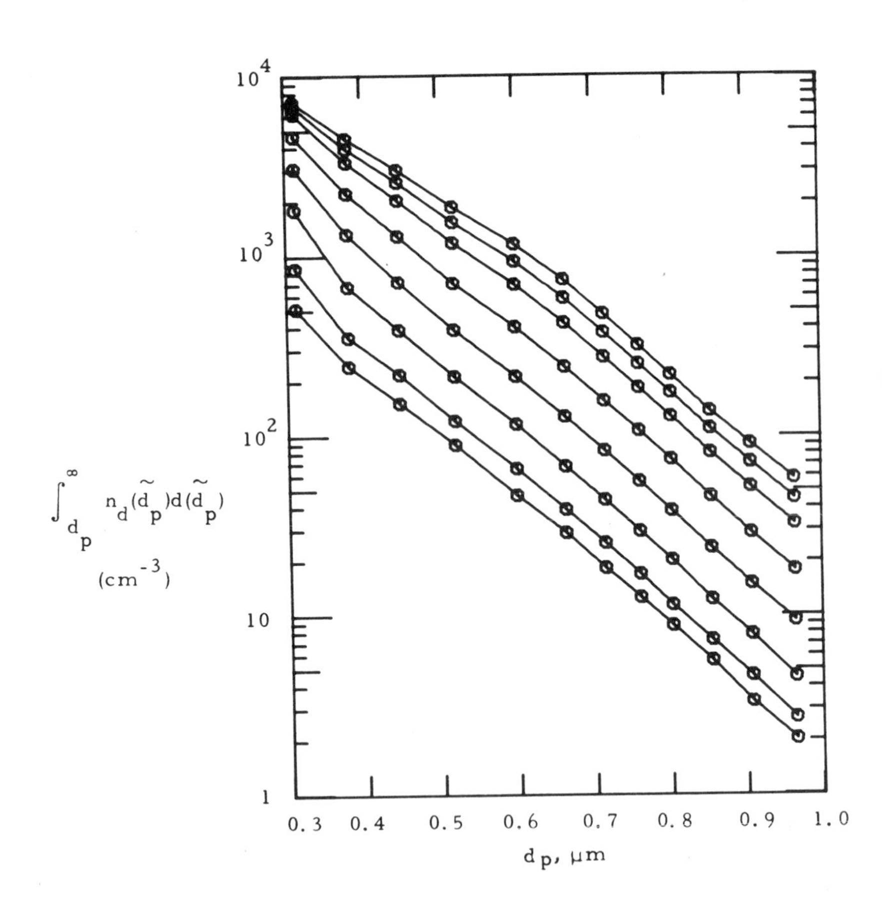

Fig. 9·5 Number concentrations of particles larger than a given diameter versus diameter at various times in a smog chamber experiment. Initial concentrations were 2.02 ppm cyclohexene, 0.34 ppm NO, and 0.17 ppm NO_2. The time between measurements was about 3 to 4 min. The first measurement shown was made 12 min after the addition of the reactants (Heisler, 1975).

The growth rate, $d(d_p)/dt \approx \Delta d_p/\Delta t$ can be obtained from adjacent distributions in Fig. 9·5 as a function of d_p and of the time. The data were then plotted with dv/dt as a function of particle diameter as shown in Fig. 9·6. For this set of data, an approximately linear relationship was found with an intercept on the positive d_p-axis.

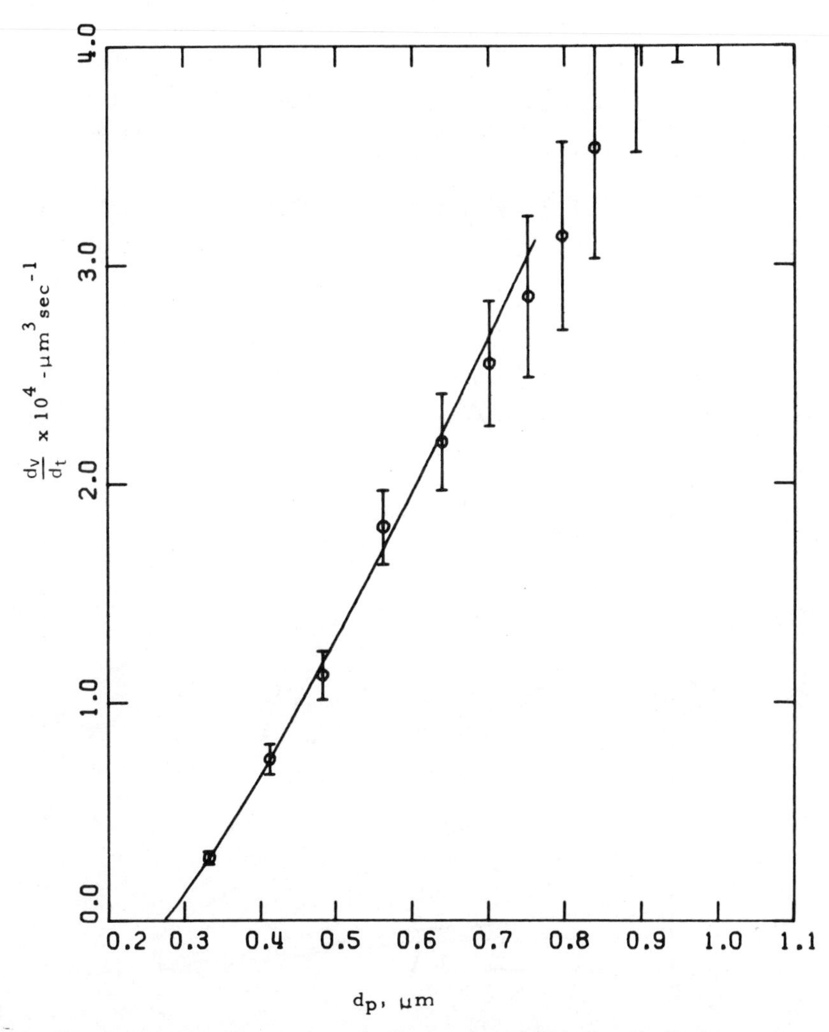

Fig. 9·6 Particle growth rates between the fourth and fifth size distribution measurements for the data of Fig. 9·5. The solid line is a least-square best-fit of the diffusional growth law, modified to include mean free path effects (9·13) and the Kelvin effect (9·15). The intercept on the size axis is the average critical size, d_p^*.

As a reasonable model for explaining these results, it was assumed that a single condensable species was formed in the gas as a result of chemical reaction, or a small group of species with similar thermodynamic properties. Molecules of these reaction products then diffused to the surfaces of existing aerosol particles. Hence a diffusion controlled growth law modified by the Kelvin effect in the small saturation ratio approximation (9·15) correlated the data as shown in Fig. 9·6. The cut-off particle diameter probably results from the Kelvin effect. For the run shown, the critical diameter was about 0.28 μm. The line is the result of a least-squares fit using (9·13) combined with (9·15) in calculating the growth law. The curvature in the line results from the form of the interpolation formula (9·13).

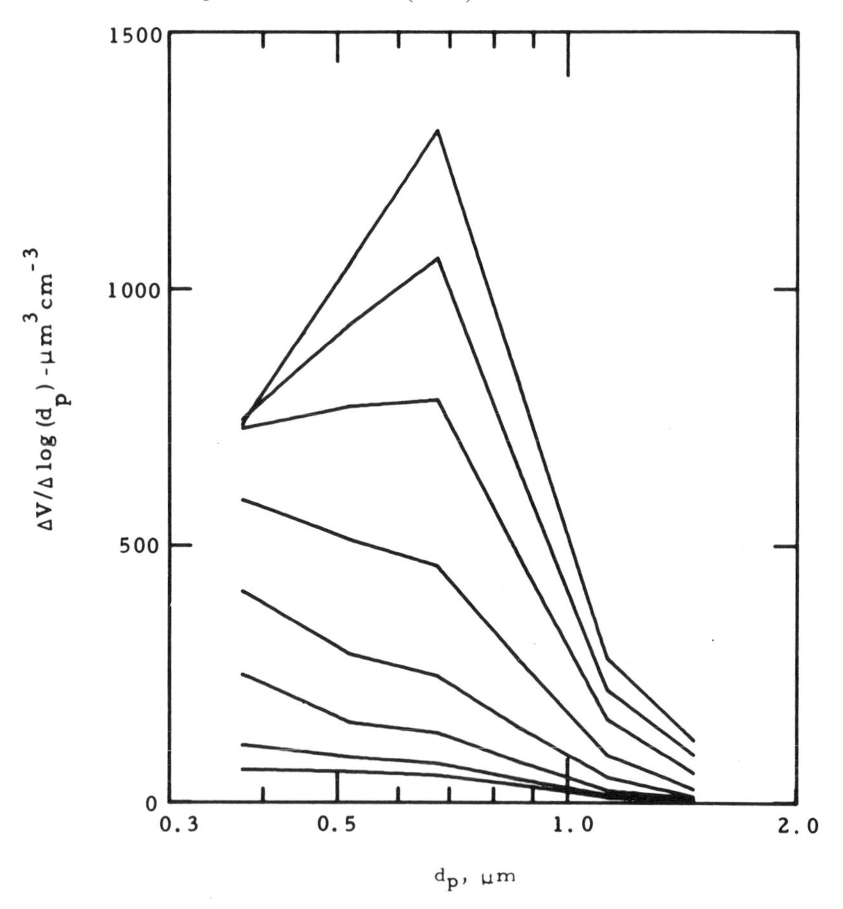

Fig. 9·7 Aerosol volume distributions from experiment Fig. 9·6. The order of the measurements is indicated by the numbers. Increases are from gas-to-particle conversion. The peak is near the size range most efficient for light scattering.

Aerosol volume distributions calculated from the data are shown in Fig. 9·7. Material accumulates in the size range near 0.6 μm, which is particularly efficient for light scattering. Small particles grow little because of the Kelvin effect.

In experiments with cyclopentene, the data showed a pronounced bend at a particle diameter larger than the lower cut-off diameter. The results were correlated using a growth law which incorporated another condensing species.

Problems

1. Estimate the time it takes for an 0.1 μm water droplet at a temperature of 25°C to evaporate completely, taking into account the Kelvin effect. Assume the vapor pressure of the water in the bulk of the gas phase is zero. Neglect the heat of evaporation in your calculation. How would evaporative effects modify your answer?

2. Determine the size of the smallest stable drop at the critical saturation ratio for toluene at 300°K. Of how many molecules are these drops composed?

3. The total surface area of the Los Angeles smog aerosol is of the order of 1000 $\mu m^2/cm^3$. Estimate the maximum rate of formation of a condensable species by chemical reaction that can be sustained *without* homogeneous nucleation taking place. Express your answer in $\mu g/m^3$ hr as a function of the saturation ratio. The molecular weight of the condensable species is 100 and its vapor pressure is 10^{-7} mm Hg.

4. Derive an expression for the form of the size distribution function as a function of time and particle size when growth is diffusion limited. Assume as a model an aerosol in a box with condensation the only process taking place.

5. Sulfur dioxide in a certain stack plume is converted to SO_4^{2-} at a rate of 1%/hr. Assume that the conversion takes place in the droplet phase and that a quasi-steady state exists between reaction in the droplets and transport through the gas phase. The local concentration of SO_2 is 0.1 ppm (260 $\mu g/m^3$) and the droplet concentration is 200 $\mu g/m^3$ and their density is 1.2 g/cm^3. Calculate $(p_1 - p_s)/p_1$ for SO_2 as a function of particle size. The temperature is 30°C and $D_{SO_2} = 0.140$ cm^2/sec.

6. As a result of a rapid expansion, one component of a gas mixture in a box becomes supersaturated and then condenses. No foreign nuclei are present. Sketch the development of the size distribution on a series of diagrams corresponding to different times from the onset of condensation. Take into account the effects of coagulation after condensation has effectively ceased. Show only a schematic representation; no detailed calculations are necessary (Dunning, W. J. (1973), *Symposia of the Faraday Society*, **7**, 7).

References

Andres, R. P. (1969) Homogeneous Nucleation in a Vapor, Zettlemoyer, A. C. (Ed.) *Nucleation*, Dekker, New York, Chap. 2.

Birks, J., and Bradley, R. S. (1949) *Proc. Roy. Soc.*, **198A**, 226.

Brock, J. R. (1971) *Atmos. Environ.*, **5**, 833.

Calvert, J. G. (1973) Interactions of Air Pollutants in *Proceedings of the Conference on Health Effects of Air Pollutants*, U. S. Gov't. Printing Office Serial No. 93-15.

Denbigh, K. G. (1966) *The Principles of Chemical Equilibrium*, Cambridge Univ. Press, Cambridge, 2nd ed.

Frenkel, J. (1955) *Kinetic Theory of Liquids*, Dover, New York.

Fuchs, N. A., and Sutugin, A. G. (1971) High-Dispersed Aerosols in Hidy, G. M., and Brock, J. R. (Eds.) *Topics in Current Aerosol Research*, Pergamon, New York.

Grosjean, D. (1975) personal communication.

Heisler, S. L., and Friedlander, S. K. (1976) Gas to Particle Conversion in Photochemical Smog: Aerosol Growth Laws and Mechanisms for Organics, to be published, *Atmos. Environ.*

Heist, R. H., and Reiss, H. (1973) *J. Chem. Phys.*, **59**, 665.

Junge, C. E., and Ryan, T. G. (1958) *Quart. J. Roy. Meterol. Soc.*, **84**, 46.

Katz, J. L., Scoppa, C. J., Kumar, N. G., and Mirabel, P. (1975) *J. Chem. Phys.*, **62**, 448.

Mason, B. J. (1971) *The Physics of Clouds*, Clarendon Press, Oxford.

Paul, B. (1962) *ARS J.*, **32**, 1321.

Reiss, H. (1950) *J. Chem. Phys.*, **18**, 840.

Roberts, P. T., and Friedlander, S. K. (1975) *Environmental Health Perspectives*, **10**, 103.

Roberts, P. T., and Friedlander, S. K. (1976) *Environ. Sci. Technol.*, **10**, 573.

Wegener, P. P., and Pouring, A. A. (1964) *Physics of Fluids*, **7**, 352.

Wilson, C. T. R. (1927) On the Cloud Method of Making Visible Ions and the Tracks of Ionizing Particles in *Nobel Lectures in Physics* 1922–1941.

CHAPTER 10

The General Dynamic Equation for the Continuous Distribution Function

Particle formation, growth, diffusion, and coagulation have been discussed in previous chapters. These processes together with convection and sedimentation determine the time rate of change of the size distribution function. A general dynamic equation (GDE) for $n(\mathbf{r}, v, t)$ is set up in this chapter. This equation is sometimes referred to as a population balance equation. By solving the equation for different initial and boundary conditions, the size distribution function can be calculated for geometries and flow conditions of practical interest. The GDE is of fundamental importance in atmospheric simulation studies, in relating particulate pollution to sources, and in the characterization of emission sources. It should also be useful in the design of industrial processes to reduce emissions, but such applications have not yet been made.

Since the GDE is a nonlinear, partial integro-differential equation, its solution is difficult, indeed. Solutions to problems of practical interest have been obtained in a few cases when two or more processes that modify the size distribution are occurring at the same time. As examples, in this chapter we consider simultaneous condensation and coagulation, turbulent diffusion and growth, and coagulation with transport to surfaces. In each case, approximate analytical or semianalytical solutions can be obtained.

Improvements in instrumentation and analytical techniques have now made it possible to follow the physical and chemical behavior of these systems. Hence research on the problems discussed in this chapter is currently very active.

10·1 General Dynamic Equation

The equation of convective diffusion (Section 3·1) represents a balance on the particle concentration in an elemental volume. Particle transport results from convective diffusion and the effects of an external force field. These may be called *external* processes since they involve movement across the walls of an elemental volume. In Chap. 7 through 9, it has been shown that particles in a given size range may also be gained or lost as a result of *internal* processes, namely, coagulation and growth by gas-to-particle conversion.

The change in the discrete distribution function with time is obtained by generalizing the equation of convective diffusion (Chap. 3) to include particle growth and coagulation:

$$\frac{\partial n_k}{\partial t} + \nabla \cdot n_k \mathbf{v} = \nabla \cdot D \nabla n_k + \left[\frac{\partial n_k}{\partial t} \right]_{growth} + \left[\frac{\partial n_k}{\partial t} \right]_{coag} - \nabla \cdot \mathbf{c} n_k \quad (10 \cdot 1)$$

where $\mathbf{c}$ is the particle velocity resulting from the external force field. The growth term refers to gas-to-particle conversion of single molecules, while the coagulation term results from particle collision and adhesion.

By (9·3) the growth term can be written

$$\left[\frac{\partial n_k}{\partial t} \right]_{growth} = I_k - I_{k+1}$$

The particle current, I_k, the flow through v space, has dimensions of particles per unit volume per unit time.

Coagulation is the result of collision and adhesion among particles for which $k \geqslant 2$. For the discrete distribution, the coagulation rate is given by (7·3):

$$\left[\frac{\partial n_k}{\partial t} \right]_{coag} = \frac{1}{2} \sum_{i+j=k} \beta(v_i, v_j) n_i n_j - n_k \sum_{i=2}^{\infty} \beta(v_i, v_k) n_i \qquad (i,j,k \geqslant 2)$$

$$(10 \cdot 2)$$

Collisions with single molecules are excluded from this expression.

The transition to the continuous distribution function requires care. For the growth term, this was shown to be (9·26)

$$\left[\frac{\partial n}{\partial t} \right]_{\text{growth}} = -\frac{\partial I}{\partial v} \qquad v \gg v_m \qquad (10 \cdot 3)$$

The particle current, I, can be expressed as the sum of diffusion and migration terms in v space:

$$I = -D_v \frac{\partial n}{\partial v} + nq$$

where $q = dv/dt$ is the migration velocity through v space. Similarly, the coagulation terms become

$$\left[\frac{\partial n}{\partial t}\right]_{coag} = \frac{1}{2}\int_0^v \beta(\tilde{v}, v-\tilde{v})n(\tilde{v})n(v-\tilde{v})\,d\tilde{v}$$

$$-\int_0^\infty \beta(v,\tilde{v})n(v)n(\tilde{v})\,d\tilde{v} \qquad (v \gg v_m) \quad (10.4)$$

Substituting (10·3) and (10·4) for growth and coagulation, respectively, in (10·1) the GDE for the continuous distribution function is obtained:

$$\frac{\partial n}{\partial t} + \nabla\cdot n\mathbf{v} + \frac{\partial I}{\partial v} = \nabla\cdot D\nabla n + \frac{1}{2}\int_0^v \beta(\tilde{v}, v-\tilde{v})n(\tilde{v})n(v-\tilde{v})\,d\tilde{v}$$

$$-\int_0^\infty \beta(v,\tilde{v})n(v)n(\tilde{v})\,d\tilde{v} - \nabla\cdot\mathbf{c}n \qquad (v \gg v_m) \quad (10\cdot5)$$

Collisions with single molecules are excluded from the coagulation terms in this expression. For the usual case of an incompressible flow, the second term on the left-hand side takes the form:

$$\nabla\cdot n\mathbf{v} = \mathbf{v}\cdot\nabla n$$

An equation of similar form has been derived by Hulburt and Katz (1964) in a different way.

To solve the GDE, expressions are needed for I and $\beta(v,\tilde{v})$ as shown later in this chapter. However, it is possible to derive useful expressions for the number and volume concentrations without assuming forms for these parameters.

10·2 The Dynamic Equation for the Number Concentration N_∞

The dynamics of the total number concentration, N_∞, and volume fraction of aerosol material, V, are moments of special interest. There is a problem in defining the total number concentration, N_∞, in an experimentally meaningful way. This parameter is usually measured with a condensation nuclei counter (CNC) (Chap. 5). The CNC detects particles larger than some minimum size that probably depends on their chemical nature and shape. Let v_d be the minimum detectable particle volume. Then

$$N_\infty = \int_{v_d}^\infty n(v)\,dv$$

and assume $v_d \geqslant v^*$ the critical particle volume for homogeneous nucleation.

The dynamic equation for the total number concentration is obtained by integrating the GDE with respect to v over all values of $v > v_d$:

$$\frac{\partial N_\infty}{\partial t} + \mathbf{v} \cdot \nabla N_\infty + \int_{v_d}^{\infty} \frac{\partial I}{\partial v} dv = \nabla^2 \int_{v_d}^{\infty} Dn\, dv$$

$$+ \frac{1}{2} \int_{v_d}^{\infty} \left[\int_0^v \beta(\tilde{v}, v - \tilde{v}) n(\tilde{v}) n(v - \tilde{v})\, d\tilde{v} \right] dv$$

$$- \int_{v_d}^{\infty} \left[\int_0^{\infty} \beta(v, \tilde{v}) n(v) n(\tilde{v})\, d\tilde{v} \right] dv$$

$$- \frac{\partial \int_{v_d}^{\infty} c_s n\, dv}{\partial z}$$

The growth term is evaluated as follows:

$$\int_{v_d}^{\infty} \frac{\partial I}{\partial v} dv = I_\infty - I_d$$

On physical grounds, $I_\infty = 0$ since there is no loss of particles by growth from the upper end of the distribution. The term I_d is the particle current flowing into the lower end of the spectrum. When homogeneous nucleation takes place, this term is important. For $v_d = v^*$, the critical particle size, I_d, is the particle current of homogeneous nucleation theory. Hence the dynamic equation for the number concentration is given by

$$\frac{\partial N_\infty}{\partial t} + \mathbf{v} \cdot \nabla N_\infty = I_d + \nabla^2 \int_{v_d}^{\infty} Dn\, dv$$

$$+ \frac{1}{2} \int_{v_d}^{\infty} \left[\int_0^v \beta(\tilde{v}, v - \tilde{v}) n(\tilde{v}) n(v - \tilde{v})\, d\tilde{v} \right] dv$$

$$- \int_{v_d}^{\infty} \left[\int_0^{\infty} \beta(v, \tilde{v}) n(v) n(\tilde{v})\, d\tilde{v} \right] dv - \frac{\partial \int_{v_d}^{\infty} c_s n\, dv}{\partial z} \tag{10·6}$$

Experiments are often carried out with the aerosol contained in a large chamber. If the surface to volume ratio is sufficiently small to neglect deposition on the walls by diffusion and sedimentation, (10·6) becomes

$$\frac{\partial N_\infty}{\partial t} = I_d + \left[\frac{\partial N_\infty}{\partial t}\right]_{coag} \tag{10·7}$$

where $[\partial N_\infty/\partial t]_{coag}$ represents the coagulation terms in (10·6). The change in N_∞ results from the competing effects of formation by homogeneous nucleation and loss by coagulation.

10·3 The Dynamic Equation for the Volume Fraction

The aerosol volume fraction, V, is closely related to the mass concentration, which is usually determined by filtration. We assume the filter is ideal, removing all particles larger than single molecules. Then

$$V = \int_0^\infty nv\,dv$$

The change in the volume fraction, V, with time is obtained by multiplying the GDE by v and integrating with respect to v:

$$\frac{\partial V}{\partial t} + \mathbf{v}\cdot\nabla V + \left[\frac{\partial V}{\partial t}\right]_{growth} = \nabla^2 \int_0^\infty Dvn\,dv + \left[\frac{\partial V}{\partial t}\right]_{coag} - \frac{\partial \int_0^\infty c_s vn\,dv}{\partial z} \tag{10·8}$$

The change in V resulting from gas-to-particle conversion can be written as follows:

$$\left[\frac{\partial V}{\partial t}\right]_{growth} = \frac{\partial \int_0^\infty nv\,dv}{\partial t} = -\int_0^\infty v\frac{\partial I}{\partial v}\,dv$$

If homogeneous nucleation is taking place, we can write $[\partial V/\partial t]_{growth}$ as the sum of two terms:

$$\left[\frac{\partial V}{\partial t}\right]_{growth} = \frac{\partial \int_0^{v^*} nv\,dv}{\partial t} + \frac{\partial \int_{v^*}^\infty nv\,dv}{\partial t} \tag{10·9}$$

where v^* is the critical particle volume.

The term $(\partial \int_0^{v^*} nv\,dv)/\partial t$ represents the accumulation of material in the cluster size range below the critical particle size range v^*. In homogeneous nucleation theory (Chap. 9), this term vanishes; there is a steady state for this portion of the distribution in which material is removed as fast as it is supplied. (This is actually true only as a quasi-steady approximation.) The second term in (10·8) can be written as follows:

$$\frac{\partial \int_{v^*}^{\infty} nv\,dv}{\partial t} = -\int_{v^*}^{\infty} v\frac{\partial I}{\partial v}\,dv = -\int_{v^*}^{\infty} \frac{\partial Iv}{\partial v}\,dv + \int_{v^*}^{\infty} I\,dv$$

but

$$\int_{v^*}^{\infty} \frac{\partial Iv}{\partial v}\,dv = [Iv]_{\infty} - [Iv]_{v^*}$$

The term $\int_{v^*}^{\infty} I dv$ represents the growth of stable particles ($v > v^*$) by gas-to-particle conversion. On physical grounds, this is clear since the particle current represents the volume of material converted per unit of v space in unit volume of gas and unit time.

Since there is no loss of material by growth to the upper end of the distribution

$$[Iv]_{\infty} = 0$$

The term $[Iv]_{v^*}$ is the volumetric rate at which material is delivered by homogeneous nucleation to the stable part of the size distribution.

The contribution of the coagulation term $[\partial V/\partial t]_{coag}$ vanishes identically no matter what the form of the collision frequency function. The coagulation mechanism only shifts matter up the distribution function from small to large sizes and does not change the local volumetric concentration of aerosol.

The balance on V (10·8) then takes the form

$$\frac{\partial V}{\partial t} + \mathbf{v}\cdot\nabla V = \underbrace{\int_{v^*}^{\infty} I\,dv}_{\substack{\text{growth}\\ \text{of}\\ \text{stable}\\ \text{particles}}} + \underbrace{[Iv]_{v^*}}_{\substack{\text{formation}\\ \text{by}\\ \text{homogeneous}\\ \text{nucleation}}} + \underbrace{\nabla^2 \int_0^{\infty} Dvn\,dv}_{\text{diffusion}} - \underbrace{\frac{\partial \int_0^{\infty} c_s v\,dv}{\partial z}}_{\text{sedimentation}} \qquad (10\cdot 10)$$

For $D \sim d_p^{-2}$ (free molecule region), the integral $\int_0^\infty Dvn\,dv$ is proportional to the average particle diameter (Chap. 1). Hence this term represents the diffusion of a quantity proportional to the average particle diameter.

10·4 Simultaneous Coagulation and Diffusional Growth: The Similarity Solution

Suppose the aerosol contained in a large chamber is composed of particles larger than the mean free path of the gas. The surface-to-volume ratio of the chamber is sufficiently small to neglect deposition on the walls, and the composition of the system is uniform. Coagulation takes place, and at the same time, the particles grow as a result of diffusion controlled condensation but sedimentation can be neglected. Homogeneous nucleation does not occur and the system is isothermal.

For growth and coagulation alone, the GDE can be written as follows:

$$\frac{\partial n}{\partial t} + \frac{\partial I}{\partial v} = \frac{1}{2}\int_0^v \beta(v-\tilde{v},\tilde{v})n(v-\tilde{v})n(\tilde{v})\,d\tilde{v} - \int_0^\infty \beta(v,\tilde{v})n(v)n(\tilde{v})\,d\tilde{v} \tag{10·11}$$

with the collision frequency function for the continuum range given by

$$\beta = \frac{2kT}{3\mu}\left(\frac{1}{v^{1/3}} + \frac{1}{\tilde{v}^{1/3}}\right)(v^{1/3} + \tilde{v}^{1/3}) \tag{10·12}$$

The particle current is assumed to be given by (Chap. 9)

$$I = n\frac{dv}{dt} \tag{10·13}$$

The diffusional growth law (9·23) can be written in the form:

$$\frac{dv}{dt} = 3^{1/3}(4\pi)^{2/3}\frac{Dp_s v_m}{kT}(S-1)v^{1/3} = B(S-1)v^{1/3} \tag{10·14}$$

where S is the saturation ratio, p_s is the saturation vapor pressure, v_m is the molecular volume in the condensed phase, and B is a constant defined by this expression. Latent heat effects in condensation are neglected, as is the Kelvin effict.

The similarity transformation, $n=(N_{\infty}^{2}/V)\psi(\eta)$ (7·26), is still applicable in this case (Pich, Friedlander, and Lai, 1970), but the volumetric concentration is no longer constant because of the condensation of material from the gas phase. Substituting the self-preserving form in (10·11) with (10·12) through (10·14), it is found that similarity is preserved provided that the saturation ratio changes with time in a special way and that the dimensionless group,

$$C=\frac{3\mu}{4kT}\left[\frac{2}{V^{2/3}N_{\infty}^{1/3}}\right]B(S-1)$$

is constant. This group is a measure of the relative rates of condensation and coagulation. When C is small, condensation proceeds slowly compared with coagulation. The time rate of change of the total number of particles is given by an expression of the same form for coagulation without condensation (Chap. 7):

$$\frac{dN_{\infty}}{dt}=-\frac{2kT}{3\mu}(1+ab)N_{\infty}^{2}$$

but the values of the moments a and b (7·32a and b) are different. The volumetric concentration increases as a result of condensation at a rate given by

$$V=V_{0}\left[1+\frac{2kT}{3\mu}(1+ab)N_{\infty}(0)t\right]^{aC/(1+ab)} \qquad (10\cdot 15)$$

where V_0 and $N_{\infty}(0)$ are the values at $t=0$. In the important special case of constant saturation ratio, it is found that the total surface area of the system is constant. The decrease of surface area by coagulation is, in this case, balanced by the formation of new surface as a result of vapor condensation. The value of ab is 1.05 and the exponent in (10·15), $aC/(1+ab)=1/2$. Calculated values of $N_{\infty}/N_{\infty}(0)$ and V/V_0 are shown in Fig. 10·1.

If the size distribution reaches a self-preserving form, a special relationship exists among the number, surface area and volume concentrations. The surface area per unit volume of gas is given by

$$A=\int_{0}^{\infty}\pi d_{p}^{2}n(v)\,dv$$

$$=(36\pi)^{1/3}\int_{0}^{\infty}v^{2/3}n\,dv$$

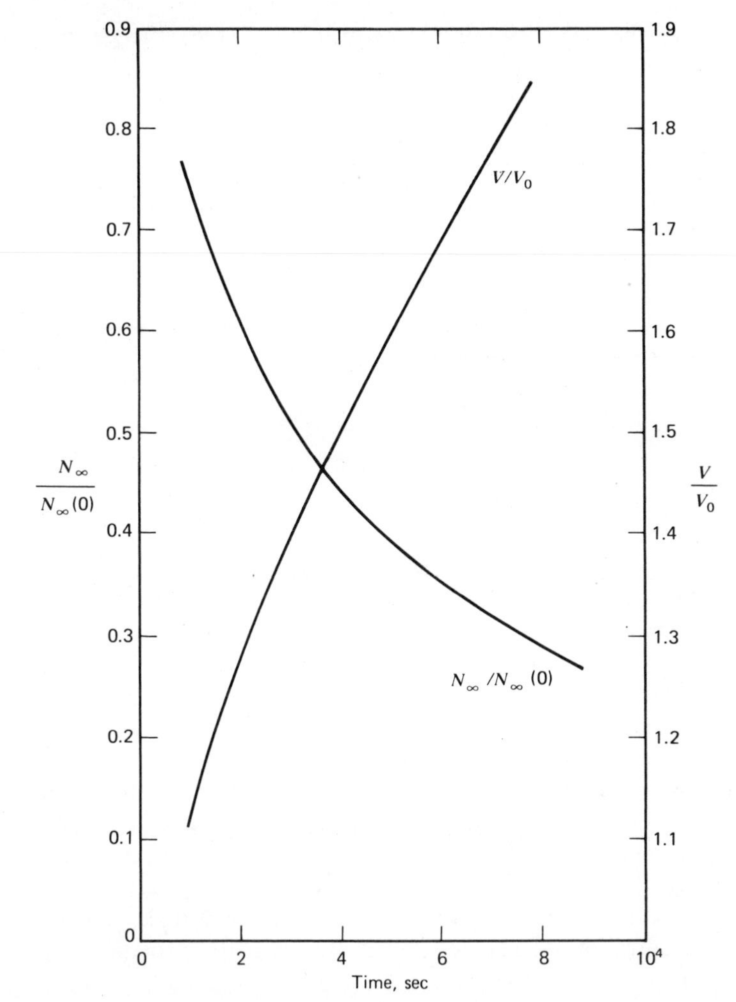

Fig. 10·1 Variations of number and volume concentration with time for a self-preserving aerosol with constant saturation ratio and constant surface area. The value of $ab = 1.05$ for this case. $N_\infty(0) = 10^6$ cm^{-3} and $T = 20$°C. The number concentration decreases as a result of coagulation, and the volume concentration increases because of condensation.

Substituting the self-preserving transformation (7·26), the result is

$$A = (36\pi)^{1/3} N_{\infty}^{\ 1/3} V^{2/3} \int_0^{\infty} \eta^{2/3} \psi \, d\eta \tag{10·16}$$

In the special case of constant A (and saturation ratio) the integral is equal to 0.951 (Pich, Friedlander, and Lai, 1970) so that in this case

$$\frac{A}{N_{\infty}^{1/3} V^{2/3}} = 4.60 \tag{10·16a}$$

10·5 Simultaneous Coagulation and Growth: Experimental Results

Experiments on simultaneous coagulation and growth have been carried out by Husar and Whitby (1973). A 90 m^3 polyethylene bag was filled with laboratory air from which particulate matter had been removed by filtration. Solar radiation penetrating the bag induced photochemical reactions among gaseous pollutants, probably SO_2 and organics, but the chemical composition was not determined. The reactions led to the formation of condensable species and photochemical aerosols. Size distributions were measured in 20 min intervals using an electrical mobility analyzer. The results of one set of experiments for three different times are shown in Fig. 10·2.

The number, surface, and volume concentrations were calculated from the size distribution function and are shown in Fig. 10·3. The variation with time of the number concentration is interpreted as follows: In the absence of foreign nuclei, particles are formed initially by homogeneous nucleation. As concentrations mount, coagulation takes place, and growth occurs on nuclei already generated. The number concentration reaches a maximum and then decays. The maximum concentration is reached when the rate of formation by self-nucleation and rate of coagulation are equal. The maximum concentration is determined by setting $\partial N_{\infty}/\partial t = 0$ in (10·7):

$$I_d = -\left[\frac{\partial N_{\infty}}{\partial t}\right]_{coag}$$

As growth continues, the aerosol surface area becomes sufficiently large to accomodate the products of gas-to-particle conversion. The saturation ratio decreases leading to a reduction in the particle formation rate. The decay in the number concentration for $t > 80$ min in Fig. 10·3 is probably due to coagulation; calculations for free molecule aerosols support this hypothesis.

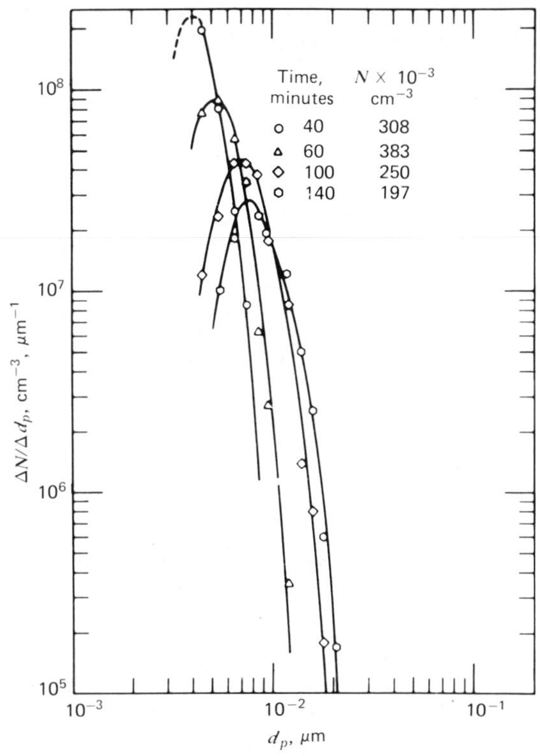

Fig. 10·2 Size distributions of an aging free modecule aerosol generated by exposing filtered laboratory air in a 90 m^3 polyethylene bag to solar radiation. Change in the distribution function results from the combined effects of coagulation and growth (Husar and Whitby, 1973).

Unlike the case of coagulation without growth, the volume fraction of dispersed material increases with time as a result of gas-to-particle conversion. The total surface area, on the other hand, tends to an approximately constant value. Coagulation tends to reduce, whereas growth tends to increase surface area, and the two effects in this case almost balanced each other.

The ratio $A/N_\infty^{1/3}V^{2/3}$ reaches a constant value after about 1 hr, indicating that the asymptotic, self-preserving stage has been reached. The relationship (10·16) holds for both the continuum and free-molecule ranges, but the value of the integral would be expected to vary somewhat. As shown in Fig. 10·3, however, the value of the ratio falls

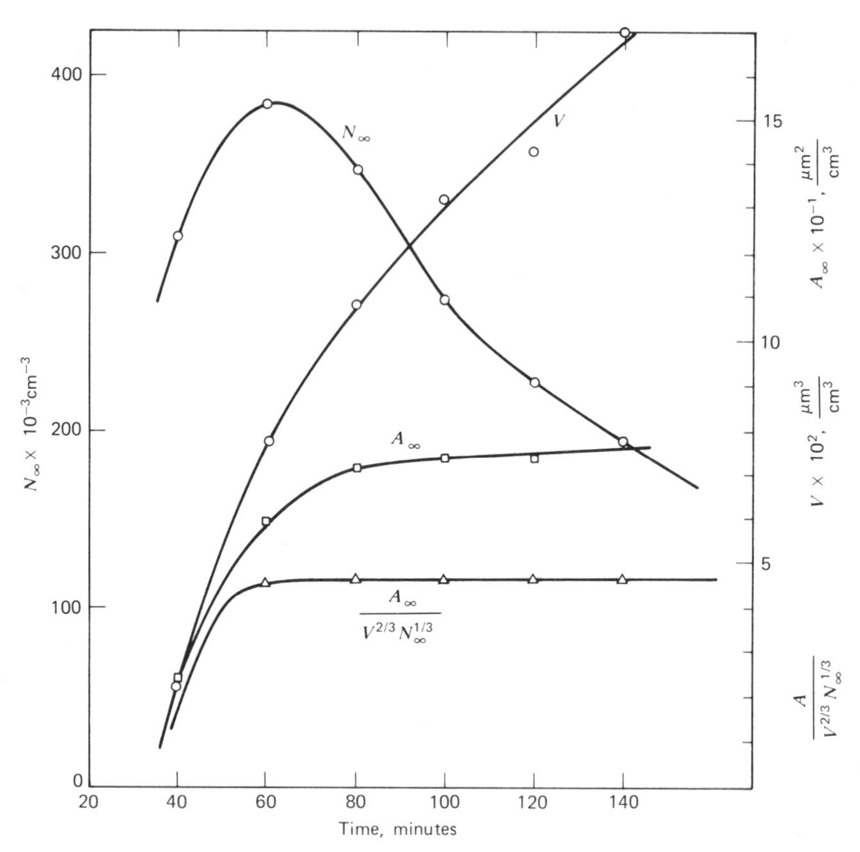

Fig. 10·3 Evolution of the moments of the size distribution function for the aerosols shown in Fig. 10·2. The peak in the number distribution probably results when formation by homogeneous nucleation is balanced by coagulation. Total aerosol volume increases with time as gas-to-particle conversion takes place. Total surface area increases at first and then approaches an approximately constant value due, probably, to a balance between growth and coagulation (Husar and Whitby, 1973). The results should be compared with Fig. 10·1.

very close to 4.60, the value for continuum range with constant A (10·16*a*).

In Fig. 10·4, the data of Fig. 10·2 have been replotted in the self-preserving form. As a good approximation, all the data fall on a single curve. The theory for the continuum range discussed in the previous section is not directly applicable, so no comparison between theory and experiment has been made.

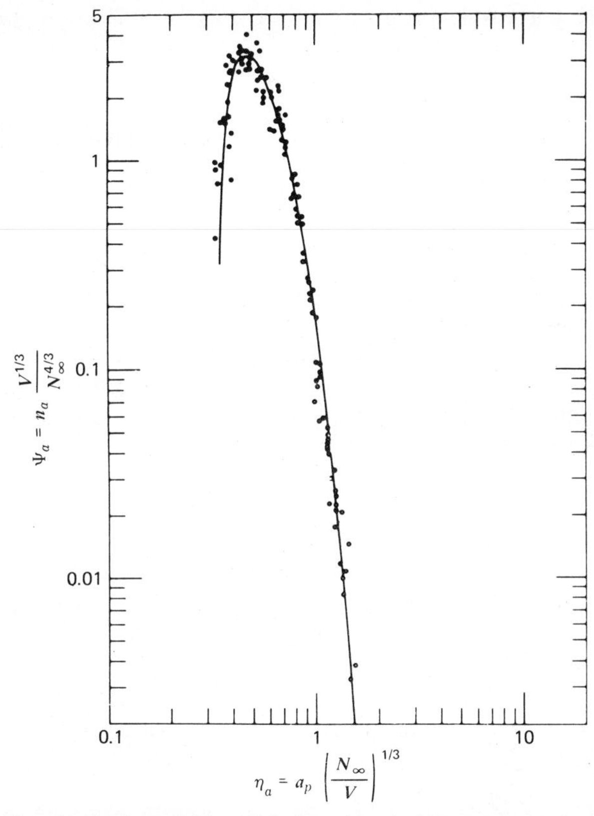

Fig. 10·4 Size distributions of Fig. 10·2 plotted in the self-preserving form (Husar and Whitby, 1973). The curve is based on the data. The self-preserving distribution for simultaneous coagulation and growth in the free-molecule range has not been calculated from theory, so no comparison is made.

10·6 The GDE for Turbulent Flow

In many if not most cases of practical interest, the fluid in which the particles are suspended is in turbulent motion. The problem is treated by making the Reynolds assumption that the fluid velocity and size distribution function can be written as the sum of mean and fluctuating components:

$$\mathbf{v} = \bar{\mathbf{v}} + \mathbf{v}' \tag{10·17}$$

$$n = \bar{n} + n' \tag{10·18}$$

It is assumed that homogeneous nucleation does not take place and that the particle current is proportional to the concentration through the growth law:

$$I = qn \tag{10·19}$$

The growth law can also be written as the sum of mean and fluctuating terms:

$$q = \bar{q} + q'$$

The fluctuations in growth rate result from local variations in the temperature and in the concentrations of the gaseous species involved in gas-to-particle transformation processes.

As an example, when the growth process is diffusion limited, $q = 2\pi D d_p p v_m / kT$, where p is the concentration of the diffusing gas at large distances from the surface and $p_s = 0$. Then for an isothermal system,

$$q = \frac{2\pi D d_p \bar{p} v_m}{kT} + \frac{2\pi D d_p p' v_m}{kT}$$

where p' is the fluctuating partial pressure of the condensing species. This form for the diffusional flux is based on a quasi-steady-state approximation and may not hold for rapid changes in concentration in the gas surrounding the particle.

Substituting (10·17), (10·18) and (10·19) in the GDE, averaging with respect to time, and making use of the equation of continuity, the result is

$$\begin{aligned}
\frac{\partial \bar{n}}{\partial t} + \mathbf{v} \cdot \nabla \bar{n} + \frac{\partial \bar{n}\bar{q}}{\partial v} + \frac{\partial \overline{n'q'}}{\partial v} = & -\nabla \cdot \overline{n'\mathbf{v}'} + D\nabla^2 \bar{n} \\
& + \frac{1}{2}\int_0^v \beta(\tilde{v}, v-\tilde{v})\bar{n}(\tilde{v})\bar{n}(v-\tilde{v})\, d\tilde{v} \\
& - \int_0^\infty \beta(v,\tilde{v})\bar{n}(v)\bar{n}(\tilde{v})\, d\tilde{v} \\
& + \frac{1}{2}\int_0^v \beta(\tilde{v}, v-\tilde{v})\, \overline{n'(\tilde{v})n'(v-\tilde{v})}\, d\tilde{v} \\
& - \int_0^\infty \beta(v,\tilde{v})\, \overline{n'(v)n'(\tilde{v})}\, d\tilde{v} - c_s \frac{\partial \bar{n}}{\partial z}
\end{aligned} \tag{10·20}$$

As a result of the time averaging, several new terms appear in the GDE. The third term on the left-hand side, the fluctuating growth term, depends on the correlation between the fluctuating size distribution function n' and the local concentrations of the gaseous species converted to aerosol. It results in a tendency for spread to occur in the particle size range—a turbulent diffusion through v space (Levin and Sedunov, 1968).

The first term on the right-hand side is a well-known form that represents the change in $\bar{n}$ resulting from turbulent diffusion. The separate components of the vector flux $n'\mathbf{v}'$ are usually assumed to follow an equation of the form:

$$\overline{n'v'_i} = -\epsilon_i \frac{\partial \bar{n}}{\partial x_i}$$

where the eddy diffusivity, ϵ_i, is a function of position and $i = 1,2,3$ refers to the components of the Cartesian coordinate system. The last two terms on the right-hand side are the contributions to coagulation resulting from the fluctuating concentrations. The importance of these terms for turbulent flows in ducts or in the atmosphere has not been evaluated in detail.

10·7 The GDE for Turbulent Stack Plumes

One of the most obvious manifestations of air pollution is the visible plume formed downwind from a stationary source. A relatively simple model for such systems is the continuous point source in a turbulent fluid with a mean velocity, $\bar{u}(x,z)$. The coordinate x is measured downwind from the source, parallel to the ground, and z is the coordinate perpendicular to the surface (Fig. 10·5). The velocity components in the y and z directions vanish, and diffusion in the x direction can be neglected compared with convection. Brownian diffusion is also neglected compared with eddy diffusion. These are the usuall simplifying assumptions made in the theory of diffusion of molecular species in turbulent stack plumes, and with them (10·20) becomes

$$\bar{u}\frac{\partial \bar{n}}{\partial x} + \frac{\partial \bar{n}\bar{q}}{\partial v} + \frac{\partial\, \overline{n'q'}}{\partial v} = \frac{\partial \epsilon_y(\partial \bar{n}/\partial y)}{\partial y} + \frac{\partial \epsilon_z(\partial \bar{n}/\partial z)}{\partial z}$$
$$+ \frac{1}{2}\int_0^v \beta(\tilde{v}, v-\tilde{v})\bar{n}(\tilde{v})\bar{n}(v-\tilde{v})\,d\tilde{v}$$
$$- \int_0^\infty \beta(v,\tilde{v})\bar{n}(v)\bar{n}(\tilde{v})\,d\tilde{v}$$
$$+ \frac{1}{2}\int_0^v \beta(\tilde{v}, v-\tilde{v})\,\overline{n'(\tilde{v})n'(v-\tilde{v})}\,d\tilde{v}$$
$$- \int_0^\infty \beta(v,\tilde{v})\,\overline{n'(v)n'(\tilde{v})}\,d\tilde{v} - c_s\frac{\partial \bar{n}}{\partial z} \qquad (10\cdot21)$$

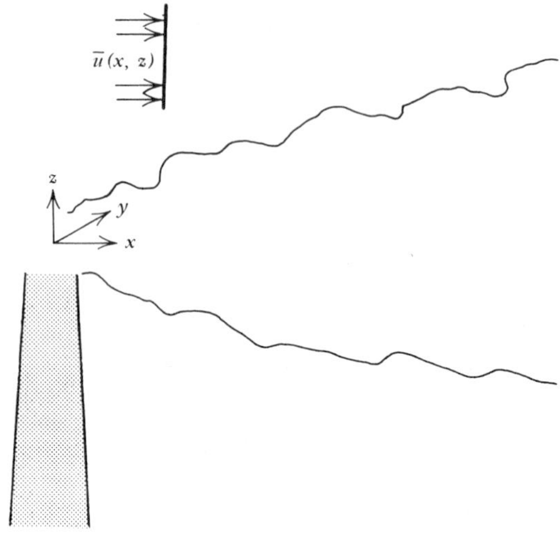

Fig. 10·5 Schematic diagram of turbulent stack plume with coordinate system employed in the text.

In general, numerical methods are necessary for the solution of (10·21).

Away from the immediate neighborhood of the source when concentrations have decreased sufficiently as a result of coagulation and dilution, additional coagulation can often be neglected. Restricting attention to particles smaller than a few micrometers for which sedimentation is not important, and neglecting the turbulent growth term in the absence of further information, (10·21) takes the following form:

$$\overline{U}\frac{\partial \bar{n}}{\partial x} + \frac{\partial \bar{n}\bar{q}}{\partial v} = \frac{\partial \epsilon_y (\partial \bar{n}/\partial y)}{\partial y} + \frac{\partial \epsilon_z (\partial \bar{n}/\partial z)}{\partial z} \tag{10·22}$$

for a constant mean velocity, $\overline{U}$.

We wish to know how the size distribution function changes with position downwind from the stack for a given form of the growth law, $\bar{q}$. This problem has a surprisingly simple solution for a growth law of the form:

$$q = V(v)X(x) \tag{10·23}$$

where V and X are arbitrary functions of v and x, respectively.

Only x and v appear as independent variables on the left-hand side, while y and z appear on the right-hand side. Hence we try as a solution

$$\bar{n} = \frac{1}{V} g(s)\overline{N}_{\infty}(x,y,z) \tag{10·24}$$

where g is an arbitrary function of $s=[\int_0 X dx/\bar{U} - \int_0 dv/V]$ and $\bar{N}_\infty(x,y,z)$ is the distribution in space of the number concentration of particles:

$$\bar{N}_\infty(x,y,z) = \int_0^\infty \bar{n}(v,x,y,z)\,dv \tag{10·25}$$

Substituting (10·24) in (10·22), the result is

$$X\bar{N}_\infty \frac{dg}{ds} + \bar{U}g\frac{\partial \bar{N}_\infty}{\partial x} - X\bar{N}_\infty \frac{dg}{ds} = g\frac{\partial \epsilon_y(\partial \bar{N}_\infty/\partial y)}{\partial y} + g\frac{\partial \epsilon_z(\partial \bar{N}_\infty/\partial z)}{\partial z}$$

Rearranging terms,

$$\bar{U}\frac{\partial \bar{N}_\infty}{\partial x} = \frac{\partial \epsilon_y(\partial \bar{N}_\infty/\partial y)}{\partial y} + \frac{\partial \epsilon_z(\partial \bar{N}_\infty/\partial z)}{\partial z}$$

This is the equation for the distribution in space of the number concentration of particles that can be derived independently from (10·22) by integrating over all values of v. Hence the form (10·24) is indeed a solution to (10·22) with the growth law (10·23).

A particularly simple solution is obtained in the case of a growth law of the form $q = vX$ corresponding to a droplet phase reaction (9·22). If the initial size distribution is of a power law form, $\bar{n} = bv^p\bar{N}_\infty$, substitution in (10·24) results in the following expression:

$$\bar{n} = bv^p\bar{N}_\infty \exp\left[-(p+1)\int \frac{X dx}{\bar{U}}\right]$$

This should be compared with the result for the growth of an aerosol in a box (Section 9·9).

The power law form cannot hold over the entire size range because singularities develop in certain integral functions. As a result, integrals of the type (10·25) must be truncated at the upper or lower ranges or both.

The growth law is, in general, a function of the local concentrations of the reactive gas phase species:

$$s = V(v)F(c_1, c_2, \ldots, c_i)$$

where the function $V(v)$ depends on the conversion mechanism (Table 9·3), and the c_i are the concentrations of the reactive gases. Since the concentrations are in general functions of x,y,z, it is clear that the form

(10·23) is an approximation. A complete solution to the problem would require simultaneous solution of the GDE and the equations of conservation of the gaseous components that participate in the reaction. While this general problem is beyond the scope of this text, a simple special case is discussed in the next section.

10·8 First-Order Reaction in a Plume

One of the most important gas-to-particle transformation processes in stack plumes is the conversion of SO_2 to particulate sulfate. The formation of particulate sulfate downwind from stationary sources contributes to visibility reduction and material damage, affects public health, and probably modifies the local climate in some cases. It is of considerable interest to be able to relate the rate of formation of particulate sulfur compounds to the composition of the stack gases and to meteorological conditions.

The forms of q and I are usually not known for the SO_2 conversion process; an alternative approach for the estimation of V is to work with the equation of conservation of SO_2. This equation is conjugate to (10·22), linked through the growth term that describes the gas-to-particle transformation process.

Field data on power plant plumes indicate that over limited distances downwind from the source, the rate of SO_2 depletion is first order in the concentration of SO_2 (Coutant et al., 1972). The steady-state equation of conservation of species for the average SO_2 concentration in the turbulent field with constant wind speed U then takes the form:

$$U\frac{\partial[SO_2]}{\partial x}=\frac{\partial\epsilon_y(\partial[SO_2]/\partial y)}{\partial y}+\frac{\partial\epsilon_z(\partial[SO_2]/\partial z)}{\partial z}-k[SO_2] \quad (10\cdot 26)$$

The first-order rate coefficient, k, describes the loss of SO_2 that is assumed to appear in the particulate phase. It is not necessary to make any assumptions concerning the mechanism of the conversion process or even the phase in which it takes place.

The solution of the turbulent diffusion equation (10·26) for a constant wind field, infinite in extent, is given by

$$[SO_2]=[S_T]e^{-kx/U} \quad (10\cdot 27)$$

where $[S_T]=[SO_2]+[S_p]$ is the total molar concentration of sulfur present in gas and particulate phases. This quantity is conserved since sulfur is assumed present either as gaseous SO_2 or in the particulate phase.

The value of $[S_T]$ is obtained from a solution to the equation of conservation of species for a nonreactive substance. A Gaussian plume model is often used (Turner, 1967) and in this case, (10·27) becomes

$$[SO_2]=\frac{Q_{SO_2}}{2\pi\sigma_y\sigma_z U}\exp\left\{-\frac{1}{2}\left[\left(\frac{y}{\sigma_y}\right)^2+\left(\frac{z}{\sigma_z}\right)^2\right]\right\}e^{-kx/U} \quad (10\cdot 28)$$

where Q_{SO_2} is the rate of SO_2 emission (moles/sec) and σ_y and σ_z are the standard deviations of the concentration distribution in the horizontal and vertical direction, respectively. The concentration of sulfur in the particulate phase (moles per unit volume of aerosol) is obtained by difference since the concentration of total sulfur in both phases is conserved:

$$[S]_p=\frac{Q_{SO_2}}{2\pi\sigma_y\sigma_z U}\exp\left\{-\frac{1}{2}\left[\left(\frac{y}{\sigma_y}\right)^2+\left(\frac{z}{\sigma_z}\right)^2\right]\right\}(1-e^{-kx/U}) \quad (10\cdot 29)$$

If the chemical form of sulfate in the particulate phase is known and local equilibrium is assumed between the water and droplet phases, the local value of the volumetric concentration can be obtained from the relation

$$V=\frac{[S]_p M}{\rho_s}$$

where ρ_s is the density of the sulfate solution in equilibrium with the local water vapor pressure in the plume and M is the molecular weight of the sulfate species.

The value of V calculated in this way should be added to the value resulting from the primary particulate matter originally in the plume. While the value of V gives no direct information on the size distribution, it is of value for approximate estimates of air pollution effects downwind from sources.

Although (10·29) for the concentration of particulate sulfur is the result of many approximations, it includes terms illustrative of most of the processes occurring in SO_2 containing stack plumes.

The assumption of a first-order rate law for the conversion process considerably simplifies the calculations. It is of interest to examine the conditions under which the first-order law is likely to be operative. This model is directly applicable if the chemical reaction leading to the production of particulate matter takes place in the gas and is first order or pseudo-first order with respect to the reacting species. It is assumed that the material generated by chemical reaction is transferred to the particulate phase as fast as it is generated.

When the conversion process is limited by reactions in a droplet phase, the situation is more complex. Suppose the rate of formation of sulfate per unit volume of solution is given by (9·8) and (8·15):

$$\frac{1}{v}\frac{dn_{SO_4^{2-}}}{dt} = k\left[SO_3^{2-}\right] = k'\frac{\left[SO_2\right]}{\left[H^+\right]^2}$$

where $[SO_2]$ refers to the gas phase and the other concentrations, to the solution phase. The constant k' includes equilibrium and Henry's law constants. The rate of loss of SO_2 by reaction per unit volume of gas is given by

$$\left[\frac{\partial\left[SO_2\right]}{\partial t}\right]_{reaction} = -\frac{k'V}{\left[H^+\right]^2}\left[SO_2\right]$$

where V is the volume of droplets per unit volume of gas. The rate of loss of SO_2 by reaction will be first order in $[SO_2]$, if the ratio $V/[H^+]^2$ is constant. The value of $[H^+]$ is related in a complex way to the gas phase concentrations of NH_3 and CO_2, and V depends on the total amount of material converted and the relative humidity. Numerical calculations are necessary to justify the adequacy of the assumption of pseudo-first-order kinetics on theoretical grounds.

10·9 Coagulation and Stirred Settling

Suppose a chamber is filled with an aerosol that is kept well mixed. The particles are coagulating and at the same time settling and diffusing to the walls. This type of model is used in analyzing the behavior of radioactive particles generated in a nuclear reactor accident and then collected in a vessel specially designed for the purpose. The contents of the containment structure are mixed as a result of natural convection induced by temperature gradients present under post-accident conditions. The effectiveness of the vessel in containing the products depends on the severity of the accident, with core melt-down representing a very severe test. Both theoretical and experimental models of such systems have been studied. The goal of the analysis is to predict the decay rate of the cloud in the vessel and the size of the particles, based on certain assumptions concerning the amount of material in the aerosol phase. Assumptions are also made concerning the leak rate from the containment vessel.

We consider only the one-dimensional problem in which the chamber is replaced by two parallel, horizontal plates a distance h apart, and sedimentation occurs only in the z direction. All three components of the mean velocity vanish. The equation for the mean concentration (10·20) takes the following form:

$$\frac{\partial \bar{n}}{\partial t} = \frac{\partial (D+\epsilon)(\partial \bar{n}/\partial z)}{\partial z} + \frac{1}{2}\int_0^v \beta(v-\tilde{v},\tilde{v})\bar{n}(v-\tilde{v})\bar{n}(\tilde{v})\,d\tilde{v}$$

$$-\int_0^\infty \beta(v,\tilde{v})\bar{n}(v)\bar{n}(\tilde{v})\,d\tilde{v} - c_s\frac{\partial \bar{n}}{\partial z} \qquad (10\cdot30)$$

The fluctuating coagulation terms can be neglected because the concentration is approximately uniform away from the walls. Moreover, since the system is well stirred, the concentration through the bulk is approximately uniform up to a small distance, δ, from the bottom of the chamber corresponding to the region where the eddy diffusion goes from its value in the bulk of the fluid to zero (at the wall). To a certain extent this distance is arbitrary and need not be defined exactly for this analysis.

The average concentration in the chamber is defined by

$$[\bar{n}] = \frac{1}{h}\int_0^h \bar{n}\,dz$$

The concentration in the bulk of the fluid is approximately equal to $[\bar{n}]$, since the volume of fluid bounded by δ and the wall is small. Outside region δ, the particle flux toward the bottom of the chamber is given by $[\bar{n}]c_s$, since the concentration gradients and, therefore, diffusion are negligible.

Assuming a quasi-stationary state, the flux of particles to the bottom of the chamber will also be $[\bar{n}]c_x$, that is,

$$\left[-(D+\epsilon)\frac{\partial \bar{n}}{\partial z} + \bar{n}c_s\right]_{z=0} = [\bar{n}]c_s \qquad (10\cdot31)$$

Integrating (10·30) term by term with respect to z over the height of the chamber with this boundary condition, the following results are obtained: The unsteady term takes the form:

$$\int_0^h \frac{\partial \bar{n}}{\partial t}\,dz = \frac{\partial \int_0^h \bar{n}\,dz}{\partial t} = h\frac{\partial [\bar{n}]}{\partial t}$$

The combined diffusion and sedimentation terms can be integrated as follows:

$$\int_0^h \frac{\partial\left[(D+\epsilon)(\partial\bar{n}/\partial z)-c_s\bar{n}\right]}{\partial z}\,dz = 0-[\bar{n}]c_s$$

Deposition on the roof of the chamber has been neglected and (10·31) has been introduced for the floor of the chamber. For the coagulation term,

$$\int_0^h \int_0^v \beta(v-\tilde{v},\tilde{v})\bar{n}(v-\tilde{v})\bar{n}(\tilde{v})\,d\tilde{v}\,dz$$

$$=\int_0^v \beta(v-\tilde{v},\tilde{v})\left[\int_0^h n(v-\tilde{v})n(\tilde{v})\,dz\right]d\tilde{v}$$

Except for a small region near the wall, $\bar{n}$ is almost independent of z at any time. Hence

$$\int_0^h \bar{n}(v-\tilde{v})\bar{n}(\tilde{v})\,dz = [\bar{n}(v-\tilde{v})][\bar{n}(\tilde{v})]h$$

with an analogous result for the other coagulation term. Thus the result of integrating (10·30) with respect to z is

$$\frac{\partial[\bar{n}]}{\partial t}=\frac{1}{2}\int_0^v \beta(v-\tilde{v},\tilde{v})[\bar{n}(v-\tilde{v})][\bar{n}(\tilde{v})]\,d\tilde{v}$$

$$-\int_0^\infty \beta(v,\,\tilde{v})[\bar{n}(v)][\bar{n}(\tilde{v})]\,d\tilde{v}-\frac{c_s[\bar{n}]}{h} \qquad (10\cdot32)$$

This is the equation that is usually solved in calculating simultaneous coagulation and settling in a well-mixed chamber. Numerical solutions for special values of the collision frequency function have been obtained by Lindauer and Castleman (1971). They report results for the decay in the mass concentration as a function of chamber height and time.

In the coagulation process, small particles from the low end of the size distribution are transferred to the large particle size range. The large particles formed in this way settle to the floor of the chamber. A quasi-steady state may develop for the upper end of the distribution in which the rate of formation in a given size range by coagulation is equal to the rate of loss by sedimentation. Equating the first and last terms on the right-hand side of (10·32), the result for the continuum regime is

$$\frac{kT}{3\mu}\int_0^v \left[1+\left(\frac{v-\tilde{v}}{\tilde{v}}\right)^{1/3}\right] n(\tilde{v})n(v-\tilde{v})\,d\tilde{v} = \frac{c_s n}{h} \qquad (10\cdot 33)$$

where $n=[\bar{n}]$. A particular solution that satisfies (10·33) and the requirement that the total aerosol volume per unit volume of gas is finite is

$$n = A_1 v^{-1/3} e^{-A_2 v}$$

where A_1 and A_2 are constants (Fig. 10·6). This can be tested by

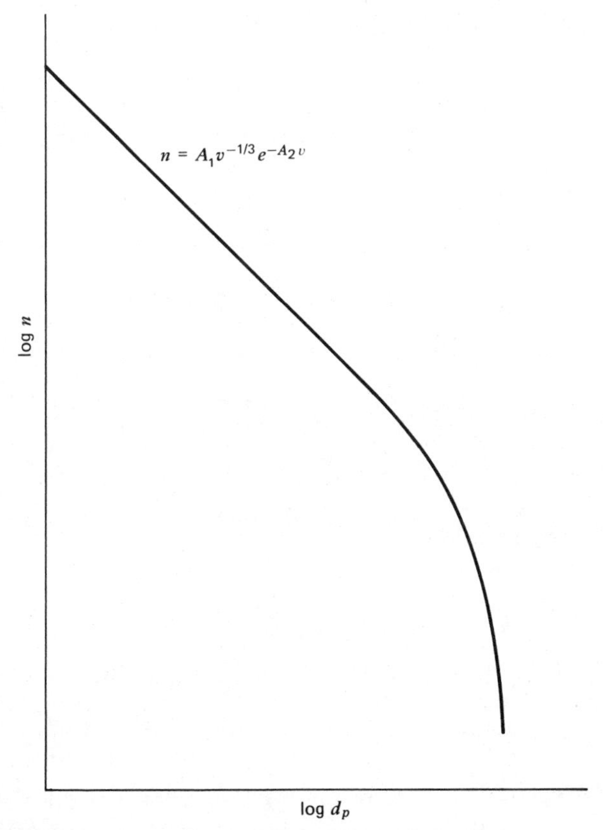

Fig. 10·6 Upper end of the size distribution for the steady state between coagulation and sedimentation. The solution breaks down for small values of the particle diameter because the second term on the right hand side of (10·32)—representing loss by coagulation—has been neglected in (10·33).

substitution in (10·33), which also shows that

$$A_1 = \left(\frac{\pi}{6}\right)^{1/3} \frac{\rho_p g}{\pi k T h [3 + B(2/3, 2/3)]}$$

where B is the beta function. The constant A_1 has dimensions L^{-5}.The total aerosol volume concentration is given by

$$V = A_1 \int_0^\infty v^{2/3} e^{-A_2 v} dv$$

The integral can be expressed in terms of a gamma function, and A_2 can be evaluated in this way

$$A_2 = \frac{[A_1 \Gamma(5/3)]^{3/5}}{V^{3/5}}$$

The rate of particle deposition (volume of particulate matter per unit surface per unit time) has the dimensions of velocity and is given by

$$\begin{aligned} \begin{matrix}\text{sedimentation}\\ \text{flux}\end{matrix} &= \int_0^\infty c_s v n(v)\, dv \\ &= \left(\frac{6}{\pi}\right)^{2/3} \frac{\rho_p g}{18\mu} A_1 \int_0^\infty v^{4/3} e^{-A_2 v}\, dv \\ &= \left(\frac{6}{\pi}\right)^{2/3} \frac{\Gamma(7/3)}{18[\Gamma(5/3)]^{7/5}} \frac{\rho_p g}{\mu A_1^{2/5}} V^{7/5} \end{aligned}$$

Hence the sedimentation flux is proportional to $V^{7/5}$ for the steady state. The analysis is approximate because the second coagulation term on the right-hand side of (10·32) is neglected.

In a quasi-steady-state situation, the concentration of suspended material changes slowly with time. The sedimentation flux then represents the rate of loss of material from the volume of the chamber above unit area of floor. The results of the calculation of the volumetric concentration compare well with the numerical computations of Lindauer and Castleman (1971) for long times.

10·10 Coagulation and Deposition by Convective Diffusion

The combustion of leaded gasoline in automobile engines is a major source of submicron particles in urban atmospheres. Ethylene bromide is added to the gasoline to combine with the combustion products of lead tetraethyl and form volatile lead salts. The salts remain in the vapor phase thus minimizing the deposition of particulate matter in the cylinders. In the tailpipe, condensation of the lead salts takes place (Fig. 10·7) as the gases cool; a secondary increase in mass concentration of particulate matter occurs closer to the exit as some of the higher molecular weight compounds condense.

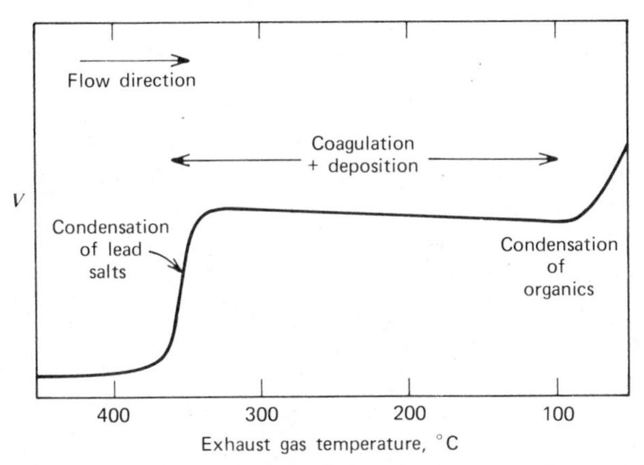

Fig. 10·7 Variation of the aerosol volume concentration with temperature along the exhaust pipe of an automobile burning leaded gasoline. The figure is based on the data of Ganley and Springer (1974). The large initial increase in concentration after the gas leaves the exhaust manifold results from the condensation of lead salt vapors, probably by homogeneous nucleation. Toward the end of the tailpipe when the gas temperature falls below 100°C, the increase in concentration probably results from the condensation of organic vapors present in the exhaust gases.

Condensation probably takes place initially by homogeneous nucleation leading to the formation of high concentrations of very small particles. These then coagulate and some deposit on the walls of the tailpipe by combined Brownian diffusion and thermophoresis since the walls are cooler than the exhaust gases. In principle, (10·20) must be solved with an additional term for transport by thermophoresis—a formidable task.

An approximate calculation can be made by adopting the following simplified model: A gas carrying many particles smaller than the mean free path flows down a straight pipe with walls at constant temperature. Particles deposit by diffusion on the walls that behave as a perfect sink ($n=0$).

If the rate of coagulation is rapid compared with the rate of loss to the walls of the pipe, the two processes—coagulation and surface deposition—can be treated separately. For the purposes of the calculation, the flow can be broken into two parts: In the turbulent core, coagulation controls the shape of the size distribution, which is then determined by the solution to the equation for coagulation under steady flow conditions.

Near the surface, the flux of particulate matter to the wall is given by the expression

$$\bar{J} = -(D+\epsilon)\frac{\partial \bar{n}}{\partial z} \tag{10·34}$$

when diffusion alone controls transport. The bars that denote time average quantities are omitted in the rest of this section to simplify the notation. If thermal gradients are present, these must also be included in the driving forces for surface deposition. Equation (10·34) can be integrated for different forms of the eddy diffusion coefficient $\epsilon(z)$ as shown in Section 3·11. The result for particle diffusion ($\nu/D \gg 1$) with $\epsilon \sim y^4$ in the viscous sublayer is given by

$$J = kn = 0.079\ Unf^{1/2}Sc^{-3/4}$$

This expression, instead of the alternate dependence on $Sc^{-2/3}$, is used because it simplifies the forms of the expressions derived in the calculations that follow. The total local volume flux of particles to the walls of the pipe is given by

$$\int_0^\infty Jv\,dv = 0.079\frac{Uf^{1/2}}{\nu^{3/4}}\int_0^\infty nD^{3/4}v\,dv \tag{10·35}$$

For spherical particles much smaller than the mean free path of the surrounding gas molecules, the diffusion coefficient is given by (2·14) and (2·15) with $\alpha = 0$:

$$D = \frac{kT}{f} = \frac{3kT}{2pd_p^2}\left[\frac{kT}{2\pi m}\right]^{1/2}$$

where m is the molecular mass of the gas molecules and mp/kT has been substituted for ρ. For gas molecules that behave as rigid elastic spheres, this expression can also be written as

$$D \approx \frac{4D_{11}}{\sqrt{2}}\left(\frac{v_m}{v}\right)^{2/3}$$

where D_{11} is the coefficient of self-diffusion for the gas and v_m is the molecular volume of the gas molecules. Substituting in (10·35), the result is

$$\int_0^\infty Jv\,dv = \frac{0.17\,Uf^{1/2}v_m^{1/2}}{(\nu/D_{11})^{3/4}} \int_0^\infty nv^{1/2}\,dv \tag{10·36}$$

The calculation is easily carried out when the size distribution is self-preserving. Substituting $n=(N_\infty^2/V)\psi(\eta)$ (where V and N_∞ are time-averaged quantities in keeping with the turbulent nature of the flow) in (10·36) and taking $\nu/D_{11}=0.7$, the value for air, the result is

$$\int_0^\infty Jv\,dv = 0.23\,Uf^{1/2}v_m^{1/2}V^{1/2}N_\infty^{1/2}\int_0^\infty \psi\eta^{1/2}d\eta \tag{10·37}$$

The integral $\int_0^\infty \psi\eta^{1/2}d\eta = 0.89$ when evaluated from the self-preserving distribution for the free molecule range. The heaviest mass deposition occurs upstream where N_∞ and V are largest.

The change in V with x is given by a mass balance on a small element of the pipe wall (Fig. 10·8):

$$\left(\frac{\pi d_{pipe}^2}{4}\right) U\,dV = -\pi d_{pipe}\left[\int_0^\infty Jv\,dv\right]dx$$

Substituting (10·37), the result is

$$-\frac{d_{pipe}}{4}\,dV = 0.20 f^{1/2}v_m^{1/2}V^{1/2}N_\infty^{1/2}\,dx \tag{10·38}$$

The variation of N_∞ with distance is obtained from (7·37), the expression for free molecule coagulation:

$$U\frac{dN_\infty}{dx} = -0.334\left(\frac{3}{4\pi}\right)^{1/6}\left(\frac{6kT}{\rho_p}\right)^{1/2} V^{1/6}N_\infty^{11/6} \tag{10·39}$$

Combining (10·39) with (10·38), the result is

$$\frac{dV}{V^{1/3}} = \frac{0.25 f^{1/2}v_m^{1/2}U}{d_{pipe}(3/4\pi)^{1/6}(6kT/\rho_p)^{1/2}}\,\frac{dN_\infty}{N_\infty^{4/3}} \tag{10.40}$$

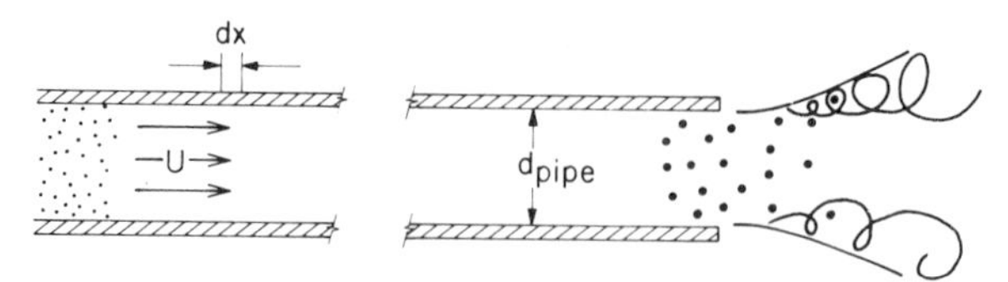

Fig. 10·8 Schematic diagram showing processes occurring in an automobile tailpipe. High concentrations of small particles form by homogeneous nucleation near the entrance. Smaller concentrations of larger particles leave at the tailpipe exit.

Integrating with the initial condition $V = V_0$ at $N_\infty = \infty$,

$$\left(V_0^{2/3} - V^{2/3}\right) = \frac{2B}{N_\infty^{1/3}} \qquad (10\cdot 41)$$

where B is the constant of (10·40).

Substituting (10·41) in (10·39), a differential equation is obtained for the variation of particulate volume along the pipe:

$$-\frac{\left[V_0^{2/3} - V^{2/3}\right]^{3/2} dV}{V^{1/2}} = \frac{0.82}{d_{pipe}} f^{1/2} v_m^{1/2} (2B)^{3/2} dx \qquad (10\cdot 42)$$

The deposition up to any point in the pipe can be obtained by numerical integration of this expression.

In practice, the surfaces over which the gas flows become roughened as a result of particle deposition, corrosion, and scaling. Hence the deposition rates are probably significantly greater than those calculated from (10·42) for smooth pipes. In addition, the tailpipe flow is unsteady because of the usual patterns of driving in traffic. These factors contribute to the reentrainment of agglomerates formed on the surface and the appearance of large mass fractions of coarse particles ($>10\ \mu$m) in the exhaust gases (Habibi, 1973). The actual size distribution of particles leaving the tailpipe is then considerably broader than the self-preserving distribution both because of reentrainment and the variation in residence times across the pipe.

10·11 Continuous Stirred Tank Reactor

An aerosol flows steadily into and out of a chamber that is kept well stirred (Fig. 10·9). In the chamber, processes that modify the size distribution take place of the type represented by the terms in the GDE. If the flow is maintained for a sufficiently long time—about five times

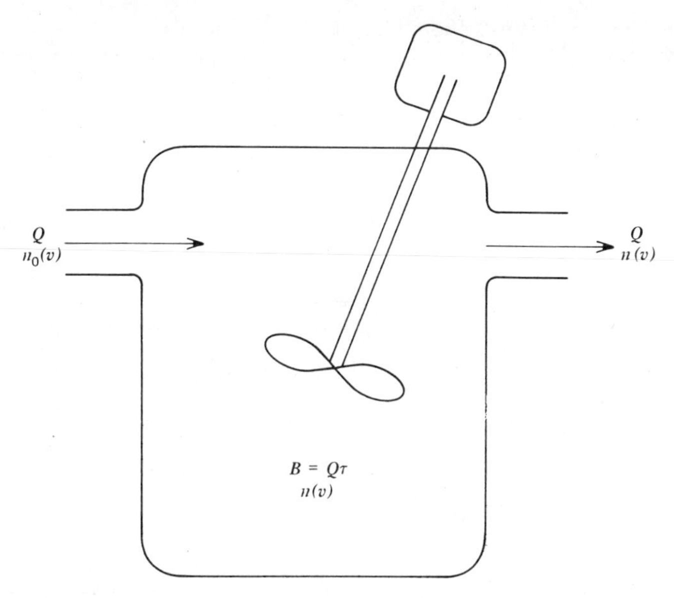

Fig. 10·9 Stirred tank reactor of volume B. The distribution at the inlet is $n_0(v)$. The distribution in the reactor and in the exit stream is $n(v)$.

the mean residence time—the chamber contents tend to approach a steady state. The steady-state distribution is determined by the size distribution of the input, the flow rate and the growth, coagulation, and deposition processes taking place within the tank. Such a system is analogous to the continuous stirred tank reactor (CSTR) often employed in modeling chemical reactors in industry or the laboratory. A basic assumption is that the concentration in the reactor is everywhere uniform and equal to the concentration at the exit. Such aerosol systems have not been carefully studied experimentally. They are of interest because of their potential use as aerosol generators. Since they represent a simple model of a chemical reactor, they may be useful in the design of processes for minimum or at least controllable pollution production.

The CSTR for aerosols is in some respects simpler to analyze than the unsteady or spatially varying systems considered previously. We consider a reactor of volume B with an aerosol entering of size distribution $n_0(v)$. In the steady state, for a volumetric flow of gas Q, a balance on the number of particles in the size range v to $v+dv$ gives

$$Qn(v) = Qn_0(v) + B\begin{pmatrix}\text{net rate of formation}\\ \text{per unit volume}\end{pmatrix} - B\begin{pmatrix}\text{rate of deposition}\\ \text{per unit volume}\end{pmatrix}$$

We consider the following simple problem that can be solved analytically: An aerosol with size distribution $n_0(v)$ enters the reactor continuously. Chemical or physical processes within the reactor produce condensable species that deposit on the aerosol particles. In the steady state, the balance on particles in the size range v to $v+dv$ becomes

$$\frac{n_0}{\tau} = \frac{n}{\tau} + \frac{d(nq)}{dv} \tag{10·43}$$

where the mean residence time $\tau = B/Q$. Particle deposition on the walls has been neglected. Equation (10·43) is an ordinary linear equation. Taking nq as the dependent variable, the integrating factor is $\exp\left(\int dv/q\tau\right)$ and the solution to (10·43) with $n=0$ at $v=0$ is given by

$$n = \frac{1}{q}\exp\left(-\int\frac{dv}{q\tau}\right)\int_0^v \exp\left(\int\frac{dv}{q\tau}\right)\frac{n_0}{\tau}\,dv \tag{10·44}$$

If the growth rate is diffusion controlled, $q = Av^{1/3}$ where A is a constant. Now suppose the aerosol entering the chamber is monodisperse, that is, $n_0 = N_0\delta(v - v_0)$ where $\delta(v-v_0)$ represents the Dirac delta function and N_0 is the number of particles per unit volume of size v_0. Then the integral on the right-hand side of (10·44) becomes $\{\exp((3/2)Av_0^{2/3}/\tau)\}N_0/\tau$ and the size distribution of the aerosol leaving the chamber is

$$n(v) = \frac{1}{Av^{1/3}}\left\{\exp\left[\frac{3}{2}\frac{A\left(v_0^{2/3} - v^{2/3}\right)}{\tau}\right]\right\}\frac{N_0}{\tau} \qquad (v > v_0) \tag{10·45}$$

Thus the result of particle growth in the chamber is to convert a monodisperse aerosol into a polydisperse aerosol. This is the reverse of what occurs in a condensation aerosol generator (Section 9·8).

Why does this spread in the distribution occur? The reason is that there is a distribution of residence times for the particles in the CSTR. Some particles stay for times longer than τ and others for times shorter so that the growth period varies among the particles leaving the reactor at any time.

In practice, deposition on the walls of the chamber must be taken into account. Small particles will be preferentially removed by diffusion, and large particles, by turbulent deposition and sedimentation. Quantitative estimates of such deposition rates are usually difficult to make. The particle size distribution in the effluent from a CSTR has been discussed by Bransom, Dunning and Millard (1949) for the case of homogeneous nucleation in the reactor.

Problems

1. Derive a general expression, based on the GDE, for the change in the total surface area of an aerosol with time and position.

2. For the size distribution of Problem 5, Chap. 1, estimate the contributions of the growth and coagulation terms to the change in the distribution function with time $\partial n/\partial t$ for particles 0.5 μm in diameter. Assume a growth law of the form corresponding to a diffusion limited process with a Kelvin cut-off particle diameter of 0.2 μm. The *total* rate of gas-to-particle conversion is 20 $\mu g/m^3$ hr and the density of converted material is 1.5 g/cm^3.

3. An aerosol issuing from a point source is dispersed in a steady turbulent plume in the atmosphere. Derive an expression for the variation of the extinction coefficient, b, (Chap. 5) with position in the plume assuming (*a*) the only mechanism affecting the light scattering portion of the spectrum is turbulent diffusion and (*b*) the only mechanisms are turbulent diffusion and growth.

4. An aerosol is injected at a point into a turbulent gas over a very short time period (an instantaneous point source). Set up the equation describing the dynamics of the size distribution. Assume that the mean flow is uniform so that the cloud, once released, spreads only radially with respect to the mean flow. Settling is not important.

5. A well-stirred vessel contains a coagulating, sedimenting aerosol. Assume that the upper end of the size distribution has reached a steady or quasi-steady state such that the rate of loss by sedimentation is equal to the rate of formation by coagulation. Consider two cases, the mass (or volumetric) concentration of aerosol in one is double that of the other. In both cases, the steady state has been attained. What are the relative rates of loss of matter by sedimentation?

6. As an idealized model for the automobile exhaust aerosol, assume the distribution is self-preserving and that deposition occurs as discussed in Section 10·10. The mass loading of aerosol at a point 15 ft upstream from the exit is 1 mg/ft^3 (STP) and the average temperature 400°F. The density of the aerosol material is 6 g/cm^3, and the gas velocity 90 ft/sec. Estimate the fraction of the aerosol that deposits on the walls of the tailpipe by diffusion.

7. A monodisperse aerosol enters a CSTR in which growth occurs only by diffusion. Show by integration of (10·45) that the total number concentration is conserved.

References

Bransom, S. H., Dunning, W. J., and Millard, B. (1949) *Faraday Soc. Discussions*, No. 5, 83.

Coutant, R. W., Merryman, E. L., Barrett, R. E., Giammar, R. D., and Levy, A. (1972) Battelle Mem. Inst., Columbus, Ohio, EPA Contract CPA 70-121, APTD-1107.

Friedlander, S. K., and Hidy, G. M. (1969) New Concepts in Aerosol Size Spectrum Theory in Podzimek, J. (Ed.) *Proceedings of the 7th International Conference on Condensation and Ice Nuclei*, Academia, Prague.

Friedlander, S. K. (1973) *Aerosol Sci.*, **1**, 295.

Friedlander, S. K. (1960) *Physics of Fluids*, **3**, 693.

Ganley, J. T., and Springer, G. S. (1974) *Environ. Sci. Technol.*, **8**, 340.
Habibi, K. (1973) *Environ. Sci. Technol.*, **7**, 223.
Hulburt, H. M., and Katz, S. (1964) *Chem. Engng. Sc.*, **19**, 555.
Husar, R. B., and Whitby, K. T. (1973) *Environ. Sci. Technol.*, **7**, 241.
Levin, L. M., and Sedunov, Y. S. (1968) *Pure Appl. Geophys.*, **69**, 320.
Lindauer, G. C., and Castleman, A. W., Jr. (1971) *Aerosol Sci.*, **2**, 87.
NOTE: There is an extensive literature in the field of containment vessel design, composed of unclassified reports issued by the Atomic Energy Commissions of various countries and their contractors. See also: International Atomic Energy Agency (IAEA) (1968) *Treatment of Airborne Radioactive Wastes*, IAEA, Vienna.
Pich, J., Friedlander, S. K., and Lai, F. S. (1970) *Aerosol Sci.*, **1**, 115.
Turner, D. B. (1967) *Workbook of Atmospheric Dispersion Estimates,* Office of Air Programs Publications No. AP-26.

CHAPTER 11

Air Quality–Emission Source Relationships

Relating air quality to emission sources is a central problem in air pollution control. In this chapter, air quality for particulate pollution is defined in terms of certain important, easily measured quantities. The rest of the chapter is concerned with methods of estimating source contributions to the degradation of air quality. If the contributions of the sources are known, a rational strategy for their control can be devised. By suitable optimization techniques, the minimum cost necessary to achieve any desired level of control can, in principle, be determined.

Estimates of source contributions to particulate pollution are often based on emission inventories of particulates. An example of such an inventory for Los Angeles is shown in Table 11 · 1, which illustrates the wide variety of sources of primary particulates. Such inventories by themselves are of limited value in determining contributions to the aerosol concentration at a given point, such as an air monitoring station. Emission inventories make no provision for natural background, or particle deposition between source and point of measurement. They give no information on atmospheric dispersion. Most important, they do not account for gas-to-particle conversion. The products of such conversion processes contribute significantly to the total mass of aerosols and affect visibility and health.

More satisfactory methods have recently been developed for relating particulate air quality to emissions: The method of *material balances* requires field data on particulate composition and emission source characteristics. This method has been applied with some success to several different sets of data for the Los Angeles and Chicago air basins. By *predictive modeling*, atmospheric concentrations can be calculated as a function of position and time based on source characteristics, growth laws, deposition velocities, and the wind field. Field data are in principle not required. Both methods, and a hybrid, are discussed in this chapter.

TABLE 11·1
1973 EMISSION INVENTORY FOR LOS ANGELES COUNTY
DAILY/AVERAGE, TONS/DAY
(Birakos, 1974)

Sources	Reactive Hydrocarbons	NO_x	Particulates (primary)	SO_2	CO
Combustion of fuel[a]	—	240	35	175	—
Petroleum refining	5	20	5	55	—
Petroleum marketing	30	10	—	—	—
Chemical processing	—	—	—	60	—
Organic solvent	40	—	15	—	—
Metallurgical	—	5	5	15	5
Mineral processing	—	5	5	15	5
Total Stationary	75	280	65	320	10
Gasoline-Powered vehicles[b]	690	775	40	30	7090
Diesel-Powered vehicles	—	20	10	—	20
Aircraft	5	15	15	5	160
Ships and railroads	—	25	—	10	20
Total transportation	695	835	65	45	7290
Grand total	770	1115	130	365	7300

Note:

Reactive hydrocarbons refers to organic gases that are photochemically reactive.

NO_x = Oxides of nitrogen.

All reported values have been rounded to the nearest 5 ton/day.

A dash (—) indicates less than 2.5 ton/day.

[a]This category includes power plants, refineries, and other industrial and residential sources. Daily power plant emissions vary due to changes in fuel type and plant operation.

[b]Motor vehicle emissions have been calculated using the ARB surveillance date (7-mode cycle).

11·1 Air Quality

Characterizing air quality is not an easy task. For scientific purposes, the particulate component can be defined fairly completely in terms of the size–composition probability density function (Chap. 1), but this quantity is not of direct use in practical applications because of the difficulty of experimental measurement and the large number of variables involved. Instead, certain relatively simple integral functions are commonly used for air quality characterization.

The total mass of particulate matter per unit volume of air is perhaps the simplest, and it is on this quantity that current federal standards for particulate pollution are based. The primary standard is 75 $\mu g/m^3$, annual geometric mean, and 260 $\mu g/m^3$, maximum 24 hr concentration not to be exceeded more than once per year. The primary standard is in principle linked to health effects.

However, total particulate mass is not an adequate measure of air quality for several reasons. Atmospheric residence times, nucleating characteristics, light scattering, and lung deposition are all sensitive functions of particle size. The cloud nucleating characteristics of one 5 and 1000 0.5 μm particles, which are of equal mass, are entirely different! These factors can be partially accounted for by adopting separate standards for particles above and below 1 μm.

Another important air quality parameter, the visibility, is closely related to the aerosol extinction coefficient. The extinction coefficient, like the total mass, is an integral function of the particle size distribution. However, for urban pollution it tends to weight the contribution of material in the 0.1 to 1.0 μm size range most heavily (Chap. 5). The concentrations of chemical species in the particulate phase are air quality parameters of public health interest. Certain trace metals, sulfates, nitrates, and carcinogens are of most interest. Their distribution with respect to particle size should be taken into account but are difficult to measure on a routine basis. Some of the species present in highest concentration in the particulate phase, including sulfates, nitrates, and certain organic compounds, result from gas-to-particle conversion.

In summary, three parameters can be used to characterize air quality with respect to particulate pollution: the total mass per unit volume of air and its contributions from different size fractions, the visibility, and the chemical composition expressed in terms of mass of species per unit volume of air, with contributions from different size fractions. These quantities provide useful (although not complete) information for estimating the effects of particulate pollution on man; at the same time, they can be measured with commercially available instrumentation. In the following sections, methods of relating the air quality variables to source contributions will be discussed.

11·2 Relationship of Mass and Chemical Composition to Sources

A standard Federal reference method has been prescribed for the determination of mass loadings of atmospheric particulates (Federal Register, 1971). Air is drawn through a filter by a blower at a rate of 40 to 60 ft^3/min. Glass fiber filters with a collection efficiency of at least 99% for $d_p = 0.3$ μm are recommended. Total mass is determined by weighing at 15 to 35°C and less than 50% relative humidity. The chemical composition of the deposited material can be determined by various analytical methods.

The total mass of aerosol per unit volume of air is the sum of the contributions of various sources:

$$\rho = \sum_j m_j \tag{11·1}$$

where ρ, the local mass concentration in mass of species per unit volume of air, is a function of position and is averaged over the sampling time. The source contribution, m_j, is the mass of material from source j per unit volume of air at the *point of measurement*. The right-hand side of (11·1) represents the sum of the source contributions to the total mass, assuming the sources can be considered discrete. In practice, it is usually convenient to lump certain classes of sources, automobiles, power plants, and so on, such that j refers to each class of source. The source contributions, m_j, are exact quantities; each type of source makes a well-defined contribution to the aerosol measured at a point. The evaluation of m_j is one of the main goals of this chapter. The concentration ρ_i of an element i in the aerosol at a given point is related to the source contributions by

$$\rho_i = \sum_j c_{ij} m_j \tag{11·2}$$

where c_{ij} is the mass fraction of element i in m_j.

When samples are obtained in discrete size fractions as with the cascade impactor, relations similar to (11·1) and (11·2) hold for the discrete fractions:

$$\Delta\rho = \sum_j \Delta m_j$$

and

$$\Delta\rho_i = \sum_j c_{ij} \Delta m_j$$

where the symbol Δ denotes the material in the discrete size range between d_{p1} and d_{p2}.

The dependence of the air quality parameters on the properties of the emission sources is thus reduced to the two quantities Δm_j, the mass contributed by source j per unit volume of air to the size range Δd_p and the mass fraction c_{ij}, the mass of chemical species i in unit mass of material from source j.

11·3 Evaluation of Source Contributions: Material Balance Method

Source contributions to particulate pollution have been estimated by balances on the chemical composition of the collected aerosol material (Heisler, Friedlander, and Husar, 1973; Gatz, 1975; and Gartrell and Friedlander, 1975). The method consists of (1) estimating certain primary source contributions to the pollution at a point using chemical elements as tracers, (2) supplementing these estimates with emission inventory data for those primary sources for which characteristic tracers are not available, and (3) determining the contributions of gas-to-particle conversion from the measured values of sulfate, nitrate, organic, and ammonium ion. In this way, data of different types are reduced to a common basis—source contributions to the mass of aerosol at a given sampling point. These calculations permit the estimation of the mass contributions, m_j, of the various sources. By (11·1), the sum of these mass contributions must equal the total measured mass. This provides a check for consistency on the overall method of calculation, although not a severe one.

The chemical tracer method of evaluating m_j depends on the inversion of (11·2), which in matrix notation can be written in terms of the vectors M and P and the inverse matrix C^{-1} of c_{ij}

$$M = C^{-1}P$$

The values of ρ_i, the components of P, are measured at a given sampling site. The matrix c_{ij} should also correspond to the point of measurement. Usually, however, it is assumed that the value of c_{ij} is equal to the value at the source, and fractionation by exchange with the gas phase or by sedimentation is neglected. Hence in carrying out the chemical element balance, it is necessary to choose elements for which fractionation is not important.

In studies of Los Angeles' particulate pollution, sodium, aluminum, calcium, lead, potassium, vanadium, and magnesium were used as the tracer elements in calculating contributions by the primary sources. Values for these elements shown in Table 11·2 were measured in Pasadena. For these elements, automobile exhaust, soil dust, sea salt, cement dust, and flyash from fuel combustion were believed to be the dominant sources. The source concentration matrix, c_{ij}, determined

TABLE 11·2
CONCENTRATIONS OF ELEMENTS USED IN CHEMICAL ELEMENT BALANCES, 24 hr averages (Gartrell and Friedlander, 1975)[a]

	Pasadena 9/20/72		Pomona 10/24/72		Riverside 9/20/72	
Na	0.716	± 0.014	2.06	± 0.20	1.13	± 0.19
Al	1.03	0.33	1.30	0.42	2.35	0.07
Ca	0.94	0.16	1.8	0.28	1.49	0.07
V	0.00482	0.00026	0.0142	0.00052	0.0088	0.0004
Pb	2.03	0.077	2.85	0.120	1.55	0.06
Mg	0.22[b]	0.22	0.22[b]	0.22	0.97	0.38
K	0.36	0.036	0.44	0.044	0.68	0.068

[a]Values in micrograms per cubic meter; errors represent analytical errors.
[b]Value assumed from lower limit of detection.

from the literature or by experimental measurement, is given in Table 11·3. Since there were seven elements and five sources, the calculation was carried out by minimizing the mean square error between measured and calculated values, weighted by the experimental error.

Convenient tracers are not available for certain sources such as aircraft emissions and the large number of relatively small industrial sources in Los Angeles. Hence the contribution of these sources to the total mass loading was obtained from the emission inventory prepared by the local air pollution control agency. This introduces errors resulting from inaccuracies associated with the emission inventory and, more important, the effects of the wind field on the dispersion of this material.

The products of secondary conversion products, sulfates, nitrates, organics, and ammonium ion, were measured. The sulfate comes primarily from the oxidation of SO_2 produced by the combustion of sulfur containing fuel oil (Table 11·1). The nitrate results from the oxidation of NO_2 emitted by automobiles with significant input also from fuel combustion. The organic compounds in the aerosol phase result largely from the oxidation of certain olefins (Chap. 9). The sources of ammonia are uncertain. Both cattle feed lots (there are many in Los Angeles county!) and combustion are probably important.

The results of the calculation of the source contributions for the Pasadena, Pomona, and Riverside aerosols are shown in Table 11·4. The importance of the contributions of the secondary conversion products, including sulfates, nitrates, and organic compounds, to the total mass is clear. Their contribution to visibility degradation is even more important as shown in the next section.

TABLE 11·3
SOURCE CONCENTRATIONS OF PARTICULATE MATTER
(Friedlander, 1973)

	Percentages					
	Sea Salt	Soil Dust	Auto Exhaust	Fuel Oil Flyash	Portland Cement	Tire Dust
C as carbon	—[a]	—	—	u[b]	—	29.3
C compounds	—	—	40.3[c]	u	—	58[d]
Na	30.6	2.5	u	5	0.4	u
Mg	3.7	1.4	u	0.06	0.48	u
Al	—	8.2	u	0.8	2.4	u
Si	—	20	u	1	10.7	u
S	2.6	—	u	u	—	u
Cl	55.0	—	6.8	—	—	u
K	1.1	1.5	u	0.2	0.53	u
Ca	1.16	1.5	u	1.3	46.0	u
Ti	—	0.4	u	0.06	0.144	u
V	—	0.006	u	7	—	u
Cr	—	—	u	0.1	—	u
Mn	—	0.11	u	0.06	—	u
Fe	—	3.2	0.4	6	1.09	u
Co	—	0.002	u	0.2	—	u
Ni	—	0.004	u	2	—	u
Cu	—	0.008	u	0.2	—	u
Zn	—	<0.01	0.14	0.02	—	1.5
Br	0.19	—	7.9	—	—	u
I	1.4×10^{-4}	—	u	—	—	u
Ba	—	0.06	u	0.1	—	u
Pb	—	0.02	40.0	0.07	—	u

[a]—Negligible.
[b]Unknown, u.
[c]As a tarry substance assumed 90% C.
[d]Mostly as a copolymer of styrene and butadiene.

Many different sources emit the SO_2, NO_x, and reactive organic compounds converted to aerosols. Their relative contributions to sulfate, nitrate, and organic compounds can be estimated from an emission inventory (Table 11·1) or, more accurately, from a diffusion model.

The industrial source contributions based on emission inventory estimates were small for Pasadena and Pomona. The contribution for Riverside, 16.4%, was relatively large and a further breakdown for control purposes would be needed. The agreement between the measured and calculated total mass is good in each case.

TABLE 11·4
24-Hour Source Breakdown[a]
(Gartrell and Friedlander, 1975)

	Pasadena 9/20/72	Pomona 10/24/72	Riverside 9/20/72
Sea salt	0.7 ± 0.06	5.7 ± 0.6	1.3 ± 0.1
Soil dust	19.8 ± 0.1	15.1 ± 0.5	28.5 ± 0.9
Auto exhaust	5.1 ± 0.15	7.2 ± 0.3	3.9 ± 0.15
Cement dust	1.4 ± 0.15	3.3 ± 0.6	2.3 ± 0.15
Flyash	0.1 ± 0.01	0.2 ± 0.01	0.1 ± 0.01
Diesel exhaust[c]	1.4	1.9	0.9
Tire dust[e]	0.5	0.7	0.4
Industrial and agricultural[c]	4.7	6.6	20.5
Aircraft[c]	1.3	1.8	7.4
SO_4^{2-} [b]	2.9 ± 0.7	19 ± 5	5.9 ± 1.5
NO_3^- [b]	4.9 ± 0.4	36.4 ± 2.7	12.9 ± 1.0
NH_4^+ [b]	2.3 ± 0.1	16.3 ± 0.8	5.7 ± 0.3
Organics[d]	29.6	29.3	24.8
Water[b]	12 ± 6	18 ± 9	unknown
Total mass (sum of above)	86.7	161.5	114.6
Measured mass	64 ± 7	180 ± 20	125 ± 14

[a]Values in micrograms per cubic meter. Errors associated with sea salt, soil dust, auto exhaust, cement dust, and flyash are standard errors from the least-squares fit for the chemical element balance; errors associated with SO_4^{2-}, NO_3^-, NH_4^+, water and measured total mass concentrations are analytical errors.
[b]Measured values.
[c]Scaled to auto exhaust based on emission inventory.
[d]Based on a carbon balance.
[e]Assumed 10% of auto exhaust component.

Table 11·5 compares measured concentrations of the elements with concentrations calculated from the concentrations of soil dust, sea salt, auto exhaust, flyash, and cement dust obtained from the chemical element balance. The species on which the calculation was based—aluminum, sodium, lead, calcium, vanadium, magnesium, and potassium—are listed first in the table and show generally good agreement between measured and calculated values. The results are not normalized to specific species so the agreement adds confidence that the sources identified are the major ones for this set of elements.

TABLE 11·5

MEASURED AND CALCULATED ELEMENT CONCENTRATIONS[a]
(Gartrell and Friedlander, 1975)

	Pasadena 9/20/72		Pomona 10/24/72		Riverside 9/20/72	
Element	Calculated	Measured	Calculated	Measured	Calculated	Measured
Na[b]	0.716	0.716 ±0.014	2.13	2.06 ±0.20	1.13	1.13 ±0.019
Al[b]	1.66	1.03 ±0.33	1.32	1.30 ±0.42	2.35	2.35 ±0.07
Ca[b]	0.94	0.94 ±0.10	1.83	0.8 ±0.28	1.49	1.49 ±0.07
V[b]	4.8×10^{-3}	$4.8\times10^{-3}\pm2.6\times10^{-4}$	0.014×10^{-4}	$0.014\pm5.2\times10^{-4}$	8.8×10^{-3}	$8.8\times10^{-3}\pm0.000$
Pb[b]	2.03	2.03 ±0.077	2.89	2.89 ±0.12	1.55	1.55 ±0.006
Mg[b]	0.34	<0.4	0.50	<0.4	0.50	0.97 ±0.38
K[b]	0.31	0.36 ±0.036	0.31	0.44 ±0.044	0.45	0.68 ±0.068
Mn	0.022	0.023 ±0.002	0.017	0.037 ±0.001	0.031	0.040 ±0.001
Cu	1.7×10^{-3}	0.043 ±0.002	1.6×10^{-3}	0.025 ±0.002	2.4×10^{-3}	0.29 ±0.01
Cl	0.73	0.44 ±0.22	3.6	0.93 ±0.035	1.0	0.76 ±0.026
Br	0.40	0.50 ±0.011	0.58	0.81 ±0.018	0.31	0.39 ±0.009
Cr	5.2×10^{-5}	0.006 ±0.002	1.9×10^{-4}	0.025 ±0.004	1.0×10^{-4}	0.013 ±0.002
Ni	1.8×10^{-3}	0.006 ±0.001	4.4×10^{-3}	0.024 ±0.002	3.2×10^{-3}	0.014 ±0.001
I	9.7×10^{-7}	0.0023 ±0.0005	7.9×10^{-6}	$0.0042\pm7.4\times10^{-4}$	1.9×10^{-6}	$6.4\times10^{-3}\pm7\times10^{-4}$
Fe	0.67	1.15 ±0.044	0.56	1.55 ±0.06	0.94	2.31 ±0.09
Zn	0.009	0.097 ±0.004	0.012	0.23 ±0.009	0.008	0.13 ±0.005

[a]Values in micrograms per cubic meter; errors represent analytical errors. Calculated values are from estimates of soil dust, sea salt, auto exhaust, cement dust, and flyash from the chemical element balances.

[b]Used in making the chemical element balance; hence close agreement is expected.

A chemical element breakdown was not carried out for the industrial sources. Hence calculated concentrations for elements in the industrial sources would be expected to be smaller than measured concentrations. This probably explains the deviations between measured and calculated concentrations for copper, chromium, and zinc in Table 11·5. The discrepancy does not appear in the material balance (Table 11·4), since these species are included on a mass basis in the industrial contribution. Chlorine was present in smaller concentrations than calculated because it is lost from the aerosol by chemical reaction.

11·4 Light Scattering–Emission Source Relationships

Visibility is reduced by particulate matter and, in some cases, by NO_2. The aerosol extinction coefficient is given by (5·12):

$$b=\int_0^{\infty}\frac{\pi d_p^2}{4}K_{ext}(x,m)n_d(d_p)d(d_p) \tag{11·3}$$

The visibility or visual range is related to the extinction coefficient by (5·32):

$$s^* = -\frac{1}{b}\ln 0.02 = \frac{3.912}{b}$$

Since b is a function of wavelength, the visibility defined in this way also depends on wavelength.

The volume of material in any given particle size range can be expressed as the sum of contributions of various sources:

$$vn_d d(d_p) = \sum_j \bar{v}_j dm_j$$

where for a solution $\bar{v}_j$ is the partial specific volume and dm_j is the mass of aerosol per unit volume of air originating from source j. Substituting in (11·3) for the extinction coefficient, assuming that a suitable average K_{ext} can be employed and that the particles are spherical ($v=(\pi/6)d_p^3$), the result is

$$b=\sum b_j = \frac{3}{2}\sum_j \int_{-\infty}^{\infty}\frac{K_{ext}}{d_p}\frac{\bar{v}_j dm_j}{d\log d_p}d\log d_p$$

Hence a contribution to the total extinction can be assigned to each source as follows:

$$b_j = \frac{3}{2}\int_{-\infty}^{\infty} \frac{K_{ext}}{d_p}\frac{\bar{v}_j\, dm_j}{d\log d_p}\, d\log d_p$$

The contribution depends on the distribution of mass with respect to size, $dm_j/d\log d_p$, and on the light scattering per unit volume $3K_{ext}/2d_p$.

To take into account the distribution of solar radiation with respect to wavelength, $3K_{ext}/2d_p$ is integrated over the wavelength to give the light scattering per unit volume of aerosol:

$$G(d_p) = \frac{3}{2d_p}\int_0^{\infty} K_{ext}(x,m) f(\lambda)\, d\lambda$$

where $f(\lambda)$ is the standard distribution of solar radiation of solar radiation at sea level. The function $G(d_p)$ is shown in Fig. 5·7 for $m=1.5$. The extinction coefficient averaged over the solar radiation distribution function is then given by

$$\bar{b}_j = \int_{-\infty}^{\infty} G(d_p)\bar{v}_j \frac{dm_j}{d\log d_p}\, d\log d_p$$

In the special case when the normalized mass distributions,

$$\frac{dm_j}{m_j d\log d_p} = \text{const} \tag{11·4}$$

the extinction coefficient is linearly related to the masses of the various source contributions:

$$b = \sum_j \gamma_j m_j \tag{11·5}$$

where

$$\gamma_j = \frac{3}{2}\int_{-\infty}^{\infty} \frac{K_{ext}}{d_p}\frac{\bar{v}_j\, dm_j}{m_j d\log d_p}\, d\log d_p$$

The coefficients γ_j differ for each source contribution depending on the distribution of the species coming from the source with respect to particle size.

In the case of total mass and chemical composition, the linear dependence (11·1) and (11·2) on m_j is exact. The linear relationship (11·5) for the extinction coefficient is an approximation that depends on the special mass distributions (11·4). It must be tested experimentally.

11·5 Visibility–Emission Source Relationships: Field Studies

As part of a large scale aerosol field study (Hidy et al., 1975), the light scattering coefficient, b_{scat}, and chemical composition of ambient aerosols were measured at several locations in the Los Angeles air basin. The relationship between b_{scat} and chemical composition was analyzed by White, Roberts, and Friedlander (1975). In agreement with previous studies, b_{scat} correlated well with the aerosol mass concentration, ρ, with an average ratio of $b_{scat}/\rho = 0.032\ (10^{-4})\ m^2/\mu g$ for sixty 2 hr samples.

Further analysis revealed that much of the variation in the ratio b_{scat}/ρ was statistically associated with variations in the sulfate and nitrate fractions of the aerosol; b_{scat}/ρ tended to be higher than average for aerosols rich in sulfates and nitrates, and lower than average for aerosols poor in these compounds. By multiple regression analysis, values of the scattering–mass ratio, γ_j, were derived for individual constituents of the aerosol assuming a relationship of the form (11·5). Sulfate compounds were the most effective scatterers with $\gamma_{SO_4} = 0.062(10^{-4})\ m^2/\mu g$. The scattering–mass ratio of nitrate compounds (same units) depended on relative humidity, ranging from 0.028 at 40% RH to 0.051 at 80% RH. The remaining constituents of the aerosol, including the organic compounds, were rather poor scatterers, with a combined scattering–mass ratio of 0.020. Possible reasons for the relatively high scattering–mass ratios of the sulfate and nitrate compounds were the preferential association of water with these compounds and, in the case of sulfates, their distribution with respect to particle size.

Because of the efficiency with which sulfate and nitrate compounds scattered light, they contributed more to the reduction of visibility than their mass concentration alone would indicate. Based on the scattering–mass ratio for sulfate compounds and the concentrations of sulfate measured, sulfate compounds contributed 17% of the aerosol mass and 32% of the light scattering measured during the sampling intervals analyzed. Nitrate compounds contributed 21% of the aerosol mass and 27% of the light scattering, leaving 62% of the aerosol mass and only 41% of the light scattering to be attributed to all other aerosol constituents combined.

Most of the particulate sulfate and nitrate result from oxidation in the atmosphere of SO_2 and NO_x. Inventories of the emission of these gases in the Los Angeles air basin (Table 11·1) indicate that they come principally from two types of sources. The combustion of gasoline in motor vehicles produces most of the NO_x and minor quantities of SO_2. The refining of crude oil and the combustion of the heavier fuels produce nearly all of the remaining NO_x and most of the SO_2.

Support for this general picture was obtained by multiple regression analysis of the relationship of ambient SO_2 and NO_x concentrations to the concentrations of tracers for the two source types. Particulate lead was used as a tracer for motor vehicle exhaust, and particulate nickel was used as a tracer for the combustion of the heavier petroleum fractions. During the intervals sampled, SO_2 correlated significantly with nickel and not at all with lead, whereas NO_x correlated significantly with lead and showed some dependence on nickel.

Based on the foregoing analyses, the contributions of motor vehicles and stationary sources to the degradation of visibility in the Los Angeles air basin were estimated. It was calculated that the particulate nitrate, sulfate, and organic compounds produced from the emissions of these sources contributed two-thirds of the light scattering observed during the sampling intervals, with the two source types contributing about equally to the reduction of visibility.

11·6 Predictive Models

Chemical tracer and material balance methods discussed in previous sections are static; they neither require nor provide information on the change with time of the aerosol properties. *Predictive models* for atmospheric aerosols start from size distributions and predict aerosol properties downwind by solution of the GDE. The information required includes source characteristics, the spatial distribution of emission sources, the wind field and associated turbulence parameters and growth laws (gas-to-particle conversion).

Size distributions have been measured for certain types of sources including the automobile, soil dust, and the marine aerosol. Data are reviewed by Gartrell and Friedlander (1975), and new data are given by Hidy (1975). Although size distributions may vary significantly from source to source even within a given class of sources, there are certain broad similarities among distributions for each type of source. For example, the automobile generates a substantial quantity of submicron particles, whereas soil dust particles are generally larger than 1 μm.

Predicted atmospheric size distributions and chemical composition can be compared with the results of experimental observation, when available. However, this is not as necessary in this method as it is in the material balance approach. Once a satisfactory model has been developed, it can be used to predict the effects of adding new sources or modifying old sources on air quality.

Predictive modeling for stack plumes starting from the GDE was discussed in Chap. 10. An approximate expression for the dynamics of the aerosol in an urban air basin can also be derived from the GDE. Neglecting the fluctuating growth and coagulation terms, (10·20) can be written as follows:

$$\bar{u}\frac{\partial \bar{n}}{\partial x} + \frac{\partial \bar{n}\bar{q}}{\partial v} = \frac{\partial F_z}{\partial z} + \frac{1}{2}\int_0^v \beta(\tilde{v}, v-\tilde{v})\bar{n}(\tilde{v})\bar{n}(v-\tilde{v})\,d\tilde{v}$$

$$-\int_0^\infty \beta(v,\tilde{v})\bar{n}(v)\bar{n}(\tilde{v})\,d\tilde{v} - c_s\frac{\partial \bar{n}}{\partial z}$$

where F_z represents the vertical diffusion flux. Integrating this expression with respect to z up to some convenient altitude, such as the inversion height, H, the result is

$$\frac{\partial[\bar{n}\bar{u}]}{\partial x} + \frac{\partial[\bar{n}\bar{q}]}{\partial v} = \frac{F_0(v,x)}{H} + \frac{1}{2}\int_0^v \beta(\tilde{v}, v-\tilde{v})[\bar{n}(\tilde{v})\bar{n}(v-\tilde{v})]\,d\tilde{v}$$

$$-\int_0^\infty \beta(v,\tilde{v})[\bar{n}(v)\bar{n}(\tilde{v})]\,d\tilde{v} \qquad (11\cdot 6)$$

where $F_0(v,x)$ is the net flux of particles at the surface by all mechanisms including sedimentation, and the brackets indicate averaging with respect to z:

$$[f] = \frac{1}{H}\int_0^H f\,dz \qquad (11\cdot 7)$$

The flux at $x = H$ is assumed to vanish. For application to atmospheric simulation, it is further assumed that

$$[\bar{n}\bar{u}] = [\bar{n}][\bar{u}]$$

$$[\bar{n}\bar{q}] = [\bar{n}][\bar{q}]$$

$$[\bar{n}(v)\bar{n}(\tilde{v})] = [\bar{n}(v)][\bar{n}(\tilde{v})]$$

Equation (11·6) then becomes

$$u\frac{\partial n}{\partial x}+\frac{\partial nq}{\partial v}=\frac{F_0(v,x)}{H}+\frac{1}{2}\int_0^v \beta(\tilde{v},v-\tilde{v})n(\tilde{v})n(v-\tilde{v})\,d\tilde{v}$$

$$-\int_0^\infty \beta(v,\tilde{v})n(v)n(\tilde{v})\,d\tilde{v} \qquad (11\cdot 8)$$

where the quantities are the average up to a fixed height H as defined by (11·7). A schematic diagram of the model system is shown in Fig. 11·1.

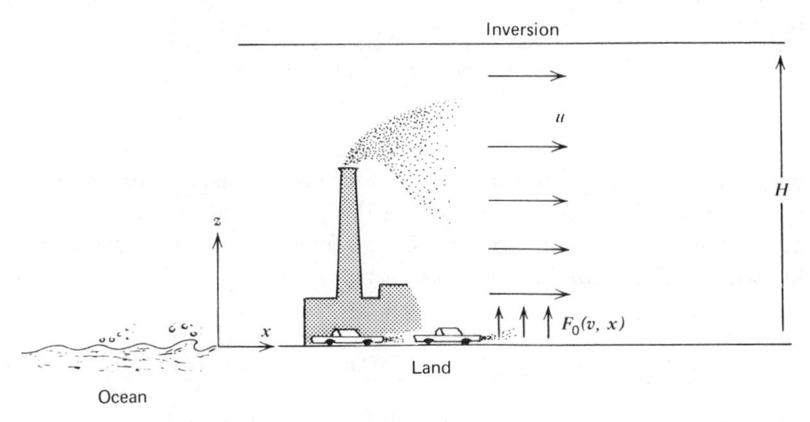

Fig. 11·1 Schematic diagram of urban air basin showing coordinate system of (11·8).

Direct application of (11·8) to aerosol simulation in urban basins has not yet been carried out. However, a closely related Lagrangian model has been employed (White and Husar, 1976; Chu and Seinfeld, 1975). In this approach, a fixed vertical element of air extending from ground to the inversion is followed. Source injections and particle growth are both taken into account.

11·7 Hybrid Models

Size distributions can be predicted starting from a set of initial distributions determined by chemical element balances (Heisler, Friedlander and Husar, 1973; Gartell and Friedlander, 1975). This approach represents a hybrid between the material balance method and predictive modeling. Coagulation must be taken into account near sources of high concentrations of small particles, such as freeways. Away from such sources, coagulation can be neglected provided high concentrations are not generated locally by homogeneous nucleation. In the absence of coagulation and homogeneous nucleation, heterogeneous growth processes shape the submicron range of the pollution aerosol. The transition region where coagulation gives way to growth is difficult to determine. If the automobile is the principal source of small particles, the ratio of the number of particles to mass of lead can be used to estimate the extent of coagulation away from sources. This ratio decreases as coagulation occurs.

Growth calculations are started from the initial primary aerosol mixture, after correcting for coagulation. For the box model, the size distribution at any time is related to the initial distribution by the relation

$$n_d(d_p) = \frac{dN}{d(d_p)} = \frac{dN}{d(d_p)_0} \Big/ \frac{d(d_p)}{d(d_p)_0}$$

where $d(d_p)/d(d_p)_0$ is the slope of a plot of d_p versus d_{p0} with time as the parameter. This quantity depends on the particle growth law, hence the mechanism of particle growth. If the mechanism of growth is not known, an empirical growth law can be used:

$$\frac{dv}{dt} = k\, d_p^{\gamma}$$

where v is the particle volume, k is a rate coefficient, and γ is a constant usually between 2 and 3 (Gartell and Friedlander, 1975). The growth calculation is continued until the total amount of material present in the aerosol phase reaches the measured value. The value of the exponent γ is adjusted to give agreement between measured and calculated particle size distributions. Such a comparison is made in Fig. 11·2 in which the calculated distribution of the primary aerosol is also shown. The area between the two curves represents the volume of secondary material. The marked effect of the secondary conversion processes on visibility results from the strong peak which develops in the volume distribution near 0.5 μm. This is the size that corresponds to the peak in the light scattering function per unit volume of dispersed matter.

Calculated distributions of chemical species with respect to size can also be checked against data; only an approximate verification is possible because size and time resolutions attainable for chemical composition in the submicron range are poor with the instrumentation currently available.

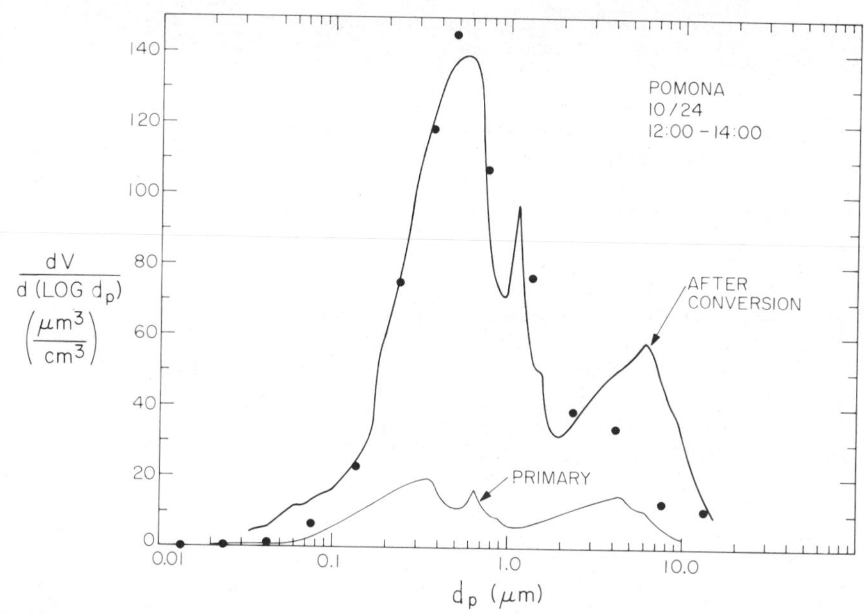

Fig. 11·2 Primary volume distribution based on chemical element balance, and final volume distribution, after secondary conversion calculations for Pomona. Points are experimental data (Gartrell and Friedlander, 1975).

Problems

1. At a sampling site near a freeway not far from the coast, an aerosol sample is collected by filtration for chemical analysis. Sodium and aluminum concentrations, measured by neutron activation analysis, are found to be 4.1 and 2.2 $\mu g/m^3$, respectively. The lead concentration, by atomic absorption, is 3.5 $\mu g/m^3$. Assuming that only sea salt, soil dust, and automobile exhaust contribute, calculate the total mass of aerosol per unit volume of air ($\mu g/m^3$) at a temperature of 20° C and determine the percentage contribution of each of the three sources. Assume that the marine aerosol component consists of a saturated salt solution.

2. On a certain day in Los Angeles, the ratio SO_4^{2-}/SO_2 is 0.20 on a mass basis averaged over a dozen monitoring stations in the region. *Estimate* the number of tons of SO_4^{2-} present in the atmosphere of that region.

3. The ratio of total particle concentration to lead concentration measured near a freeway was 8.4×10^4 (particles/cm^3)/(μg lead/m^3) while at a point in the surrounding urban atmosphere it was 2.2×10^4. Explain why this ratio might be expected to decrease.

References

Birakos, J. N. (1974) 1974 *Profile of Air Pollution Control*, County of Los Angeles Air Pollution Control District.

Chu, K. J., and Seinfeld, J. H. (1975) *Atmos. Environ.*, **9**, 375.

Federal Register (1971) Rules and Regulations, Title 42-Public Health Chap. IV, Part 410, National Primary and Secondary Ambient Air Quality Standards, 8191.

Friedlander, S. K. (1973) *Environ. Sci. Technol.*, **7**, 235.

Gartrell, G., Jr., and Friedlander, S. K. (1975) *Atmos. Environ.*, **9**, 279.

Gatz, D. F. (1975) *Atmos. Environ.*, **9**, 1.

Heisler, S. L., Friedlander, S. K., and Husar, R. B. (1973) *Atmos. Environ.*, **7**, 633.

Hidy, G. M. (Ed.) (1975) Characterization of Aerosols in California (ACHEX) Final Report to the Air Resources Board, State of California, ARB Contract No. 358 (available through NTIS).

This collection of reports provides the most detailed set of data available on the chemical and physical properties of aerosols in polluted air basins.

White, W. H., and Husar, R. B. (1976) *J. Air Poll. Control Assoc.*, **26**, 32.

White, W. H., Roberts, P. T., and Friedlander, S. K. (1975) On the Nature and Origins of Visibility-Reducing Aerosols in the Los Angeles Air Basin, unpublished report to the California Air Resources Board.

Index